Thinkwell Geometry
Course Notes

This Course Notes book is designed to be used alongside Thinkwell's online course. For more information, please contact us at support@thinkwell.com.

To the Student

Congratulations! You are about to add a new dimension to your homeschool learning experience. *Thinkwell Geometry* will give you more engaging, thorough, and effective access to content than any traditional textbook can provide.

Use this Book to Help You Succeed

This book is designed to help you take maximum advantage of the online resources by providing a hardcopy, streamlined version of the multimedia content from the Thinkwell website: **www.thinkwellhomeschool.com**. It also serves as a useful place to record your own notes, questions, and ideas.

This book summarizes the concepts, vocabulary, and examples presented in the Video Lectures, and thus serves as an effective tool for navigating the course and referencing the multimedia content. Useful as it is, however, it is not intended to serve as a substitute for viewing the Video Lectures.

Take Advantage of Thinkwell's Online Resources

Video Lectures

All of the course's objectives are presented in the Video Lectures. These Video Lectures are your main source of content.

The right-hand side of the Video Lecture screen contains text and graphics that build as the lecture progresses. Pay special attention to the headings, charts, bulleted lists, and diagrams that appear in this area, as well as any definitions and other important concepts related to the topic. Video Lectures also include a key concepts area that lists the main ideas presented in the lecture.

Online Video Lecture Notes

Each Video Lecture is summarized in the accompanying online notes, which comprise the digital version of the contents of this book.

Interactive Exercises

The Thinkwell website contains hundreds of interactive exercises. These are automatically graded, providing you with instant feedback on your knowledge of the Video Lecture content. These exercises may also include a detailed explanation of how to arrive at the correct answer.

To the Student, *continued*

View the Thinkwell Video Lectures Actively

Research shows that students who interact with course content, rather than just passively reading, listening, or viewing it, are much more likely to gain a superior understanding of that content.

Allow yourself about 30 minutes to actively view a typical 10- to 15-minute Video Lecture. Pause the video whenever you have a question or an idea, and write down your thoughts in this book. Alternatively, try watching each Video Lecture once without stopping. Then watch it a second time and use this book to take your own notes. You may also want to highlight the key concepts as they are discussed in the Video Lectures, noting any important points or details that will help you remember and understand this critical content.

Try to find some time immediately after viewing the Video Lecture to review and flesh out your notes. Most students' memories of even the major points of a lecture begins to fade after about 30 minutes. This exercise is an extremely good use of time and one that you'll appreciate when you review your notes prior to an exam.

For more help with the online components of this course, contact us at **technicalsupport@thinkwell.com**.

Finally, from all of us at Thinkwell:
We are committed to engaging your interest and ensuring your success. We wish you all the best!

Table of Contents

1

1 Fundamentals of Geometry

Understanding Points, Lines, and Planes

Key Objectives

- Identify, name, and draw points, lines, segments, rays, and planes.
- Apply basic facts about points, lines, and planes.

Key Terms

- A basic geometric figure that cannot be defined in terms of other figures is an **undefined term**.
- A **point** names a location and has no size. It is represented by a dot.
- A **line** is a straight path that has no thickness and extends forever.
- A **plane** is a flat surface that has no thickness and extends forever.
- Points that lie on the same line are **collinear**.
- Points that lie in the same plane are **coplanar**.
- A **segment** or line segment is the part of a line consisting of two points and all points between them.
- An **endpoint** is a point at the end of a segment or the starting point of a ray.
- A **ray** is a part of a line that starts at an endpoint and extends forever in one direction.
- **Opposite rays** are two rays that have a common endpoint and form a line.
- A **postulate**, or axiom, is a statement that is accepted as true without proof.
- An **intersection** is the set of all points that two or more figures have in common.

Theorems, Postulates, Corollaries, and Properties

- **Postulate** Through any two points there is exactly one line.
- **Postulate** Through any three noncollinear points there is exactly one plane.
- **Postulate** If two points lie in a plane, then the line containing those points lies in the plane.
- **Postulate** If two unique lines intersect, then they intersect at exactly one point.
- **Postulate** If two unique planes intersect, then they intersect at exactly one line.

Geometry is an area of mathematics concerned with the study of two-dimensional and three-dimensional figures. Before a figure can be studied in geometry, even the most basic parts of the figure must be identified and named. For example, consider a triangle. A triangle is a very simple figure with three sides where each pair of sides meets at a corner. Even though a triangle is a simple figure, its parts must be identified using geometric terms before a triangle can be explicitly defined. The most elementary geometric figures are discussed in the examples below.

Example 1 Naming Points, Lines, and Planes

The three fundamental geometric figures are points, lines, and planes. These three figures cannot be defined in terms of other figures, so they are referred to as undefined terms.

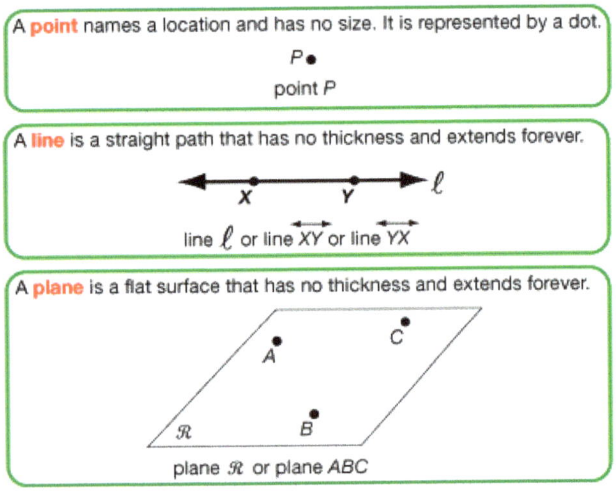

A point names a location and is represented by a dot. The dot must have some thickness or size when drawn, but an idealized point has no size because it is only a location. A point is named with a capital letter, such as P.

A line is a straight path of points that extends infinitely in two opposite directions. An idealized line contains an infinite number of points and has length, but no thickness. A line is named with a lowercase letter or by drawing a double-headed arrow over the names of two points on that line. For example, the line here, which passes through the points X and Y, can be named $\overleftrightarrow{XY}$ or ℓ.

A plane is a flat surface that extends infinitely. Points and lines are contained on a plane. A plane is named with a capital script letter, such as $\mathcal{R}$, or with any three points on that plane, and these points are not on a single line. For example, ABC names a plane that contains points A, B, and C if A, B, and C are not contained on a single line. The points used for naming a plane can be listed in any order.

Points that lie on the same line are collinear points, and points that lie in the same plane are coplanar points. There are also terms to describe points that do not lie on the same line or plane: noncollinear and noncoplanar.

Name four coplanar points.

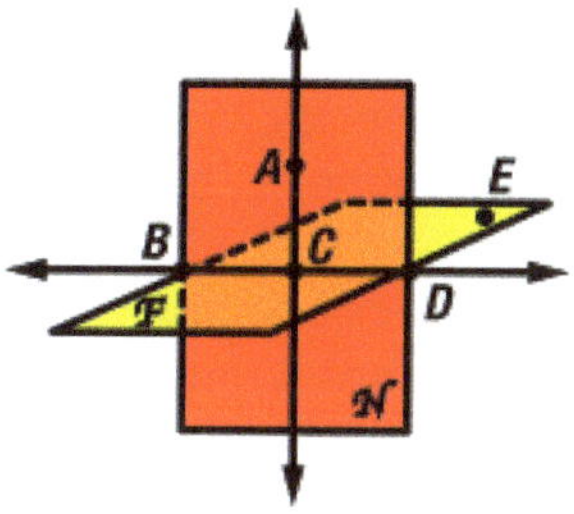

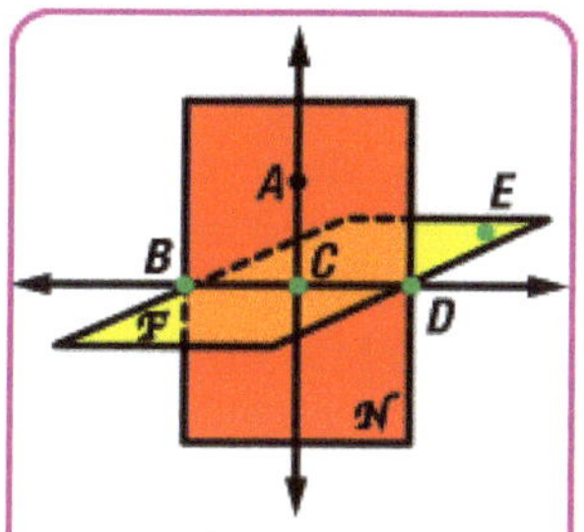

B, *C*, *D*, and *E* are four coplanar points.

The given figure shows two planes, $\mathcal{F}$ and $\mathcal{N}$; five points, *A*, *B*, *C*, *D*, and *E*; and two lines.

Points *B*, *C*, *D*, and *E* are contained on $\mathcal{F}$. So, *B*, *C*, *D*, and *E* are four coplanar points.

Furthermore, $\mathcal{N}$ contains points *A*, *B*, *C*, and *D*, so these four points are also coplanar.

Name two lines.

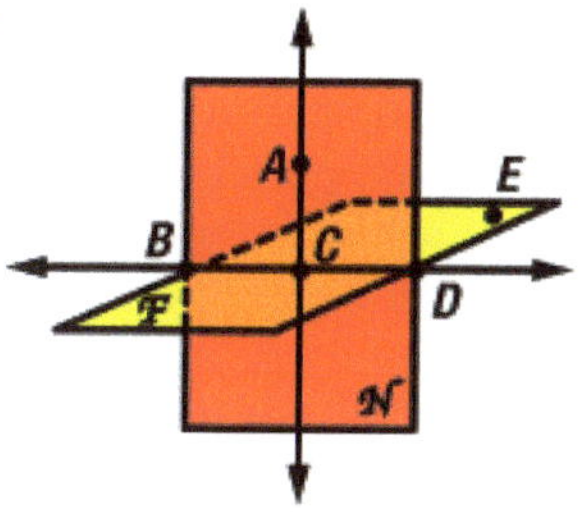

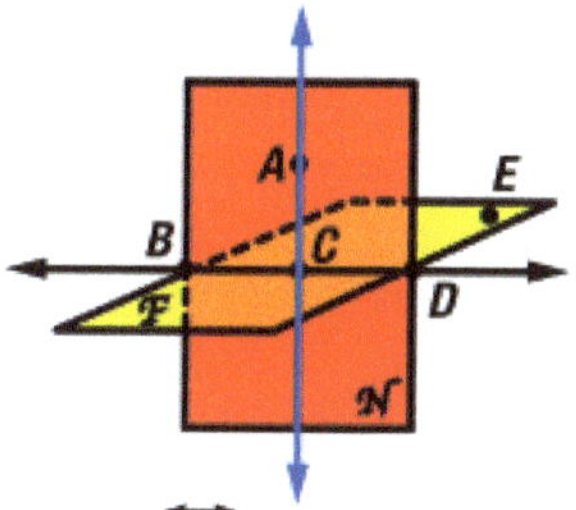

$\overleftrightarrow{AC}$ is one line.

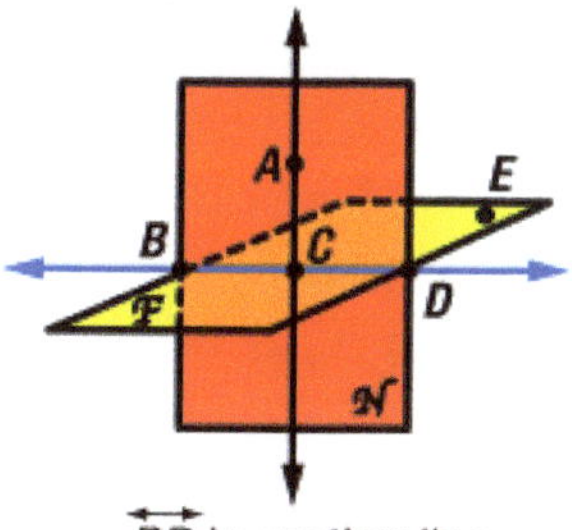

$\overleftrightarrow{BD}$ is another line.

Lines are drawn with arrows at each end. There are only two lines drawn in this figure; one line is a vertical line and the other is a horizontal line. (Note that lines do *not* have to be horizontal or vertical.)

A line is named using any two points that the line passes through. Since the vertical line passes through *A* and *C*, it can be named $\overleftrightarrow{AC}$.

The horizontal line passes through *B*, *C*, and *D*. So, the horizontal line can be named $\overleftrightarrow{BC}$, $\overleftrightarrow{CD}$, or by using any pair of the points it passes through.

Example 2 Drawing Segments and Rays

A **segment**, or **line segment**, is the part of a line consisting of two points and all points between them.

segment $\overline{AB}$ or $\overline{BA}$

A and B are the endpoints of $\overline{AB}$.

A **ray** is a part of a line that starts at an endpoint and extends forever in one direction.

ray $\overrightarrow{AB}$

A is the endpoint of $\overrightarrow{AB}$.

Opposite rays are two rays that have a common endpoint and form a line.

opposite rays $\overrightarrow{BA}$ and $\overrightarrow{BC}$

B is the common endpoint of opposite rays $\overrightarrow{BA}$ *and* $\overrightarrow{BC}$.

A section of a line defined by two endpoints that does not extend infinitely in either direction is called a segment. A segment is named by drawing a line (with no arrows) over the names of the two endpoints. For example, $\overline{AB}$ names a segment where *A* and *B* are the endpoints. Either endpoint can be listed first in the name of a segment.

A section of a line defined by one endpoint that extends infinitely in one direction is called a ray. A ray is named by drawing a single-headed arrow over the name of the endpoint and any other point on that ray. For example, $\overrightarrow{AB}$ names a ray where *A* is the endpoint and *B* is a point on the ray.

When two rays have a common endpoint and form a line, they are called opposite rays.

Draw a segment with endpoints *A* and *B*.

The name of this line segment is $\overline{AB}$.

To draw a segment with endpoints A and B, draw a straight line with a point at the beginning and another point at the end. Identify these endpoints as A and B. The segment can be named $\overline{AB}$ since its endpoints are A and B.

Draw opposite rays with a common endpoint *W*.

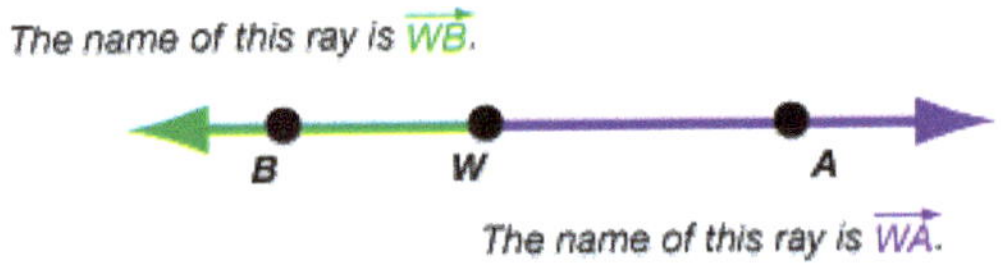

To draw opposite rays with a common endpoint W, begin by drawing a point and labeling it W, the endpoint for both rays. Opposite rays extend in opposite directions from a common endpoint to form a line. So, if the first ray extends from W to the left, then in order for the two rays to form a line, the second ray must extend from W to the right. (Note that opposite rays do *not* have to form a horizontal line; any line can be formed.)

Example 3 Identifying Points and Lines in a Plane

Many facts regarding geometric figures will be presented as a postulate. A postulate, or axiom, is a statement that is accepted as true without proof. Three postulates regarding the relationship between points, lines, and planes are listed below.

- Through any two points there is exactly one line.
- Through any three noncollinear points there is exactly one plane.
- If two points lie in a plane, then the line containing those points lies in the plane.

Name a line that passes through two points.

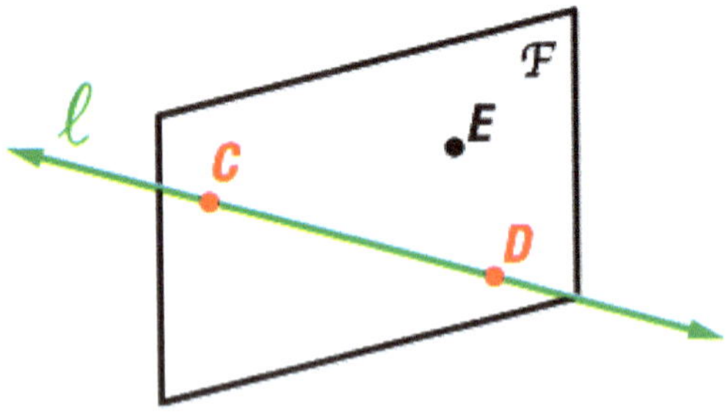

The line ℓ passes through the two points C and D.

Other names for ℓ are $\overleftrightarrow{CD}$ and $\overleftrightarrow{DC}$.

The given figure shows one plane, $\mathcal{F}$; three points, C, D, and E; and one line.

The green line passes through two points, C and D. This line can be named ℓ or it can be named with a reference to the two points it passes through, C and D.

Example 4 Representing Intersections

When two lines or planes cross each other, they are said to intersect. Consider two intersecting lines. The point at which the two lines intersect is the point that the two lines have in common. This common point is called the intersection of the lines. Generally, the intersection of two figures is the set of all points that two figures have in common. An intersection may contain one point, a limited number of points, or an infinite number of points, depending on the figures. The following postulates describe the intersections of lines and planes.

- If two unique lines intersect, then they intersect at exactly one point.
- If two unique planes intersect, then they intersect at exactly one line.

Sketch two lines that are intersecting. 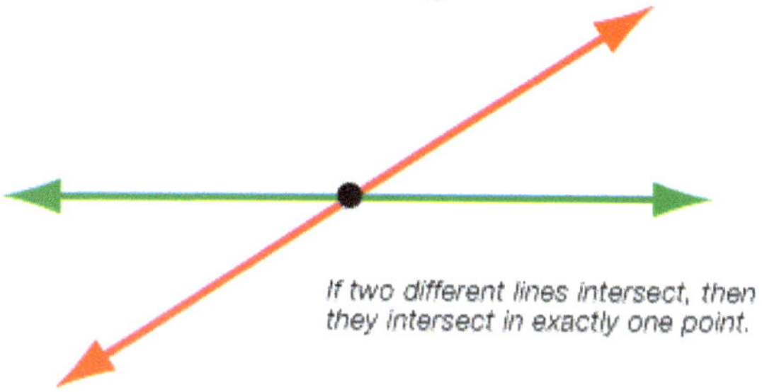	In order for two lines to intersect, they must not have the same slant. So, to draw two intersecting lines, draw two lines (each with arrows at the ends) such that the degree of slant in the first line is different from that of the second line. It is possible to draw two lines that do not intersect. Non-intersecting lines will have the same degree of slant. For example, two horizontal lines will never intersect.
Sketch two planes that intersect in one line. 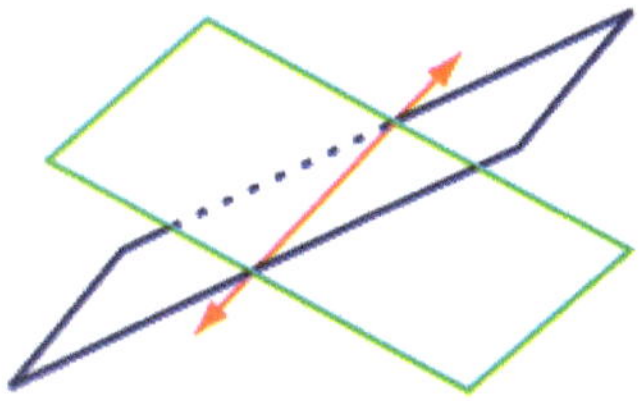	Here, draw two planes that intersect in one line. So, draw two unique planes that cross each other. The line formed by the intersection of the two planes is contained on both planes.
Sketch a second line that intersects the two planes but does not lie in either plane. 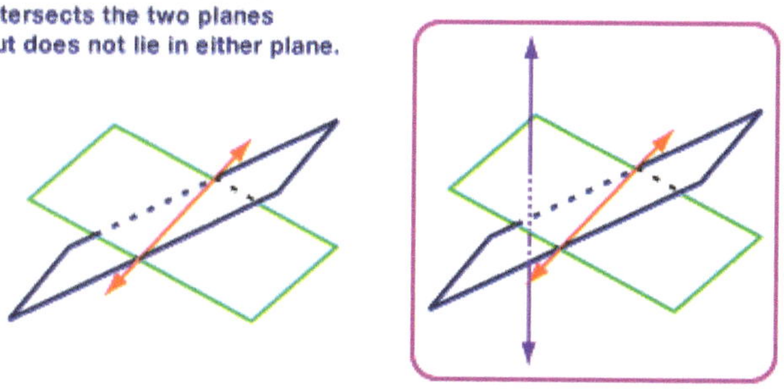	To draw a line that intersects the two planes, but does not lie in either plane, draw a line that intersects with each plane at a single point. The vertical line drawn here intersects with both of the planes, but it does not lie in either plane.

Measuring and Constructing Segments

Key Objectives

- Use length and midpoint of a segment.
- Construct midpoints and congruent segments.

Key Terms

- A **coordinate** is a number used to identify the location of a point. On a number line, a point corresponds to one number and this number is called a coordinate.
- The **distance** between any two points is the absolute value of the difference of the coordinates.
- The distance between two points A and B is also called the **length** of $\overline{AB}$, or AB.
- **Congruent segments** are segments that have the same length.
- A **construction** is a way of creating a figure that is more precise than a sketch.
- In order for you to say that a point B is **between** two points A and C, all three of the points must lie on the same line, and $AB + BC = AC$.
- The **midpoint** M of a segment AB is the point that bisects, or divides, the segment into two congruent segments. If M is the midpoint of segment AB, then $AM = MB$.
- To **bisect** is to divide into two congruent parts.
- A **segment bisector** is any ray, segment, or line that intersects a segment at its midpoint.

Theorems, Postulates, Corollaries, and Properties

- **Ruler Postulate** The points on a line can be put into a one-to-one correspondence with the real numbers.
- **Segment Addition Postulate** If B is between A and C, then $AB + BC = AC$.

Example 1 Finding the Length of a Segment

When a segment is drawn on a number line, each of the segment's endpoints corresponds with exactly one real number on a number line, called a coordinate. The relationship between points and real numbers is described in the Ruler Postulate. The points on a line can be put into a one-to-one correspondence with the real numbers.
In other words, when a segment is on a number line, there is exactly one number on that number line that corresponds with each endpoint.

A segment is a part of a line that has a defined length. The length of a segment on a number line is the distance between its two endpoints, and the distance between any two points on a number line is the absolute value of the difference of the coordinates. So, to find the length of a segment on a number line, find the difference between the coordinates (the numbers at the endpoints), and then take the absolute value of that difference.

Find the length of $\overline{XW}$ and $\overline{WY}$.

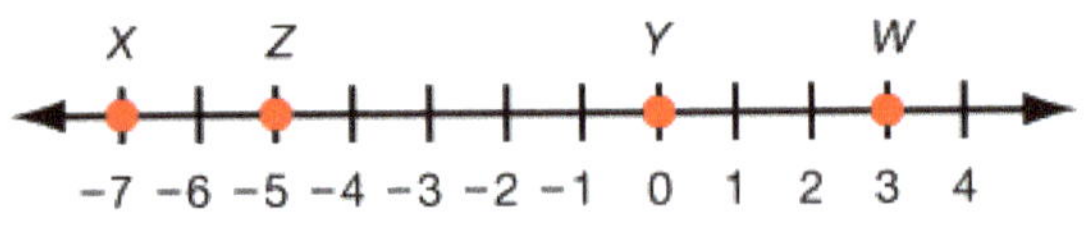

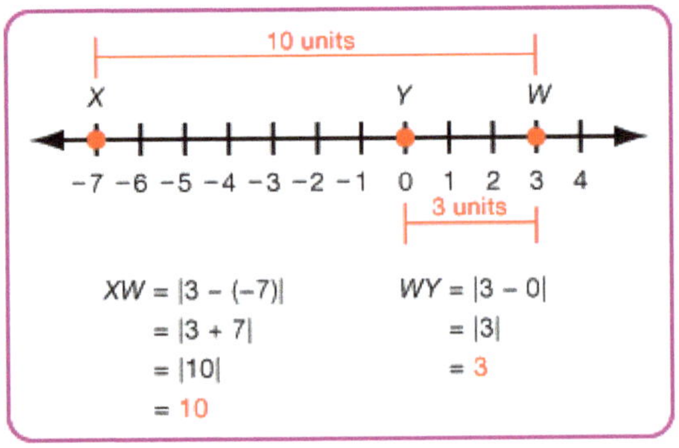

In this example a number line is given with the points X, Z, Y, and W, named on the number line. The coordinate of each point is the number that corresponds with that point. So, the coordinate of W is 3.

To find XW, first identify coordinates of its endpoints: –7 and 3. Then subtract these coordinates (in either order) and take the absolute value of the difference. Here, the length of the segment was found by subtracting –7 from 3. Note that subtracting in the reverse order results in the same length.
$|-7 - 3| = |-10| = 10$

Notice that the segment from X to W is represented here as XW and as $\overline{XW}$. The notation XW refers to the length of the segment, so XW is equal to a number (in this case $XW = 10$). The notation $\overline{XW}$ refers to the segment itself.

Example 2 Copying a Segment

Congruent segments are segments that have the same length.

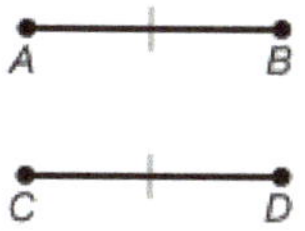

$\overline{AB} \cong \overline{CD}$ *Segments AB and CD are congruent.*

$AB = CD$ *The lengths of segments AB and CD are equal.*

Tick marks are used to indicate that segments are congruent.

When two segments have the same length, they are congruent. The symbol for congruent is similar to the symbol =, but the congruent symbol has a squiggle over the =.

The difference between congruent and equal is that segments themselves can be congruent, while the *lengths* of segments can be equal.

Small dashes, called tick marks, are drawn in figures to indicate that segments are congruent.

A sketch of a figure is an approximate version of a figure. A construction is a way of creating a figure that is more precise than a sketch. One way to make a geometric construction is to use a compass and straightedge.

Construct a segment congruent to $\overline{PQ}$.

Use a straightedge and a compass to construct a segment that is congruent to the given segment, $\overline{PQ}$.

Step 1 Using a straightedge, draw a segment that is longer than the given segment. Then, near one end of the drawn segment, draw and label a point. This is one of the constructed segment's endpoints.

Here, the drawn segment is longer than $\overline{PQ}$ and the endpoint is labeled A.

Step 1

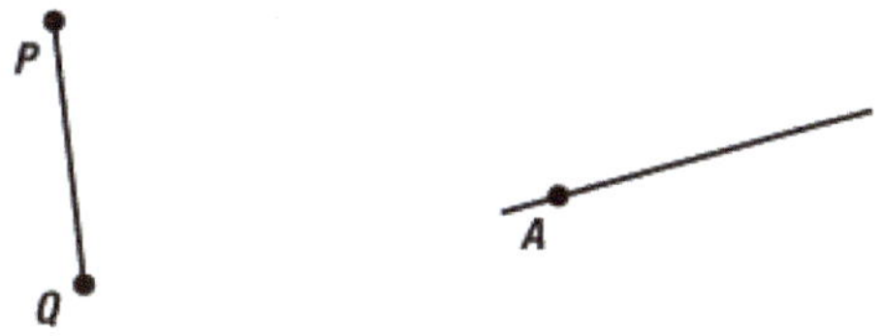

Step 2

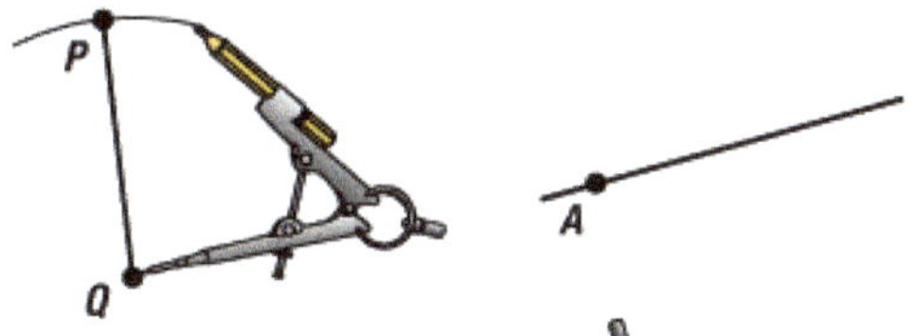

Step 3

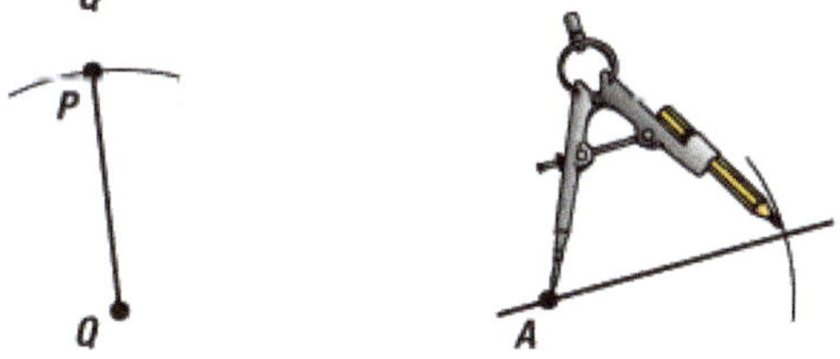

Step 4

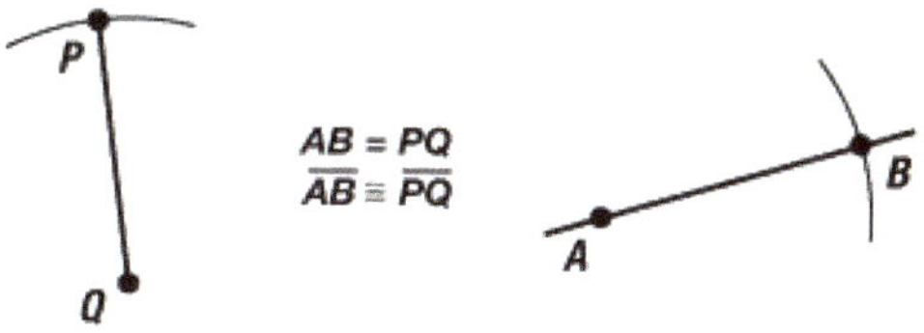

Step 2 Place the compass so that its ends are at the given segment's endpoints. Then, confirm that the compass is placed correctly by drawing a small arc that passes through the given segment's endpoint.

The compass is open to Q and P in this example. The drawn arc passes through P, so the compass is placed correctly.

Step 3 *Without moving the arms of the compass*, place it on the drawn segment so that its tip is on the labeled point. Then use the compass to sketch an arc that intersects the drawn segment.

Here, the tip of the compass is placed at A and the arc intersects the drawn segment at the opposite end.

Step 4 Label the point where the arc intersects the drawn segment. This is the constructed segment's second endpoint.

In this example, the point where the arc intersects the drawn segment is labeled B. So, the constructed segment is $\overline{AB}$. Since $\overline{PQ}$ and $\overline{AB}$ are congruent segments, they have the same length. So, $AB = PQ$.

Example 3 Using the Segment Addition Postulate

Segment Addition Postulate
If B is between A and C, then $AB + BC = AC$.
(Seg. Add. Post.)

B is between A and C.

$AB + BC = AC$

In order for you to say that a point B is between two points A and C, all three of the points must lie on the same line, and $AB + BC = AC$.

Suppose that A, B, and C are collinear points, B is between A and C, $AC = 25$, and $BC = 7.2$. Find AB.

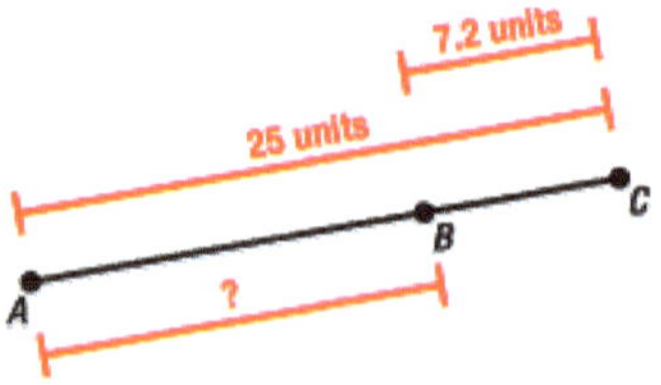

$AC = 25$	*Given*
$BC = 7.2$	*Given*
$AB + BC = AC$	*Segment Addition Postulate*
$AB + 7.2 = 25$	*Substitute the given values.*
$-7.2 \quad -7.2$	*Subtract 7.2 from both sides.*
$AB = 17.8$	*Simplify.*

In this example, the fact that B is between A and C is given. So, by the Segment Addition Postulate, it must be true that $AB + BC = AC$.

Use the given information to sketch a segment. Since B is between A and C, the segment's endpoints must be A and C. Now label the segment with the given lengths $AC = 25$ and $BC = 7.2$.

From the sketched segment it is easy to see that $AB + BC = AC$, or $AB + 7.2 = 25$. Solve this equation by subtracting 7.2 from both sides to find AB.

The length of $\overline{AB}$ is $25 - 7.2$, or 17.8.

Find PQ.

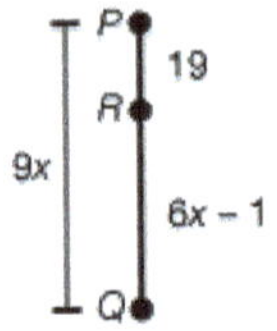

$PR = 19$	*Given*
$RQ = 6x - 1$	*Given*
$PQ = 9x$	*Given*
$PQ = PR + RQ$	*Segment Addition Postulate*
$9x = 19 + 6x - 1$	*Substitute the given values.*
$9x = 18 + 6x$	*Combine like terms.*
$-6x \quad -6x$	*Subtract 6x from both sides.*
$\frac{3x}{3} = \frac{18}{3}$	*Simplify.*
	Divide both sides by 3.
$x = 6$	*Simplify.*
$PQ = 9x = 9(6) = 54$	*Substitute 6 for x.*

The figure is given in this example. Begin by identifying the given information from the figure. The bar to the left of the segment extending from P to Q indicates that the corresponding expression, $9x$, represents PQ (read "PQ" as "the length of line segment PQ"), or $PQ = 9x$. To the right of the segment is 19 and $6x - 1$. The placement of 19 between P and R indicates that $PR = 19$, and the placement of $6x - 1$ between R and Q indicates that $RQ = 6x - 1$.

Furthermore, since P and Q are the endpoints of a segment containing R, R must be between P and Q. Therefore, by the Segment Addition Postulate, $PR + RQ = PQ$. Substitute the given lengths into this equation and solve for x.

Note that x is not equal to PQ, but that $PQ = 9x$. So, once the value of x is found, substitute that value into the expression $9x$ to find PQ.

Example 4 Travel Application

The distance from Dr. Burger's home to the math party is 900 m. The gas station is located at the midpoint between the math party and where Dr. Burger works. The distance from where Dr. Burger works to his home is 3 km. How far is the gas station from Dr. Burger's home?

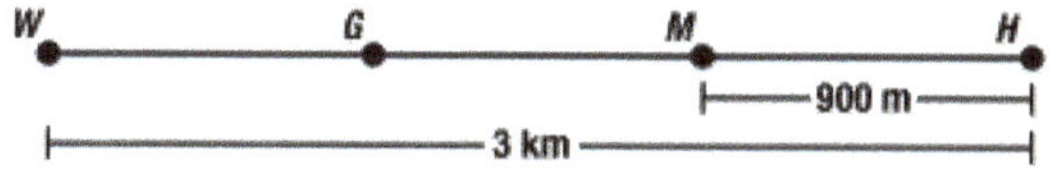

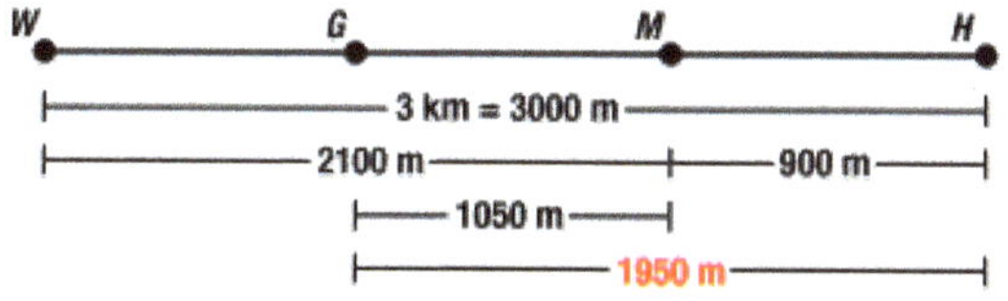

$WH = WM + MH$	*Segment Addition Postulate*
$3000 = WM + 900$	*Substitute 3000 for WH and 900 for MH.*
$-900 \quad -900$	*Subtract 900 from both sides.*
$2100 = WM$	*Simplify.*
$1050 = GM$	*G is the midpoint of $\overline{WM}$, so GM = 1/2 WM.*
$GH = GM + MH$	*Segment Addition Postulate*
$= 1050 + 900$	*Substitute 1050 for GM and 900 for MH.*
$= 1950$ m	

The given information is organized on a series of segments, where the outermost endpoints represent Prof. Burger's workplace, W, and his home, H. It is given that the distance from W to H is 3 kilometers and the distance from H to the party, M, is 900 meters.

Notice that units of these two distances are different: kilometers (km) and meters (m). Only one unit can be used in calculations, so convert either 3 km to meters by multiplying, or 900 m to kilometers by dividing. Because multiplication is usually easier than division, convert the kilometers to meters using the fact that there are 1000 meters in 1 kilometer.
$WH = 3$ km $= 3(1000$ m$) = 3000$ m.

Since M is between W and H in the segment, by the Segment Addition Postulate $WM + MH = WH$. Substitute the given lengths $MH = 900$ and $WH = 3000$ into the equation and solve to find WM.

Now use the fact that G is the midpoint of WM to find GM.
$GM = WM/2 = 2100/2 = 1050$ m

The distance from the gas station, G to H, is
$GH = GM + MH = 1050 + 900 = 1950$ m.

Example 5 Using Midpoints to Find Lengths

B is the midpoint of $\overline{AC}$. Find AB, BC, and AC.

Step 1 Write an equation and solve for x.

$AB = BC$	*B is the midpoint of $\overline{AC}$.*
$5x - 8 = 3x$	*Substitute 5x − 8 for AB and 3x for BC.*
$-3x \quad -3x$	*Subtract 3x from both sides.*
$2x - 8 = 0$	*Simplify.*
$+8 \quad +8$	*Add 8 to both sides.*
$2x = 8$	*Simplify.*
$\frac{2x}{2} = \frac{8}{2}$	*Divide both sides by 2.*
$x = 4$	*Simplify.*

Step 2 Use $x = 4$ to find AB.

$AB = 5x - 8$	*Given*
$= 5(4) - 8 = 12$	*Substitute 4 for x and simplify.*

Step 3 Use $x = 4$ to find BC.

$BC = 3x$	*Given*
$= 3(4) = 12$	*Substitute 4 for x and simplify.*

Step 4 Use $AB = 12$ and $BC = 12$ to find AC.

$AC = AB + BC$	*Segment Addition Postulate*
$= 12 + 12 = 24$	*Substitute 12 for AB and 12 for BC and simplify.*

It is given that B is the midpoint of the segment from A to C. Therefore, the segment is divided into two equal parts, where $AB = BC$.

Begin by using the fact that $AB = BC$ to write an equation using the expressions for AB and BC given in the figure. Solve this equation for x and then substitute the value of x into the given expressions to find AB and BC.

Once AB and BC are known, the Segment Addition Postulate can be used to find the length of AC.

Measuring and Constructing Angles

Key Objectives

- Name and classify angles.
- Measure and construct angles and angle bisectors.

Key Terms

- An **angle** is a figure formed by two rays, or sides, with a common endpoint called the **vertex**.
- The set of all points between the sides of the angle is the **interior of an angle**.
- The **exterior of an angle** is the set of all points outside the angle.
- The **measure** of an angle is usually given in degrees.
- Since there are 360° in a circle, one **degree** is 1/360 of a circle.
- An **acute angle** measures greater than 0° and less than 90°.
- A **right angle** measures 90°.
- An **obtuse angle** measures greater than 90° and less than 180°.
- A **straight angle** is formed by two opposite rays and measures 180°.
- **Congruent angles** are angles that have the same measure.
- An **angle bisector** is a ray that divides an angle into two congruent angles.

Theorems, Postulates, Corollaries, and Properties

- **Protractor Postulate** Given $\overrightarrow{AB}$ and a point O on $\overrightarrow{AB}$, all rays that can be drawn from O can be put into a one-to-one correspondence with the real numbers from 0 to 180.
- **Angle Addition Postulate** If S is in the interior of $\angle PQR$, then $m\angle PQS + m\angle SQR = m\angle PQR$.

Example 1 Naming Angles

An angle is a figure formed when two rays meet at a common endpoint. The rays are called the angle's sides, and the common endpoint is called the angle's vertex.

There are three ways to name an angle. If an angle's vertex is not a point on any other angle, then the angle can be named using only its vertex. For example, if A is the vertex of an angle and A is not a point on any other angle (i.e., A is not a vertex of another angle and A is not on the side of another angle), then the angle can be named $\angle A$, where $\angle$ is the symbol for angle.

The second method for naming an angle is with a number. For example, if 1 is used to label an angle in a figure, then the angle can be named $\angle 1$.

The third method for naming an angle is explained in the example below.

Name three angles in the diagram.

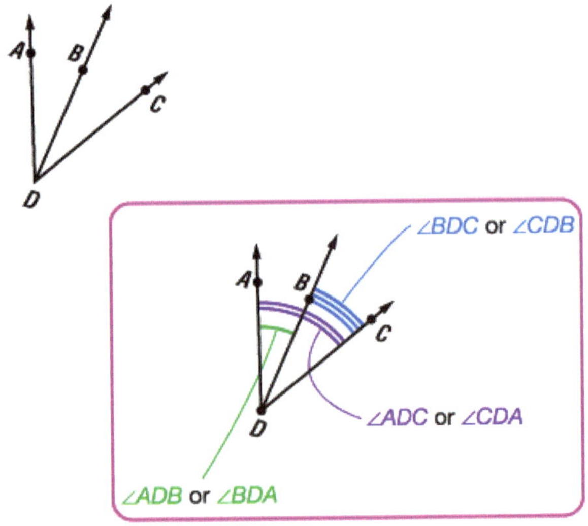

The third method for naming an angle is with its vertex and a point on each side, where the vertex is listed between the two points. This method is used when the angle's vertex is either shared with another angle or it is a point on the side of another angle.

D is the vertex of three angles in the given figure (one large angle and two smaller angles). Since D is the vertex of more than one angle, none of these three angles can be named simply $\angle D$. And since none of these angles is labeled with a number, the third method for naming angles must be used. So, each of the three angles must be named with a point from each side and the vertex, D.

To name the largest angle, first identify the sides of this angle. The sides are formed by rays DA and DC. So, A and C are each points on one of the angle's sides. And since the vertex is D, the largest angle can be named either $\angle ADC$ or $\angle CDA$.

Example 2 Measuring and Classifying Angles

An angle is measured by relating the angle to a circle that is divided into 360 parts, shaped like pie pieces. Each one of these 360 pieces is called a degree. Imagine that an angle's vertex is the center of a circle and that the circle itself intersects each side of the angle. The number of pie pieces, or degrees, that would be in the interior of the angle (i.e., between the two sides) is equal to the measure of the angle.

Protractor Postulate

Given $\overleftrightarrow{AB}$ and a point O on $\overleftrightarrow{AB}$, all rays that can be drawn from O can be put into a one-to-one correspondence with the real numbers from 0 to 180. (Prot. Post.)

By the Protractor Postulate, a protractor can be used to find the measure of an angle that is between 0° and 180°.

Classifications of Angles by their Measures

A **straight angle** measures 180° and is formed by two opposite rays.

An **obtuse angle** measures greater than 90° and less than 180°.

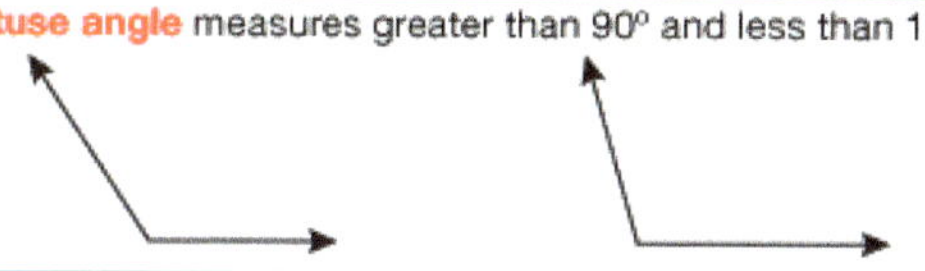

A **right angle** measures exactly 90°.

An **acute angle** measures greater than 0° and less than 90°.

Angles can be classified by their measures. There are four types of angles with measures between 0° and 180°: straight, obtuse, right, and acute angles.

A straight angle is also a line; its measure is 180°.

An obtuse angle measures between 90° and 180°.

The measure of a right angle is exactly 90°. A small box placed in an angle's interior, near its vertex, is used to indicate that the angle is a right angle.

All angles with a measure that is less than that of a right angle (so, less than 90°) are called acute angles.

Find the measure of $\angle BOD$. Then classify it as acute, right, or obtuse.

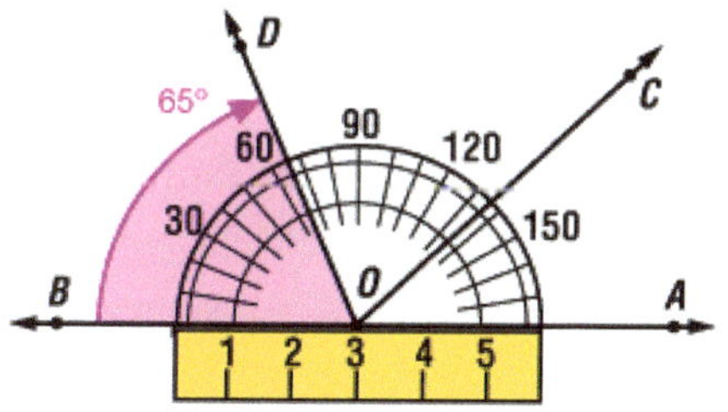

$m\angle BOD = 65°$

$\angle BOD$ is acute since its measure is less than 90°.

In this example, a protractor is placed over a given figure and then used to find $m\angle BOD$ (read as "the measure of angle BOD"). The vertex of $\angle BOD$ is O. So, the center of the protractor is placed on O.

To use the protractor to find $m\angle BOD$, place it so that 0° is on one of the angle's sides, ray OB. Then identify the degree measure on the protractor that corresponds with placement of the angle's other side, ray OD. Since OD passes through the protractor at 65°, $m\angle BOD = 65°$. The measure of $\angle BOD$ is less than 90°. Thus, $\angle BOD$ is an acute angle.

Angle Addition Postulate

If S is in the interior of $\angle PQR$, then $m\angle PQS + m\angle SQR = m\angle PQR$. ($\angle$ Add. Post.)

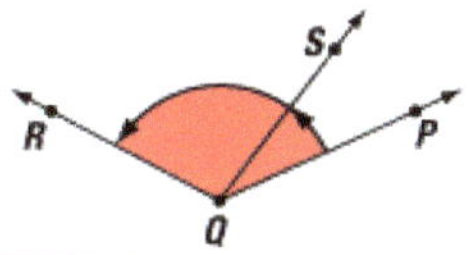

The Segment Addition Postulate states that when a segment is divided into two parts, the sum of the lengths of those two parts is equal to the length of the whole segment. The Angle Addition Postulate relates the parts of an angle to each other in the same way that the Segment Addition Postulate relates the parts of a segment to each other. By the Angle Addition Postulate, when an angle is divided into two parts, the sum of the measures of those two parts is equal to the sum of the whole angle.

Find the measure of $\angle COD$. Then classify it as acute, right, or obtuse.

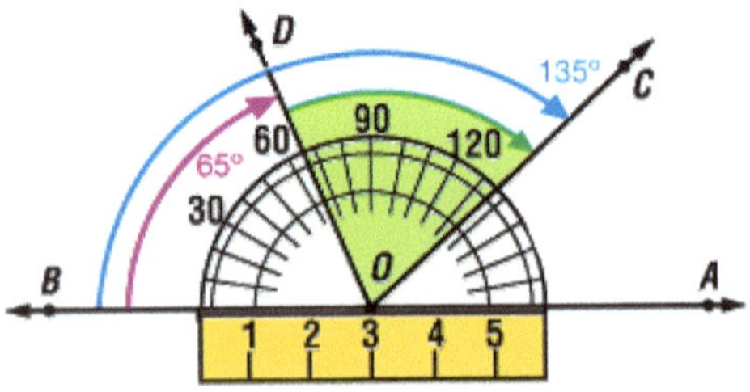

$m\angle BOD = 65°$ $m\angle COB = 135°$

$m\angle COB = m\angle COD + m\angle DOB$
$135° = m\angle COD + 65°$
$-65° \qquad -65°$ *Subtract 65° from both sides.*
$70° = m\angle COD$

$\angle COD$ is acute since its measure is less than 90°.

In the given figure, $\angle COD$ is one part of the larger angle $\angle COB$, where the second part is $\angle DOB$. So, by the Angle Addition Postulate, $m\angle COB = m\angle COD + m\angle DOB$.

Use the Angle Addition Postulate to find $m\angle COD$. From the previous example we know that $\angle DOB = 65°$. (Remember, $\angle DOB$ and $\angle BOD$ are the same angle.) Use the protractor to find $m\angle COB$: $m\angle COB = 135°$. Substitute the two degree measures into the equation and solve for $m\angle COD$. Since $135° - 65° = 70°$, $m\angle COD = 70°$. This measure is less than 90°, so the angle is acute.

Example 3 Using the Angle Addition Postulate

Given $m\angle PMR = 65°$ and $m\angle QMR = 38°$, find $m\angle PMQ$.

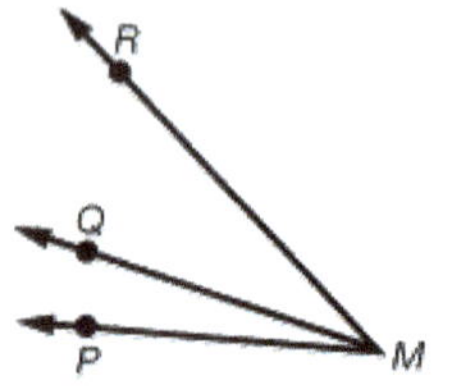

$m\angle PMR = m\angle QMR + m\angle PMQ$ *∠ Addition Postulate*
$65° = 38° + m\angle PMQ$ *Substitute the given values.*
$-38° \quad -38°$ *Subtract 38° from both sides.*
$27° = m\angle PMQ$

The measure of the large angle, $\angle PMR$, is given.
$m\angle PMR = 65°$
And, the measure of one of the two angles that make up $\angle QMR$ is also given.
$m\angle QMR = 38°$
Since $\angle QMR$ and $\angle PMQ$ make up $\angle PMR$, use the Angle Addition Postulate to find $m\angle PMQ$.

Example 4 Finding the Measure of an Angle

Two angles that have equal measures are **congruent angles**.

$m\angle 1 = m\angle 2$

$\angle 1 \cong \angle 2$

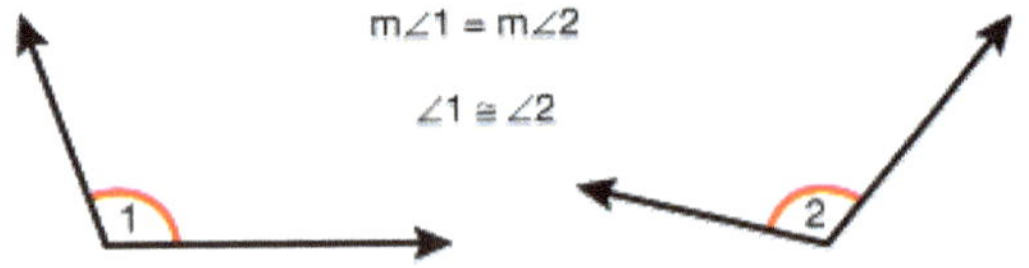

Arc marks indicate that angles are congruent.

An **angle bisector** is a ray that divides an angle into two congruent angles.

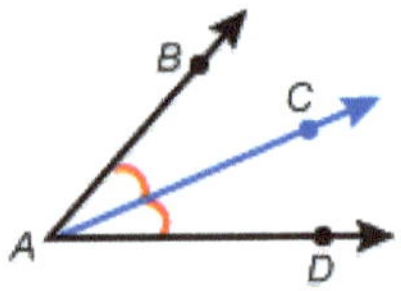

$\overrightarrow{AC}$ bisects $\angle BAD \Rightarrow$
$\angle BAC \cong \angle CAD$

When angles have the same measure, they are called congruent angles. For example, if the measure of $\angle 1$ is 100° and the measure of $\angle 2$ is 100°, then the angle *measures* are *equal*.
$m\angle 1 = m\angle 2$
But the *angles themselves* are *congruent*.
$\angle 1 \cong \angle 2$

When a ray is an angle bisector, it divides an angle into two congruent angles. For example, if $\overrightarrow{AC}$ bisects $\angle BAD$, then the two angles created by this ray are congruent.
$\angle BAC \cong \angle CAD$

$\overrightarrow{ZY}$ bisects $\angle XZW$, $m\angle XZY = (7x + 1)°$ and $m\angle YZW = (5x + 11)°$. Find $m\angle XZW$.

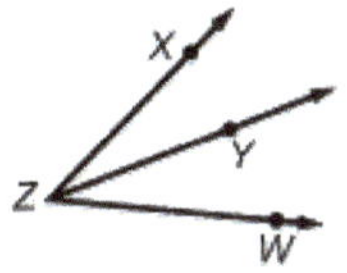

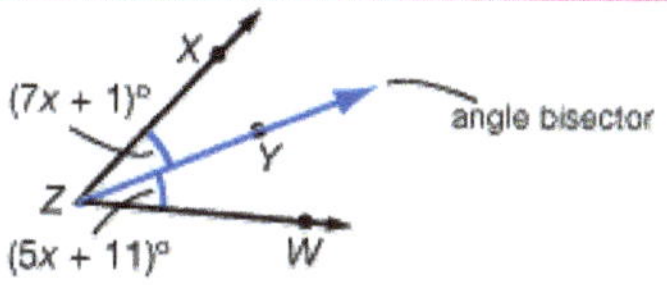

$\overrightarrow{ZY}$ bisects $\angle XZW$.	*Given*
$\angle XZY \cong \angle YZW$	*Def. of ∠ Bisector*
$m\angle XZY = m\angle YZW$	*Def. of ≅ ∠s*
$7x + 1 = 5x + 11$	*Substitute the given expressions.*
$-5x \quad -5x$	*Subtract 5x from both sides.*
$2x + 1 = 11$	*Simplify.*
$-1 \quad -1$	*Subtract 1 from both sides.*
$2x = 10$	*Simplify.*
$\frac{2x}{2} = \frac{10}{2}$	*Divide both sides by 2.*
$x = 5$	*Simplify.*

$m\angle XZY = (7x + 1)°$	*Given*
$= 7(5) + 1 = 36°$	*Substitute 5 for x.*

$m\angle XZW = 36° + 36° = 72°$

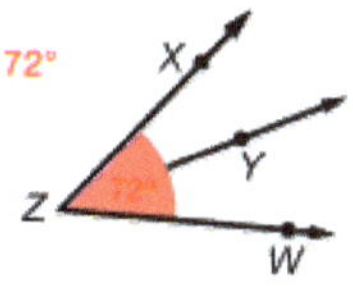

In this example, the fact that $\angle XZW$ is bisected by a ray is given. Therefore, by the definition of angle bisector, $\angle XZW$ is divided into two congruent angles.

$\angle XZY \cong \angle YZW$

So by the definition of congruent angles, $m\angle XZY = m\angle YZW$.

Algebraic expressions are given for the measures of $\angle XZY$ and $\angle YZW$. Substitute these expressions into the equation and solve for x.

$m\angle XZY = m\angle YZW$

$(7x + 1)° = (5x + 11)°$

Thus, $x = 5$.

Note that the value of x is not the answer to the question. The value of x must be used to find $m\angle XZW$.

First, substitute 5 for x into either expression given for the small angles. Then, double that result to get $m\angle XZW$.

Pairs of Angles

Key Objectives

- Identify adjacent, vertical, complementary, and supplementary angles.
- Find measures of pairs of angles.

Key Terms

- **Adjacent angles** are two angles in the same plane with a common vertex and a common side, but no common interior points.
- A **linear pair** of angles is a pair of adjacent angles whose noncommon sides are opposite rays.
- **Complementary angles** are two angles whose measures have a sum of 90°.
- **Supplementary angles** are two angles whose measures have a sum of 180°.
- **Vertical angles** are two nonadjacent angles formed by two intersecting lines.

Example 1 Identifying Angle Pairs

Adjacent angles are two angles in the same plane with a common vertex and a common side, but no common interior points. 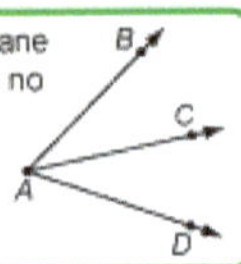 ∠*BAC* and ∠*CAD* are adjacent angles. A **linear pair** of angles is a pair of adjacent angles whose noncommon sides are opposite rays. 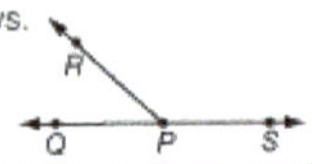 ∠*QPR* and ∠*RPS* form a linear pair.	Two angles that have a common vertex and side, but do not overlap, are called adjacent angles. The figure to the left contains a pair of adjacent angles, ∠*BAC* and ∠*CAD*. Notice that ∠*BAD* and ∠*BAC* share a common vertex, point *A*, and a common side, ray *AB*, but those two angles are not adjacent angles because they overlap. A linear pair is a pair of adjacent angles that combine to form a straight angle. In a linear pair, the noncommon sides of the two adjacent angles are opposite rays.
Tell whether the angles are only adjacent, adjacent and form a linear pair, or not adjacent. ∠1 and ∠2 ∠4 and ∠5 ∠1 and ∠3 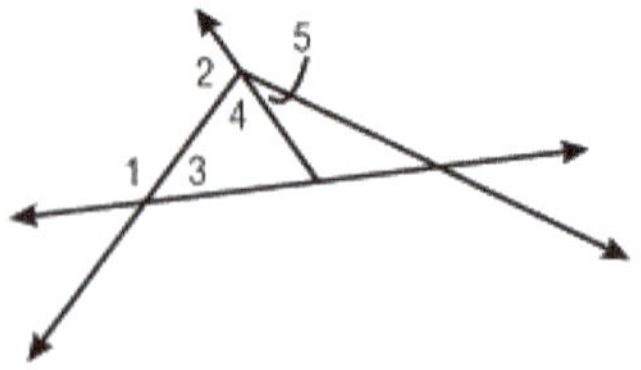 ∠1 and ∠2 share a side, but do not have a common vertex. ∠1 and ∠2 are not adjacent. ∠4 and ∠5 have a common vertex and a common side, and no common interior points. ∠4 and ∠5 are adjacent. Since their noncommon sides are not opposite rays, ∠4 and ∠5 are not a linear pair. ∠4 and ∠5 are adjacent only. ∠1 and ∠3 have a common vertex and a common side, and no common interior points. ∠1 and ∠3 are adjacent. Their noncommon sides are opposite rays. ∠1 and ∠3 form a linear pair.	Five angles are named with numbers in this figure. For each given pair of angles, determine the type of pair by first considering whether they are adjacent. If they are adjacent, then consider whether they are a linear pair. ∠1 and ∠2 do not have a common vertex, so they are not adjacent. ∠4 and ∠5 have a common vertex and a common side. Furthermore, ∠4 and ∠5 do not overlap (i.e., they do not have any common interior points), so they must be adjacent. However, they are not a linear pair because their noncommon sides are not opposite rays. Therefore, ∠4 and ∠5 are adjacent only. ∠1 and ∠3 have a common vertex, a common side, and they do not overlap. So, they are adjacent. The noncommon sides of ∠1 and ∠3 are opposite rays. Therefore, ∠1 and ∠3 form a linear pair.

Example 2 Finding the Measure of Complements and Supplements

The relationship between adjacent angles is based on the position of the angles. There are also special relationships between angles based on their measures. Complementary and supplementary angles are pairs of angles that are related to each other by their measures. Complementary angles are two angles whose measures have a sum of 90°. So, the complement of an angle can be found by subtracting its measure from 90°. Supplementary angles are two angles whose measures have a sum of 180°. So, the supplement of an angle can be found by subtracting its measure from 180°.

Find the measure of the complement of ∠*M*.

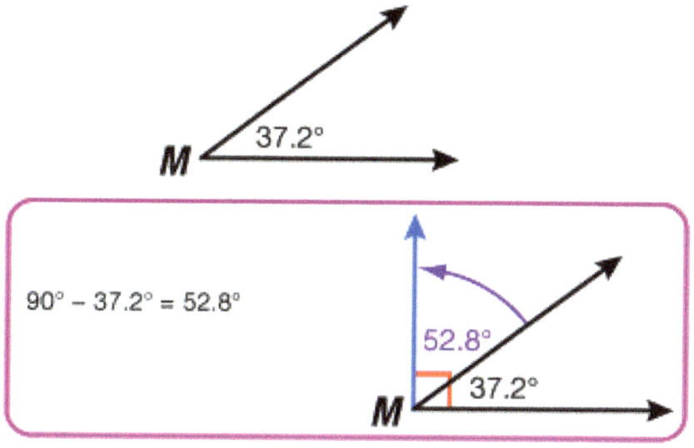

The measure of $\angle M$ is given in the figure.
$m\angle M = 37.2°$

The complement of an angle is found by subtracting the measure of the angle from 90°. So, the measure of the complement of $\angle M$ is $90° - 37.2°$, or 52.8°.

You can check your work by adding the angles. If the sum of two angles is equal to 90°, then the angles are complementary.
$37.2° + 52.8° = 90°$
So, the angles are complementary and 52.8° is the complement of 37.2°.

Find the measure of the supplement of ∠*N*.

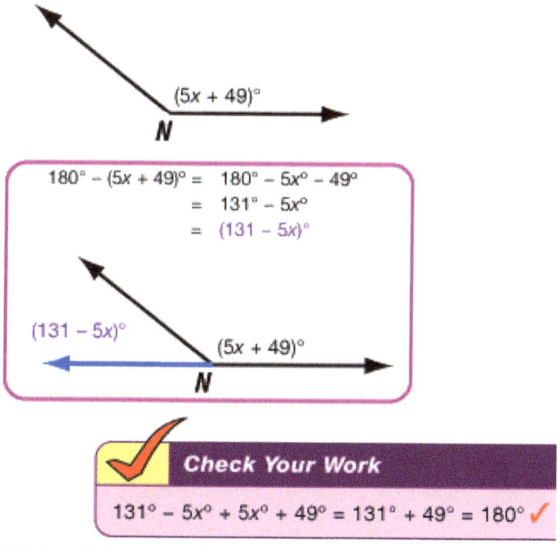

An expression for the measure of $\angle N$ is given in the figure.

$m\angle N = (5x + 49)°$

The supplement of an angle is found by subtracting the measure of the angle from 180°. So, the measure of the supplement of $\angle N$ is $180° - (5x + 49)°$, or $(131 - 5x)°$.

Add the expressions for the two angles to confirm that they are supplementary angles. The sum is 180°, so the angles are supplementary.

Example 3 Applying Complements and Supplements

If $x°$ is the measure of an angle, then the measure of that angle's complement can be represented by the expression $(90 - x)°$. Similarly, the angle's supplement can be represented by the expression $(180 - x)°$.

Twice an angle measures 9° more than the measure of its complement. Find the measure of its complement.

Let x be the unknown angle.

$2x$	=	9 +	$(90 - x)$
Twice x	is	9° more than	the measure of its complement.

$2x = 9 + (90 - x)$

$2x = 99 - x$ *Simplify.*

$+x \quad +x$ *Add x to both sides.*

$3x = 99$ *Simplify.*

$\frac{3x}{3} = \frac{99}{3}$ *Divide both sides by 3.*

$x = 33°$ *Simplify.*

$90° - 33° = 57°$

The measure of the complement of the angle is 57°.

Translate this sentence into an equation to find the measure of the complement.

Dividing the sentence into parts can make the translation easier. First, identify a variable for the unknown. If the unknown is represented by x, then "twice an angle" can be translated to "$2x$," because "twice" a number means to multiply the number by 2. The next word in the sentence, "measures," can be translated to "=" because if an angle "measures" some number of degrees, that means the angle's measure is *equal* to some number of degrees. The next part of the sentence is "9° more than." The phrase "9 more than" some number is equivalent to *9 plus* some number. So, "9° more than" can be translated to "9° + ." The last part of the sentence, "the measure of its complement," can be translated to $(90 - x)°$.

Put these parts together to form an equation and then solve the equation for x.

Remember, x is the measure of the unknown angle, but the answer to this example is the "measure of its complement," or the measure of the complement of x. So, subtract the value of x, 33°, from 90° to find the answer.

Example 4 Science Application

Light passes through a tunnel and reflects off the walls in such a way that ∠2 ≅ ∠3, ∠1 and ∠2 are complementary, and ∠3 and ∠4 are complementary. If m∠4 = 52°, find m∠1, m∠2, and m∠3.

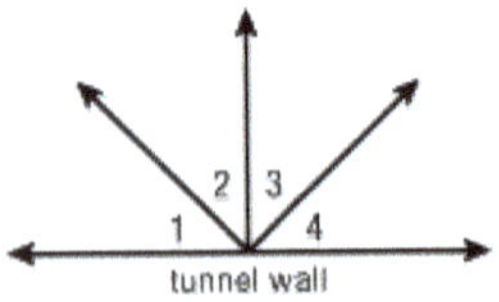

List the important information.

- ∠2 ≅ ∠3
- ∠1 and ∠2 are complementary.
- ∠3 and ∠4 are complementary.
- m∠4 = 52°

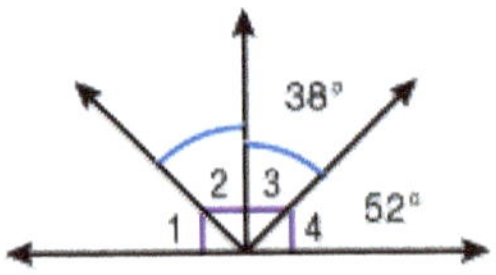

Make a plan and solve.

If ∠3 and ∠4 are complementary, then m∠3 = 90° − 52° = 38°.

If ∠2 ≅ ∠3, then m∠2 = m∠3 = 38°.

If ∠1 and ∠2 are complementary, then m∠1 = 90° − 38° = 52°.

m∠1 = 52°

m∠2 = 38°

m∠3 = 38°

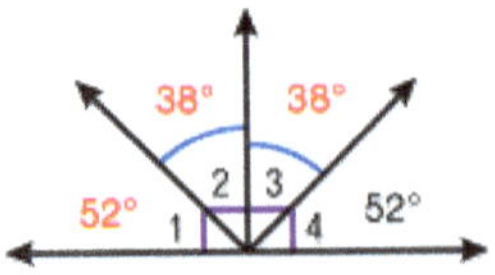

Four angles are named in this figure. Angles may appear to be congruent in a figure, but it cannot be assumed that angles (or any figures) are congruent unless that fact is given information. Begin by listing the given information. Then note the given information in the figure. Since ∠2 and ∠3 are congruent, each of those angles can be marked in the figure with a single curve to denote their congruency. It is also given that there are two pairs of complementary angles. ∠1 and ∠2 are complementary and ∠3 and ∠4 are complementary. Remember, the sum of two complementary angles is 90° and 90° is the measure of a right angle. So, in the figure, the angle composed of ∠1 and ∠2 can be marked as a right angle, as can the angle composed of ∠3 and ∠4. The last piece of given information is the measure of an angle.

m∠4 = 52°

So, in the figure ∠4 can be labeled with 52°. Now use all of this information to find m∠1, m∠2, and m∠3. Since ∠3 and ∠4 are complementary and m∠4 = 52°, m∠3 = 90° – 52° = 38°. Now that m∠3 is known, m∠2 can easily be found since ∠3 and ∠2 are congruent. So, m∠2 = 38°. Now m∠1 can be found by using the fact that ∠1 and ∠2 are complementary. So, m∠1 = 90° – 38° = 52°.

Example 5 Identifying Vertical Angles

Vertical angles are nonadjacent angles formed when two lines intersect. Vertical angles have a common vertex, but no common sides.

∠*BAC* and ∠*DAE* are vertical angles.
And, ∠*CAD* and ∠*EAB* are vertical angles.

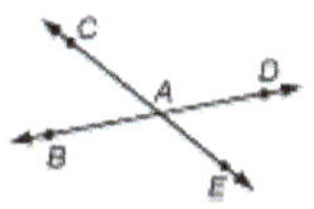

When two lines intersect, four angles are formed where the point of intersection is the vertex for all four angles. The angles that are opposite from one another are called vertical angles. Vertical angles have a common vertex, but no common sides.

Name four pairs of vertical angles in the given diagram.

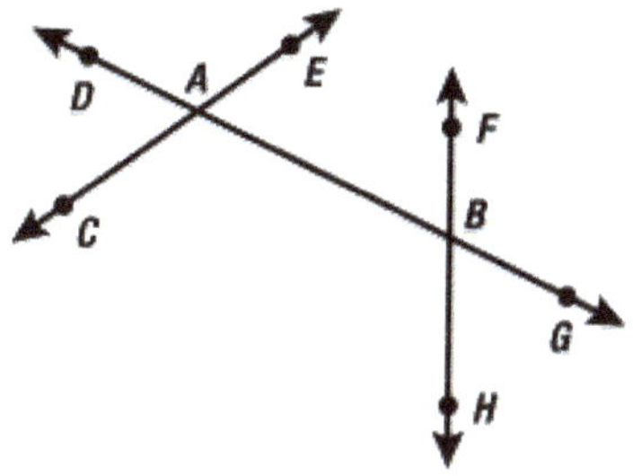

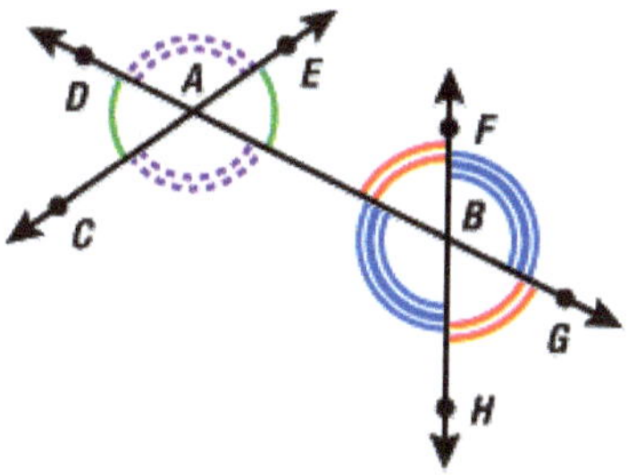

In this figure there are three lines. Two of these lines intersect the third line, creating two points of intersection. So, there are four pairs of vertical angles in this figure.

Using Formulas in Geometry

Key Objectives

- Apply formulas for perimeter, area, and circumference.

Key Terms

- The **perimeter** P of a plane figure is the sum of the side lengths of the figure.
- The **area** A of a plane figure is the number of nonoverlapping square units of a given size that exactly cover the figure.
- The **base** b of a triangle can be any side of the triangle.
- The **height** h of a triangle is a segment from a vertex that forms a right angle with a line containing the base of the triangle.
- In a circle, a **diameter** is a segment that passes through the center of the circle and whose endpoints are on the circle.
- A **radius** of a circle is a segment whose endpoints are the center and a point on the circle.
- The **circumference** of a circle is the distance around a circle.
- **Pi** (π) is the ratio of a circle's circumference to its diameter.

Formulas

- **Perimeter of a Rectangle** $P = 2l + 2h$, where l is the length and h is the height of the rectangle.
- **Area of a Rectangle** $A = lh$, where l is the length and h is the height of the rectangle.
- **Perimeter of a Triangle** $P = a + b + c$, where a, b, and c are the lengths of the sides of the triangle.
- **Area of a Triangle** $A = 1/2\ bh$, where b is the length of the base and h is the height of the triangle.
- **Area of a Square** $A = s^2$, where s is the length of one side of the square.
- **Circumference of a Circle** $C = 2\pi r$, where r is the radius of the circle.
- **Area of a Circle** $A = \pi r^2$, where r is the radius of the circle.

Example 1 Finding Perimeter and Area

The perimeter P of a plane figure is the sum of the side lengths of the figure. So, if a figure has five sides, then add the lengths of those five sides to find the figure's perimeter. Multiplication can be used to find the perimeter of a figure when that figure has sides of equal length. For example, if the sides of a five-sided figure all have equal length s, then the perimeter of that figure is equal to $s + s + s + s + s$, or $5s$.

The area A of a plane figure is the number of nonoverlapping square units of a given size that exactly cover the figure. Area can be visualized as the number of squares on a checkerboard. If a checkerboard has five squares across the top and nine squares on the side, then there are 5(9) or 45 squares on the board and 45 squares is the area of the checkerboard. In this lesson, the area of a rectangle, square, triangle, and circle will be found using given formulas.

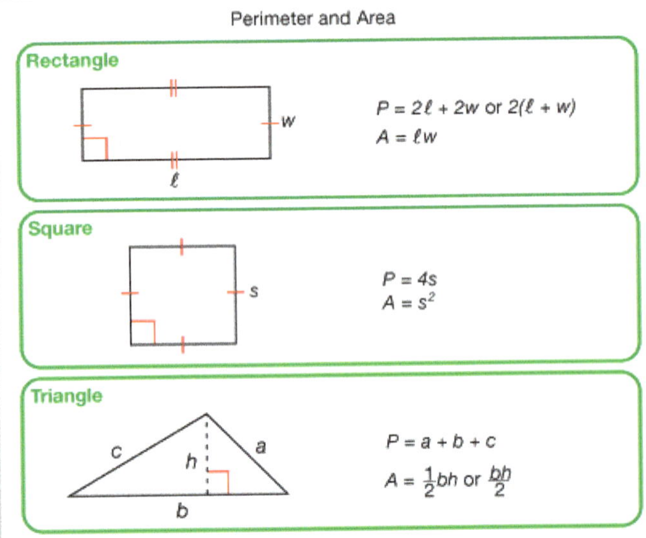

A rectangle has two pairs of equal sides. So, if the length and width of a rectangle are represented by l and w respectively, then the rectangle has two sides of length l and two sides of length w. So, the perimeter of the rectangle is $l + l + w + w$. Or, multiplication and addition can be used to find the perimeter of the rectangle.

$P = l + l + w + w = 2l + 2w$

The area of a rectangle can be found by multiplying the length by the width.

The perimeter of a square can also be found by using multiplication because all four sides of a square are always equal. The area of a square can be found by multiplying the length by the width. However, the length and width of a square are always equal and so the two dimensions can be represented by the same variable.

$A = (s)(s) = s^2$

The perimeter of a triangle is found by adding the lengths of the three sides. The area of a triangle is found by taking one half of the product of the base and height.

Find the perimeter and area of the figure.

rectangle with length 3 cm and height 11 cm

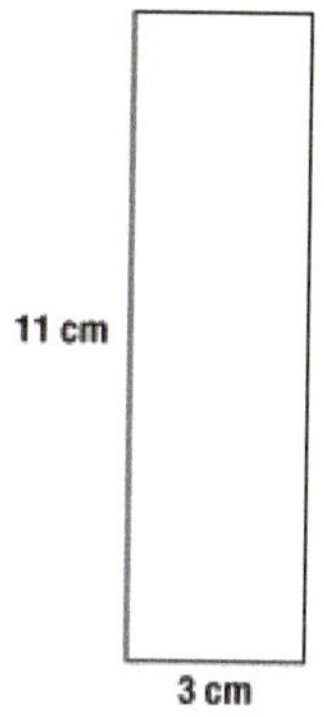

$P = 2\ell + 2h$ — *Use the formula for perimeter of a rectangle.*

$= 2(3) + 2(11)$ — *Substitute 3 for ℓ and 11 for h.*

$= 6 + 22$ — *Simplify.*

$= 28$ cm

$A = \ell h$ — *Use the formula for area of a rectangle.*

$= (3)(11)$ — *Substitute 3 for ℓ and 11 for h.*

$= 33\text{ cm}^2$ — *Simplify.*

The figure in this example is a rectangle with length 3 cm and height 11 cm. Note that the width of a rectangle can also be referred to as the rectangle's height and the length of a rectangle can also be referred to as the rectangle's base.

Substitute the given dimensions, 3 cm and 11 cm, into the formula for the perimeter of a rectangle and simplify to find the perimeter of the rectangle.

To find the area, multiply the given dimensions. Since area is measured in square units, the units for the area of this rectangle are cm^2.

Find the perimeter and area of the figure.

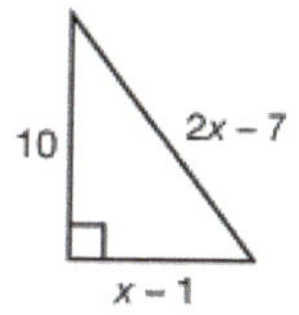

$P = a + b + c$ — *Use the formula for perimeter of a triangle.*

$= 10 + x - 1 + 2x - 7$ — *Substitute 10 for a, x – 1 for b, and 2x – 7 for c.*

$= 2 + 3x$ — *Combine like terms.*

$A = \frac{1}{2}bh$ — *Use the formula for area of a triangle.*

$= \frac{1}{2}(x - 1)10$ — *Substitute x – 1 for b and 10 for h.*

$= 5(x - 1)$ — *Multiply $\frac{1}{2}$ and 10.*

$= 5x - 5$ — *Distribute 5.*

The lengths of the sides of this triangle are 10, $x - 1$, and $2x - 7$. To find the perimeter, simply add these three lengths and simplify.

The formula for the area of a triangle is $A = 1/2\ bh$, where b is the length of the base and h is the height of the triangle. The base b of a triangle can be any side of the triangle. The height h of a triangle is a segment from a vertex that forms a right angle with a line containing the base of the triangle.

In the triangle given here, the base is represented by the expression $x - 1$ and the height is 10 units. So, to find the area of the triangle, multiply 1/2 by $(x - 1)$ and $(2x - 7)$, and then simplify.

Example 2 Crafts Application

The Texas Treasures quilt block includes 8 white squares. A side of each square is about 3 inches. Find the approximate amount of fabric used to make the 8 squares.

Area of one square

$A = s^2$
$= 3^2$
$= 9\text{ in}^2$

Total area of 8 white squares

$8(9\text{ in}^2) = 72\text{ in}^2$

To find the approximate amount of fabric used to make the 8 squares in the quilt, first find the amount of fabric needed for 1 square, which is equivalent to the area of 1 square.

Once you have the area of 1 square, multiply that area by 8 to find the area of 8 squares. The total area of the 8 squares is the amount of fabric needed to make the 8 squares on the quilt.

Example 3 Finding the Circumference and Area of a Circle

The circumference of a circle is the distance around a circle. The circumference of a circle is similar to the perimeter of a figure since both are the measure of the total distance around the figure. A circle's circumference depends on the length of its radius (a segment whose endpoints are the center and a point on the circle) or on the length of its diameter (a segment that passes through the center of the circle and whose endpoints are on the circle). The length of a circle's diameter is always twice the length of its radius, $d = 2r$.

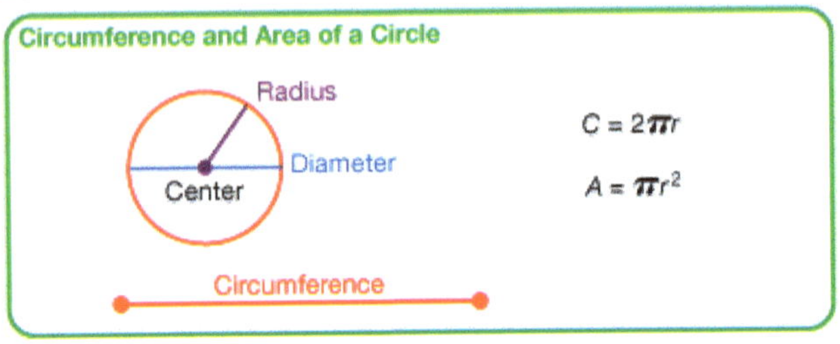

The formula for the circumference of a circle is $C = 2\pi r$, where r is the radius of the circle. Alternatively, since the diameter d of a circle is equal to twice the radius ($d = 2r$), the circumference of a circle can also be found using the formula $C = \pi d$.

The formula for the area of a circle is $A = \pi r^2$. Pi (π) is the ratio of a circle's circumference to its diameter. The value of π is approximated as 3.14, unless otherwise stated.

Find the circumference and area of the circle.

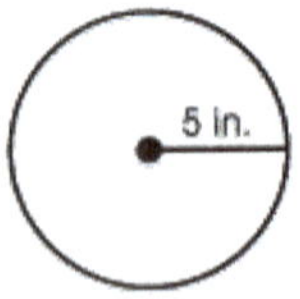

$C = 2\pi r$	*Use the formula for the circumference of a circle.*
$= 2\pi(5)$	*Substitute 5 for r.*
$= 10\pi$	*Multiply 2 and 5.*
$\approx 10(3.14)$	*Use 3.14 for π.*
≈ 31.4 in.	
$A = \pi r^2$	*Use the formula for the area of a circle.*
$= \pi(5)^2$	*Substitute 5 for r.*
$= \pi 25$	*Simplify.*
$\approx 25(3.14)$	*Use 3.14 for π.*
$\approx 78.5\ \text{in}^2$	

In this example, the circumference and area are found for a circle with radius 5 inches.

To find the circumference and the area, simply substitute 5 into each formula for r and simplify.

When the circumference is found, the first step after substituting 5 is to multiply 2 and 5, and the result is 10π. This answer is in terms of π. To find an approximate decimal value of the circle's circumference, substitute the approximate value of π, 3.14, into the expression 10π and multiply.

If the directions had said to find the circle's area in terms of π, the answer would have been $A = 25\pi$. As with the calculation of circumference, to find the approximate decimal value of the circle's area, substitute 3.14 into the expression 25π and multiply.

Midpoint and Distance in the Coordinate Plane

Key Objectives

- Develop and apply the formula for midpoint.
- Use the Distance Formula and the Pythagorean Theorem to find the distance between two points.

Key Terms

- A **coordinate plane** is a plane that is divided into four regions by a horizontal line (x-axis) and a vertical line (y-axis).
- In a right triangle, the two sides that form the right angle are the **legs**.
- In a right triangle, the side across from the right angle that stretches from one leg to the other is the **hypotenuse**.

Formulas

- **Midpoint Formula** $M\left(\frac{x_1 + x_2}{2}, \frac{y_1 + y_2}{2}\right)$, where (x_1, y_1) and (x_2, y_2) are two points
- **Distance Formula** $d = \sqrt{(x_2 - x_1)^2 + (y_2 - y_1)^2}$, where (x_1, y_1) and (x_2, y_2) are two points

Theorems, Postulates, Corollaries, and Properties

- **Pythagorean Theorem** In a right triangle, the sum of the squares of the lengths of the legs is equal to the square of the length of the hypotenuse.

Example 1 Finding the Coordinates of a Midpoint

The Midpoint Formula

The midpoint M of $\overline{AB}$ with endpoints $A(x_1, y_1)$ and $B(x_2, y_2)$ is found by

$$M\left(\frac{x_1 + x_2}{2}, \frac{y_1 + y_2}{2}\right)$$

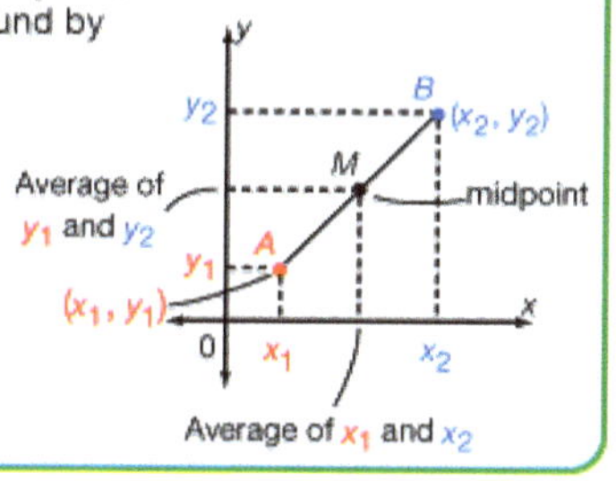

Given two points (x_1, y_1) and (x_2, y_2), the x-coordinate of the midpoint between those two points is the average of the x-coordinates, $(x_1 + x_2)/2$. The y-coordinate of the midpoint between those two points is the average of the y-coordinates, $(y_1 + y_2)/2$.

Find the coordinates of the midpoint of $\overline{CD}$ with endpoints $C(-3, 4)$ and $D(1, -2)$.

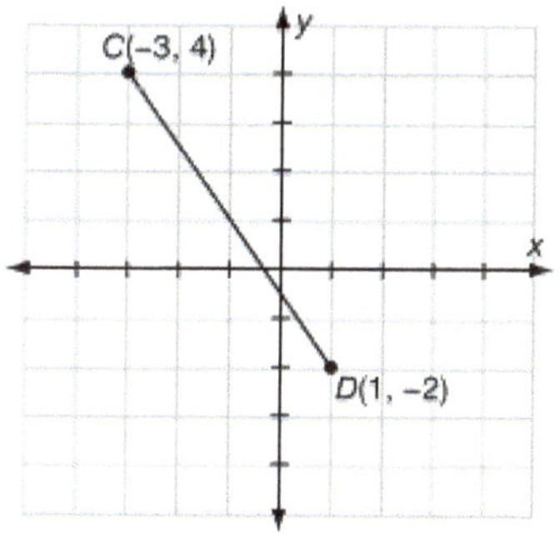

In this example, a segment is given with endpoints $C(-3, 4)$ and $D(1, -2)$. Use the midpoint formula to find the midpoint of this segment.

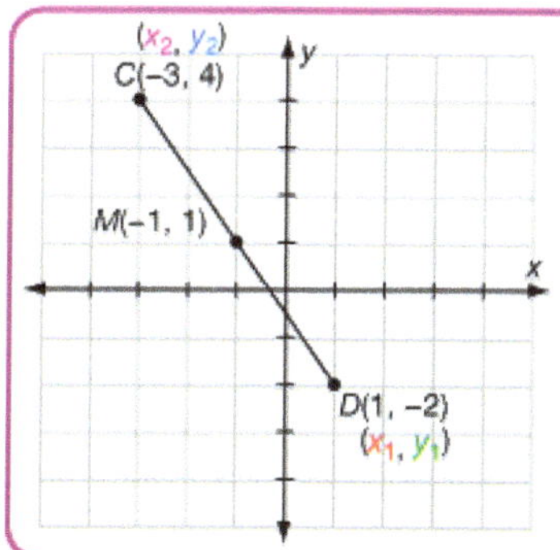

Substitute the coordinates into the midpoint formula and simplify.

$$M\left(\frac{x_1 + x_2}{2}, \frac{y_1 + y_2}{2}\right)$$

$$M\left(\frac{1 + -3}{2}, \frac{-2 + 4}{2}\right)$$

$$= \left(\frac{-2}{2}, \frac{2}{2}\right)$$

$$= (-1, 1)$$

First identify the values of x_1, x_2, y_1, and y_2. It doesn't matter which point, C or D, is picked to be (x_1, y_1). Here, $D(1, -2)$ is (x_1, y_1) and $C(-3, 4)$ is (x_2, y_2). So $x_1 = 1$, $x_2 = -3$, $y_1 = -2$, and $y_2 = 4$. Substitute these values into the midpoint formula and simplify to find the x- and y-coordinates of the midpoint.

Example 2 Finding the Coordinates of an Endpoint

M is the midpoint of $\overline{AB}$. A has coordinates (3, 5), and M has coordinates (4, −3). Find the coordinates of B.

Given: $A(3, 5)$ $M(4, -3)$

Step 1 Let the coordinates of B equal (x, y).

Step 2 Use the Midpoint Formula: $\left(\frac{3+x}{2}, \frac{5+y}{2}\right) = (4, -3)$.

Step 3 Find the x-coordinate.

$\frac{3+x}{2} = 4$ *Set the coordinates equal.*

$2\left(\frac{3+x}{2}\right) = 2(4)$ *Multiply both sides by 2.*

$3 + x = 8$ *Simplify.*

$-3 \quad -3$ *Subtract 3 from both sides.*

$x = 5$

Find the y-coordinate.

$\frac{5+y}{2} = -3$ *Set the coordinates equal.*

$2\left(\frac{5+y}{2}\right) = 2(-3)$ *Multiply both sides by 2.*

$5 + y = -6$ *Simplify.*

$-5 \quad -5$ *Subtract 5 from both sides.*

$y = -11$

$B(5, -11)$

The midpoint formula can be used to find one of the endpoints of a segment when the other endpoint and the midpoint of the segment are known.

In this example, the endpoints are A and B, and the midpoint is M. The coordinates of A and M are known, and the coordinates of B are unknown. So, let the coordinates of B be (x, y). Now, substitute the coordinates into the midpoint formula where $(x_1, y_1) = (3, 5)$ and $(x_2, y_2) = (x, y)$. The first coordinate from the formula is equal to the x-coordinate from the midpoint. So, set the first coordinate from the formula equal to 4. Then solve the equation for x, which is the x-coordinate of B. Use a similar process to find the y-coordinate of B.

Example 3 Using the Distance Formula

Distance Formula

In a coordinate plane, the distance d between two points (x_1, y_1) and (x_2, y_2) is

$$d = \sqrt{(x_2 - x_1)^2 + (y_2 - y_1)^2}.$$

If two points, (x_1, y_1) and (x_2, y_2), lie on a coordinate plane, the Distance Formula can be used to find the distance between the two points.

Find AB and CD. Then determine if $\overline{AB} < \overline{CD}$.

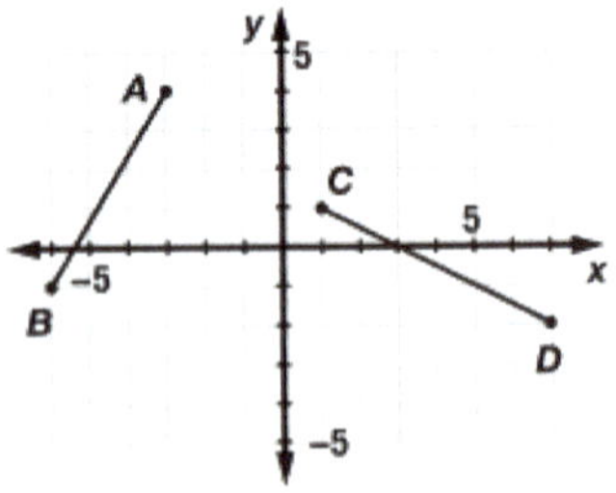

Step 1 Find the coordinates of each endpoint.

$A(-3, 4)$ $B(-6, -1)$ $C(1, 1)$ $D(7, -2)$

Step 2 Use the distance formula $d = \sqrt{(x_2 - x_1)^2 + (y_2 - y_1)^2}$ to find the length of each line segment.

$AB = \sqrt{(-3 - (-6))^2 + (4 - (-1))^2}$
$= \sqrt{(3)^2 + (5)^2}$
$= \sqrt{9 + 25}$
$= \sqrt{34}$

$CD = \sqrt{(7 - 1)^2 + (-2 - 1)^2}$
$= \sqrt{(6)^2 + (-3)^2}$
$= \sqrt{36 + 9}$
$= \sqrt{45}$

Step 3 Determine if $\overline{AB} < \overline{CD}$.

Since $\sqrt{34} < \sqrt{45}$, the inequality $\overline{AB} < \overline{CD}$ is true.

In this example, the lengths of two line segments are found and then compared. Since these two segments lie in a coordinate plane, use the Distance Formula to find the distance between the endpoints, which is equivalent to the length of the segment.

First identify the coordinates of each endpoint, then substitute these coordinates into the Distance Formula. As with the Midpoint Formula, it doesn't matter which of the two points is identified as (x_1, y_1). However, do be consistent. For example, if x_1 is the x-coordinate of A then y_1 must be the y-coordinate of A.

Example 4 Finding Distances in the Coordinate Plane

The Pythagorean Theorem

The sum of the squares of the lengths of the legs of a right triangle equals the square of the length of the hypotenuse. (Pyth. Thm.)

$a^2 + b^2 = c^2$

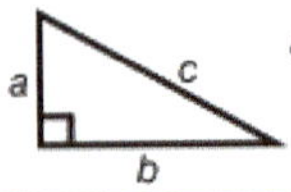

The Pythagorean Theorem states that in any right triangle, $a^2 + b^2 = c^2$, where a and b are the triangle's legs and c is the hypotenuse (the side opposite of the right angle). The distance between two points in a coordinate plane can be found using $a^2 + b^2 = c^2$.

Use the Pythagorean Theorem to find the distance, to the nearest tenth, from *P* to *R*.

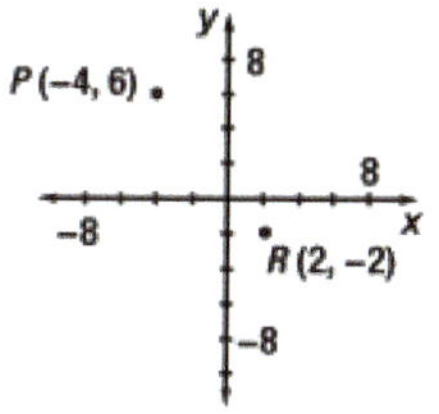

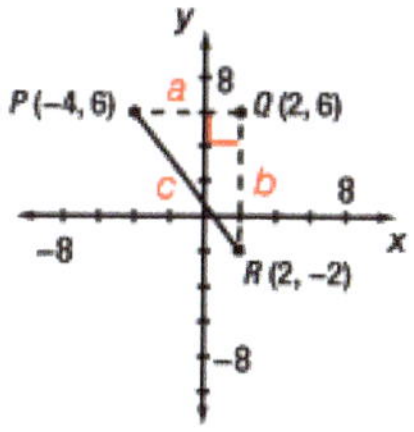

$a^2 + b^2 = c^2$

$6^2 + 8^2 = c^2$

$36 + 64 = c^2$

$100 = c^2$

$c = \sqrt{100}$

$c = 10$

The distance from *P* to *R* is 10 units.

The Distance Formula could be used to find the distance from *P* to *R*. Alternatively, $a^2 + b^2 = c^2$ could also be used. At first glance it may appear that the Pythagorean Theorem does not apply, but a right triangle can be drawn where the hypotenuse is the line segment from *P* to *R*. The legs of the right triangle are horizontal and vertical line segments. Since the legs are horizontal and vertical, the lengths of the legs can be found by subtracting their coordinates (or by counting along the axes).

To draw the right triangle, draw a segment to the right from *P* and up from *R*. The point at which these two segments meet is the third vertex of the right triangle, marked here as *Q*. The horizontal leg, named *a*, extends along the *x*-axis from –4 to 2. So, *a* is 6. The vertical leg, named *b*, extends along the *y*-axis from –2 to 6. So, *b* is 8. Substitute these values into $a^2 + b^2 = c^2$ and solve for *c*, which is the distance from *P* to *R*.

Example 5 Sports Application

The four bases on a little league baseball field form a square with 50 ft sides. When a player throws the ball from first base to third base, what is the distance of the throw, to the nearest tenth?

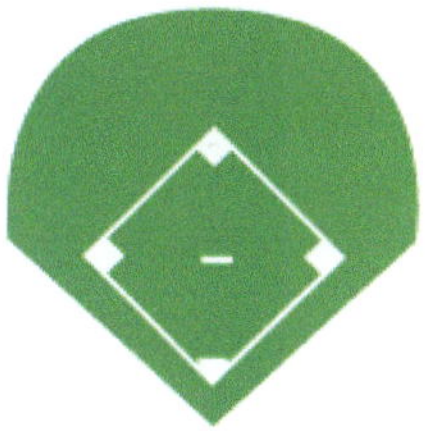

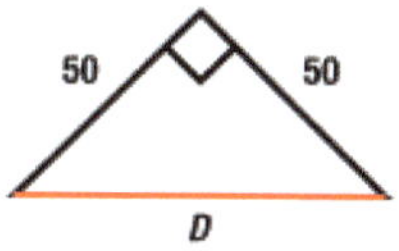

$a^2 + b^2 = c^2$	*Use the Pythagorean Theorem.*
$50^2 + 50^2 = D^2$	*Substitute 50 for a, 50 for b, and D for c.*
$D = \sqrt{50^2 + 50^2}$	*Solve for D.*
$= \sqrt{2(50)^2}$	*Simplify.*
$= \sqrt{2}\sqrt{(50)^2}$	*Product Property of Square Roots*
$= \sqrt{2}\,(50)$	*Simplify.*
≈ 70.7 ft	*Use a calculator to approximate.*

The distance from first base to third base is approximately 70.7 feet.

In this example, $a^2 + b^2 = c^2$ is used to find the hypotenuse of a right triangle.

The baseball field is a square, so there are four right angles, one at each base. It is given that each side of the square is 50 feet.

The distance from first base to third base is the hypotenuse of a right triangle where one leg is the length from first base to second base and the second leg is the length from second base to third base. Therefore, the length of each leg is 50.

Substitute 50 into $a^2 + b^2 = c^2$ for *a* and for *b* and let the length of the hypotenuse be represented by *D*. Then, simplify to find *D*, the distance from first base to third base.

Transformations in the Coordinate Plane

Key Objectives

- Identify reflections, rotations, and translations.
- Graph transformations in the coordinate plane.

Key Terms

- A **transformation** is a change in position, size, or shape of a figure.
- A **reflection** (or *flip*) is a transformation across a line, called the line of reflection. Each point and its image are the same distance from the line of reflection.
- A **translation** (or *slide*) is a transformation in which all the points of a figure move the same distance in the same direction.
- A **rotation** (or *turn*) is a transformation about a point *P*, called the center of rotation. Each point and its image are the same distance from *P*.
- The original figure in a transformation is called the **preimage**.
- The resulting figure in a transformation is called the **image**.

Transformations

A **transformation** is a change in the position, size, or shape of a figure.

The original figure is called the **preimage**.
The resulting figure is called the **image**.

A **reflection** (or *flip*) is a transformation across a line, called the line of reflection. Each point and its image are the same distance from the line of reflection.

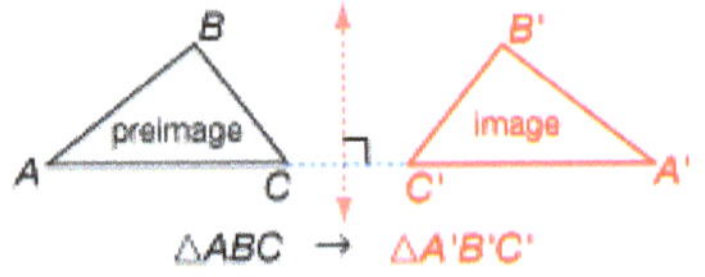

A **translation** (or *slide*) is a transformation in which all the points of a figure move the same distance in the same direction.

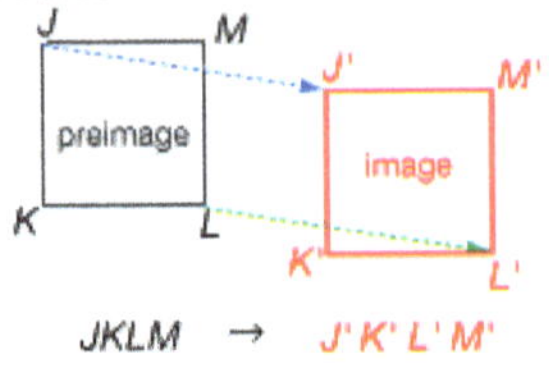

A **rotation** (or *turn*) is a transformation about a point *P*, called the center of rotation. Each point and its image are the same distance from *P*.

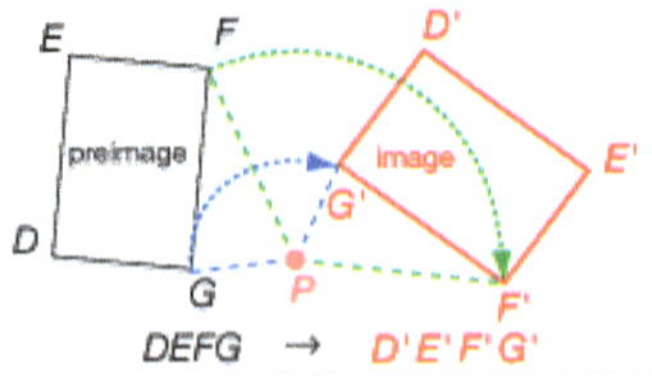

When a figure is changed in terms of its position, size, or shape, this change is called a transformation. Three types of transformations are reflections, translations, and rotations. In a transformation, the original figure, called the preimage, is mapped to a new figure, called the image. A transformation from a preimage to an image can be described using arrow notation where the image and preimage are named using the same point names, but with a prime symbol (') added after each point for the image name.

Example 1 Identifying Transformations

Reflections, translations, and rotations are three types of transformations where the position of the preimage is changed, but the size and shape of the preimage are not changed. In a reflection, the preimage is reflected, or flipped, over a line, called the line of reflection. You can think of the line of reflection as a mirror. When a figure is reflected, each point and its image are the same distance from the line of reflection. In a translation, the preimage is translated, or slid, such that all of the points on the preimage move the same distance in the same direction. In a rotation, the preimage is rotated, or turned, about a point, called the center of rotation. When a figure is rotated, each point and its image are the same distance from the center of rotation.

Identify the transformation. Then use arrow notation to describe the transformation.

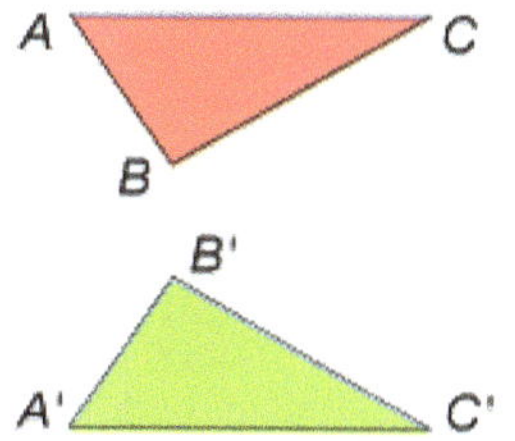

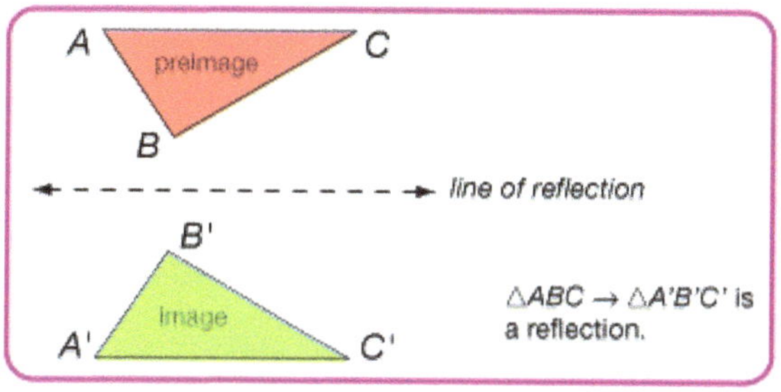

In the transformation shown here, the preimage is $\triangle ABC$ and the image is $\triangle A'B'C'$. Notice that if a mirror were placed between the two figures so that $\triangle ABC$ was reflected in this mirror, the reflection in the mirror would be in the same position as $\triangle A'B'C'$. Therefore, the transformation is a reflection and the line of reflection is a horizontal line between the two figures.

Identify the transformation. Then use arrow notation to describe the transformation.

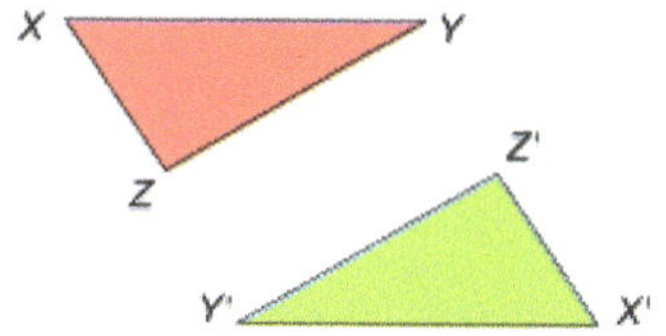

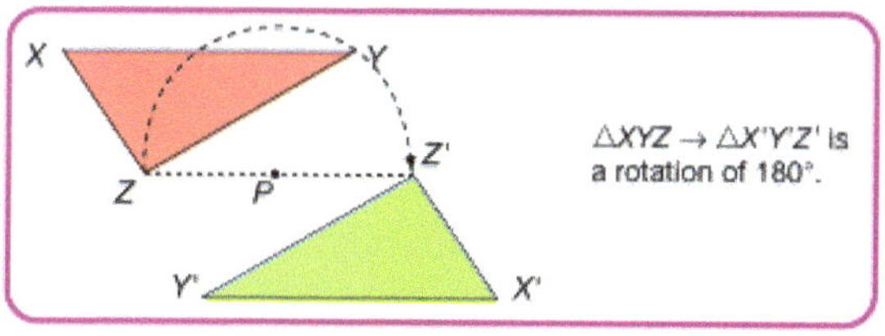

The transformation shown here may at first appear to be a reflection. However, notice that if a mirror were placed diagonally between the two figures, point Y would map to the position of point Z', not Y'.

If $\triangle ABC$ is rotated 180° about the point P shown here, the result is $\triangle A'B'C'$. Therefore, the transformation of $\triangle ABC \rightarrow \triangle A'B'C'$ is a rotation of 180°.

Example 2 Drawing and Identifying Transformations

A figure has vertices at $A(2, 4)$, $B(4, -1)$, and $C(2, -1)$. After a transformation, the image of the figure has vertices at $A'(-2, 4)$, $B'(-4, -1)$, and $C'(-2, -1)$. Draw the preimage and the image. Then identify the transformation.

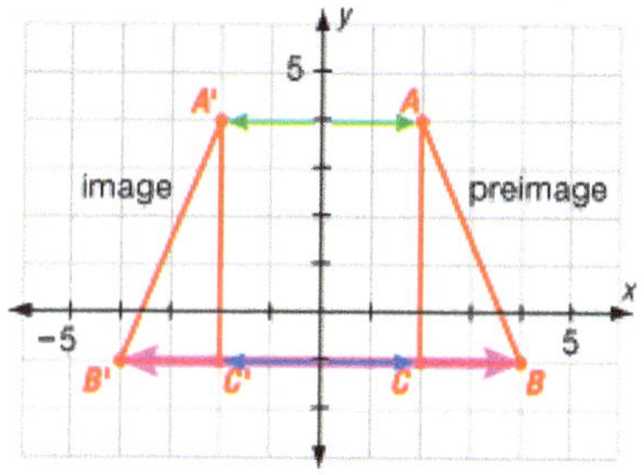

$\triangle ABC \rightarrow \triangle A'B'C'$ is a reflection across the y-axis.

In this example, the coordinates of the preimage's vertices, as well as the image's vertices, are given. Since three points are given for each figure, the preimage and image must be triangles.

Begin by plotting the points to identify the vertices. Then, draw lines to connect the vertices. These are the sides of the figures.

Now determine if the transformation of $\triangle ABC \rightarrow \triangle A'B'C'$ is a reflection, translation, or rotation. Notice that the horizontal distance from the y-axis to A is equal to the horizontal distance from the y-axis to A'. Furthermore, the horizontal distance is the same from the y-axis to B and B', and the same is true for C and C'. So, each point and its image is the same distance from the y-axis (which is a line). The transformation where each point and its image is the same distance from a line is a reflection. Therefore, the transformation $\triangle ABC \rightarrow \triangle A'B'C'$ is a reflection and the line of reflection is the y-axis.

Example 3 Translations in the Coordinate Plane

All of the points on the preimage move the same distance in the same direction to form the image in a translation.

Given a series of points, such as the vertices of a figure, the result of adding some number a to each of the x-coordinates and adding some number b to each of the y-coordinates is that all of the points will be moved the same distance in the same direction, which is a translation. So, a translation can be generalized as $(x, y) \rightarrow (x + a, y + b)$, which means that for any preimage point (x, y), the corresponding image point is $(x + a, y + b)$.

Find the coordinates for the image of $\triangle ABC$ after the translation $(x, y) \rightarrow (x - 3, y + 2)$. Draw the image.

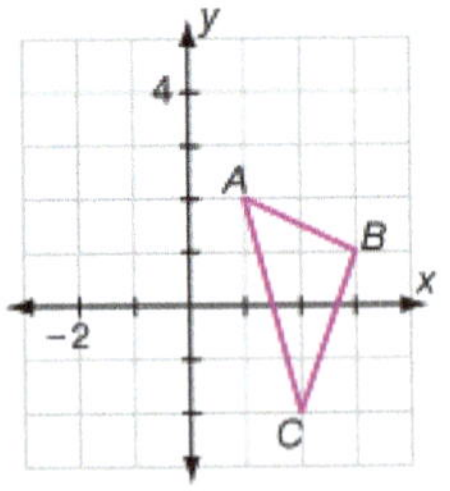

Rule: $(x, y) \rightarrow (x - 3, y + 2)$

$A(1, 2) \rightarrow A'(1 - 3, 2 + 2) = A'(-2, 4)$

$B(3, 1) \rightarrow B'(3 - 3, 1 + 2) = B'(0, 3)$

$C(2, -2) \rightarrow C'(2 - 3, -2 + 2) = C'(-1, 0)$

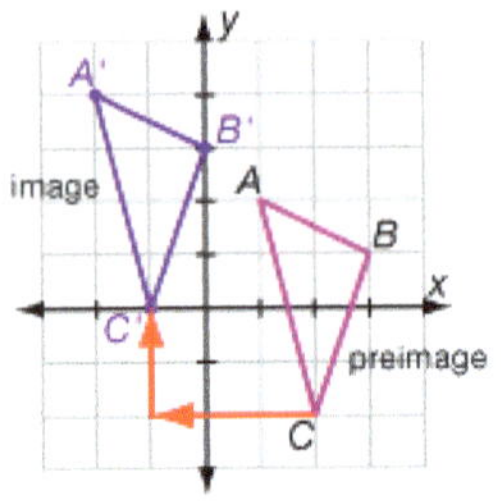

In this example, the translation is defined by the rule $(x, y) \rightarrow (x - 3, y + 2)$. This rule means that each preimage point (x, y) is translated 3 units left and 2 units up to get the corresponding image point.

The translation rule does not need to be applied to every single point on the figure. Translating the figure's vertices is sufficient. So, begin by identifying the coordinates of each vertex. Then apply the translation to the coordinates of each point. Specifically, subtract 3 from each x-coordinate and add 2 to each y-coordinate.

Once the translation has been applied to all three of the preimage's vertices, plot the three image points and connect the points with lines to draw the image.

Example 4 Art History Application

The pattern shown is similar to a pattern on a wall of the Alhambra. Write a rule for the translation of square 1 to square 2.

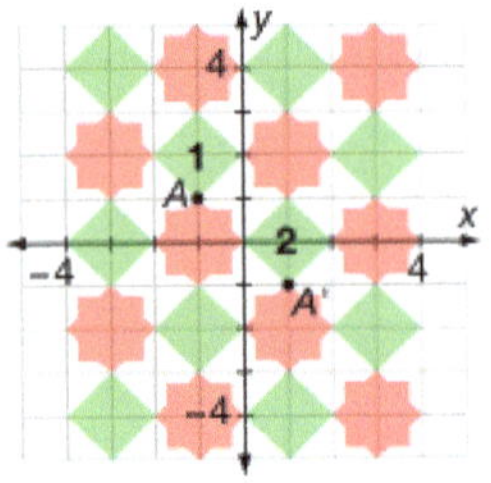

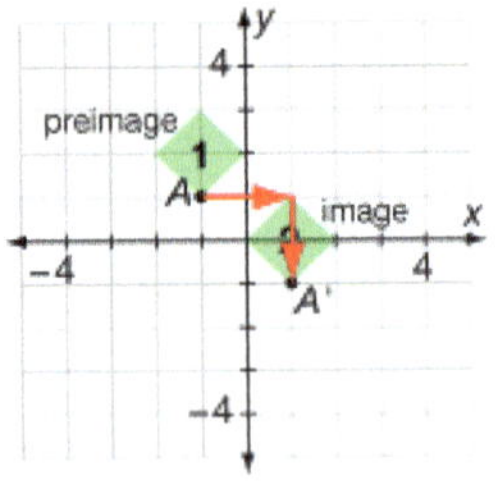

Translation Rule

$(x, y) \rightarrow (x + 2, y - 2)$

In this example the preimage and image of a translation are given and the translation rule must be written. To identify the rule, determine the distance and direction that a preimage point is moved.

First, identify a point on the preimage and its image. Here, point $A(-1, 1)$ on the preimage is mapped to $A'(1, -1)$ on the image. A is translated 2 units to the right and 2 units down. In terms of the coordinates, "2 units to the right" means to add 2 to the x-coordinate and "2 units down" means to subtract 2 from the y-coordinate. So, the translation rule is $(x, y) \rightarrow (x + 2, y - 2)$.

2

Using Inductive Reasoning to Make Conjectures

Key Objectives

- Use inductive reasoning to identify patterns and make conjectures.
- Find counterexamples to disprove conjectures.

Key Terms

- **Inductive reasoning** is the process of reasoning that a rule or statement is true because specific cases are true.
- A **conjecture** is a statement believed to be true based on inductive reasoning.
- A **counterexample** is a specific example that shows a statement of conjecture is false.

Example 1 Identifying a Pattern

A pattern in a list of items is a description of what the items have in common. To identify a pattern in a given list, first determine something that the first two items have in common. Then, consider whether the third item also has this commonality. If the commonality holds for each of the items in the list, then the pattern has been identified.

Find the next item in the pattern.

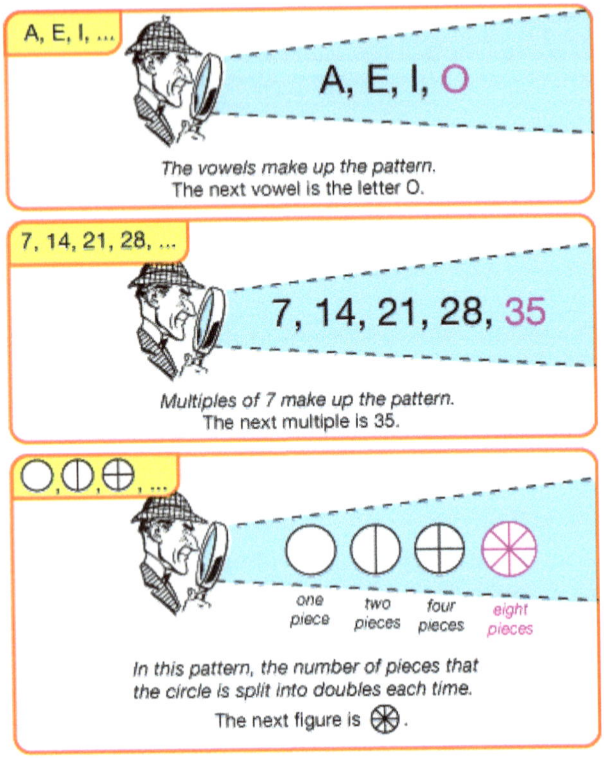

In this example, three lists are given. For each list, describe the pattern and find the next item in each list.

Each item in the first list is a letter. So, the next item in the list must also be a letter. Since the first item is the letter A, the pattern might be simply the letters in the alphabet. In this case the second item in the list must be the letter B. However, the second item is E, not B. So, determine what, besides the fact that the items are letters in the alphabet, the items in the list have in common. The letters A and E are the first two vowels in the alphabet, so that is a possible pattern. And since the next item, I, is the third vowel in the alphabet, this pattern holds. So, the next item in the list must be the vowel that follows I in the alphabet, which is the letter O.

The second list is composed of numbers. Determine what the numbers have in common to identify the pattern in the list. Each number in the list is a multiple of 7; the first item is 7(1), the second is 7(2), the third is 7(3), and the fourth is 7(4). So, the next item in the list must be 7(5), or 35, since it is the fifth item in the list.

The third list is composed of circles and each subsequent circle is divided into a greater number of pieces. The number of parts in each circle is twice the number of parts in the previous circle. In other words, the number of parts is doubled in each item. The third item has 4 parts. Therefore, the fourth item must have 8 parts, since $4(2) = 8$.

Example 2 Making a Conjecture

Inductive reasoning is the process of reasoning that a rule or statement is true because specific cases are true. Inductive reasoning was used in the previous example to determine the next item in each list. When a statement is believed to be based on inductive reasoning, that statement is called a conjecture.

Complete the conjecture.

The product of two odd numbers is _?_.

(1)(3) = 3 *List some examples and look for a pattern.*

(5)(7) = 35

(11)(9) = 99

odd

The product of two odd numbers is an odd number.

The given statement is a conjecture. So, inductive reasoning will be used to complete each conjecture. Using inductive reasoning means to consider specific examples. So, consider some specific examples that correspond to the given part of the conjecture and then use the pattern from those examples to complete the conjecture.

This conjecture begins with "The product of two odd numbers." So, each example should show the product of any two odd numbers. Three possible products are listed here. In each case, the result is an odd number. Therefore, based on the result of these examples, the complete conjecture will be "The product of two odd numbers is an odd number."

Complete the conjecture.

The number of circle parts formed by n diameters is _?_.

Diameter	Circle Parts	
1	2	2(1) = 2
2	4	2(2) = 4
3	6	2(3) = 6
4	8	2(4) = 8
n	$2n$	$2(n) = 2n$

The number of circle parts formed by n diameters is $2n$.

In this example, the conjecture begins with "The number of circle parts formed by n diameters." So, each example should show a circle with some number of diameters. When choosing the examples, it is often easiest to begin with a small number and increase by 1 in each example. So, the first example shows a circle with 1 diameter, which divides the circle into 2 parts. The second example shows a circle with 2 diameters. This circle is divided into 4 parts. Based on these two examples, the pattern appears to be that the number of parts is equal to twice the number of diameters. Test this result with the third example. In the third example, there are 3 diameters and the circle is divided into 6 parts. Since 6 = 3(2), the pattern "the number of parts is equal to twice the number of diameters" holds. Now apply this pattern to the case of n diameters. If the number of diameters is n, then the number of parts is $2n$. Therefore, the complete conjecture is "The number of circle parts formed by n diameters is $2n$."

Example 3 Biology Application

In Beth's neighborhood, there are equal numbers of Mockingbirds, Cardinals, and Blue Jays. Each morning, for five days, Beth counted the number of birds at the feeder outside her kitchen window. Make a conjecture based on the data.

	Day 1	Day 2	Day 3	Day 4	Day 5
Mockingbirds	5	4	5	3	3
Cardinals	7	10	12	8	7
Blue Jays	1	2	0	1	0

More Cardinals were viewed eating the food each day. The data supports the conjecture that the feed Beth is leaving out is most desired by the Cardinals.

In this example you are asked to make a conjecture based on the data. Since a conjecture is based on inductive reasoning, base a statement upon the results of specific examples in this table and that statement will be a conjecture.

Look at the table and consider a statement that is true about the number of birds observed on each day. On Day 1 and Day 2, more cardinals were seen than either mockingbirds or blue jays. This pattern holds for Days 3, 4, and 5. So, there are equal numbers of each type of bird in the neighborhood and more cardinals were observed eating the food each day. The data suggests that the food is most desired by the cardinals.

Example 4 Finding a Counterexample

Steps of Inductive Reasoning

1. Look for a pattern.
2. Make a conjecture.
3. Prove the conjecture or find a counterexample.

A conjecture cannot be proven true using specific examples. However, a conjecture can be proven false using only one specific example. An example that shows a conjecture is false is called a counterexample.

Show that the conjecture is false by finding a counterexample.

For all positive numbers x, $x^2 \geq x$.

Pick positive values for x.

Let $x = 3$.

Since $x^2 = 9$ and $9 \geq 3$, the conjecture holds.

Let $x = \frac{1}{2}$.

Since $x^2 = \frac{1}{4}$ and $\frac{1}{4} \ngeq \frac{1}{2}$, the conjecture is false.

$x = \frac{1}{2}$ is a counterexample.

This conjecture states that for all positive numbers, a number's square is greater than or equal to the number. Squaring a number means to multiply the number by itself. Consider specific examples to find a counterexample. If the example involves a number greater than 1, then the conjecture holds. However, the conjecture includes all positive numbers, so numbers between 0 and 1 are included and must be considered. When the example includes a fraction (a number between 0 and 1), the conjecture does not hold. Therefore, any example involving a fraction is a counterexample.

Show that the conjecture is false by finding a counterexample.

Four points on a plane always form a quadrilateral.

Draw 4 coplanar points.

Since the 4 coplanar points form a quadrilateral, the conjecture holds.

Draw 4 coplanar points, where 3 are collinear.

Since the 4 coplanar points form a triangle, the conjecture is false.

Counterexample

This conjecture may appear to be true since it concerns four points and only a quadrilateral is defined by four points. However, the conjecture is that four coplanar points *always* form a quadrilateral, so if a case can be found where four coplanar points form some figure other than a quadrilateral, then this case will be the counterexample to the conjecture. Notice that in the first figure shown here, all four points are noncollinear. Consider the case where some of the points are collinear. If three of the four points are collinear, then the figure formed is a triangle, which is not a quadrilateral. Therefore, four coplanar points such that three of those points are collinear is a counterexample.

Show that the conjecture is false by finding a counterexample.

The temperature in Lubbock, Texas, never exceeds 107°F during the spring months (March, April, May).

Monthly High Temperatures (°F) in Lubbock, Texas

Jan	Feb	Mar	Apr	May	Jun	Jul	Aug	Sep	Oct	Nov	Dec
83	84	91	97	109	109	111	114	103	102	91	82

Counterexample

Since the temperature in May was 109°F, the conjecture is false.

Here, the conjecture states that the temperature never exceeds 107° F in Lubbock, Texas, during March, April, and May. So, a counterexample to this conjecture would be any example of a temperature reaching above 107° F during March, April, or May. The table gives the monthly high temperatures in Lubbock for some year. Notice that the high temperature in June was 109° F. However, this is not a counterexample because June is not March, April, or May. So, consider only the high temperatures in March, April, and May. The months of March and April are not a counterexample because the high temperature in those months was not higher than 107° F. But, the high temperature in May was 109° F, which is higher than 107° F. Thus, the month of May is a counterexample.

Conditional Statements

Key Objectives

- Identify, write, and analyze the truth value of conditional statements.
- Write the inverse, converse, and contrapositive of a conditional statement.

Key Terms

- A **conditional statement** is a statement that can be written in the form "if p, then q."
- The **hypothesis** is the part p of a conditional statement following the word *if.*
- The **conclusion** is the part q of a conditional statement following the word *then.*
- A conditional statement has a **truth value** of either true (T) or false (F).
- The **negation** of a statement p is "not p," written as $\sim p$.
- The **converse** is the statement formed by exchanging the hypothesis and conclusion.
- The **inverse** is the statement formed by negating the hypothesis and conclusion.
- The **contrapositive** is the statement formed by both exchanging and negating the hypothesis and conclusion.
- Related conditional statements that have the same truth value are called **logically equivalent statements**.

Theorems, Postulates, Corollaries, and Properties

- **Law of Contrapositive** A conditional and its contrapositive are logically equivalent.

Example 1 Identifying the Parts of a Conditional Statement

A conditional statement is a statement that can be written in the form "if p, then q," where p is called the hypothesis and q is called the conclusion. If a statement or conjecture does not include the words "if" and "then," the statement can be rewritten as a conditional statement by rephrasing and adding "if" and "then."

Identify the hypothesis and conclusion of the conditional.

If Ed has a test, then he will study.

If Ed has a test, then he will study.

Hypothesis: Ed has a test.
Conclusion: Ed will study.

Remember, the hypothesis of a conditional statement is the part that follows "if" and the conclusion of a conditional statement is the part that follows "then."

The first conditional given here is written in "if/then" form, so the hypothesis and conclusion are easily identified.

A number is a real number if it is a rational number.

If a number is a rational number, then it is a real number.

Rephrase the conjecture as an if-then statement.

Hypothesis: A number is a rational number.
Conclusion: A number is a real number.

The second statement given here is not written in "if/then" form, but it does include the word "if." So, rephrase the statement so that the part that follows "if" comes first and add the word "then" before the new second part.

Example 2 Writing a Conditional Statement

Write a conditional statement from the following information.

A cat is an animal that can climb a tree.

Hypothesis: An animal is a cat.
Conclusion: A cat can climb a tree.

Identify the hypothesis and the conclusion.

Conditional: If an animal is a cat,
then it can climb a tree.

A right angle measures 90°.

Hypothesis: An angle is a right angle.
Conclusion: The angle measures 90°.

Identify the hypothesis and the conclusion.

Conditional: If an angle is a right angle,
then the angle measures 90°.

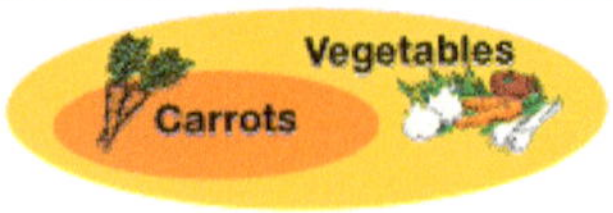

Hypothesis: You are a carrot.
Conclusion: You are a vegetable.

Identify the hypothesis and the conclusion.

Conditional: If you are a carrot,
then you are a vegetable.

In each of these three examples, the given information must be used to write a conditional statement. So, first identify the two parts from the given information, then determine which part depends on the other, and then phrase the two parts in the "if/then" form.

In the first example, the two parts are "a cat is an animal" and "that can climb a tree." The first part is the hypothesis and the second part is the conclusion.

The second example may appear to have only one part. But, the two parts are "an angle is a right angle" and "an angle measures 90°."

The information in the third example is given in a Venn diagram. The Venn diagram shows two parts, the group of vegetables and the group of carrots. When writing a conditional using a Venn diagram, the hypothesis is always gathered from the information given in the inner oval and the conclusion is gathered from the information given in the outer oval.

Example 3 Analyzing the Truth Value of a Conditional Statement

A conditional statement has a truth value of either true (T) or false (F), which depends upon the truth value of the hypothesis and the conclusion. A conditional statement's truth value is *false* only when the hypothesis is true and the conclusion is false. Otherwise, the conditional's truth value is true. So, the conditional statement's truth value is true in all other cases, including when the hypothesis is false. If the hypothesis is false, the conditional is true and the truth value of the conclusion is irrelevant.

Determine the truth value of a conditional by examining the truth value of the hypothesis and the conclusion. Showing that a conditional statement is false requires only one counterexample, which is an example where the hypothesis is true but the conclusion is false.

Determine if the conditional is true. If false, provide a counterexample.

If this month is April, then the next month is May.

The conditional is true.

If a quadrilateral has four 90° angles, then it is a square.

The conditional is false.
The hypothesis is true, but the conclusion is false.

Counterexample:

Find an example of a quadrilateral with four 90° angles that is not a square.

A rectangle is a quadrilateral with four 90° angles that is *not* a square.

If 121 is a prime number, then 10 is a multiple of 3.

If 121 is a prime number, then 10 is a multiple of 3.

121 is ***not*** a prime number, so the hypothesis is false.
10 is ***not*** a multiple of 3, so the conclusion is false.

However, the conditional is true because the hypothesis is false.

TIP
If the hypothesis is false, the conditional is true regardless of the truth of the conclusion.

The hypothesis in the first example is true because a month can be April. The conclusion is true because the month that follows April is always May; there is no counterexample. So, since the hypothesis and the conclusion are true, the conditional is also true.

For the second example, consider the truth value of the hypothesis. The hypothesis is true because a quadrilateral can have four 90° angles. Now consider the conclusion. Is it always true that a quadrilateral with four 90° angles is a square? A square does have four 90° angles, so the conclusion can be true, but it is not always true. A counterexample is the case of a rectangle because a rectangle is a quadrilateral with four 90° angles and is not a square. Thus, because the hypothesis is true and the conclusion is false, the conditional is false.

In the third example, again begin by considering the truth value of the hypothesis. Is the statement "121 is a prime number" true? No, the hypothesis is false because 121 is not a prime number (since 11(11) = 121). So, the conditional is true because the hypothesis is false. There is no need to consider the truth value of the conclusion in this case.

Example 4 Writing Related Conditionals

Remember, a conditional is a statement that can be written in the form "if p, then q," where p is the hypothesis and q is the conclusion. Related conditionals are statements based on a given conditional. In a related conditional, the hypothesis and conclusion of the given conditional are either reversed, negated, or both reversed and negated. The negation of a statement p is "not p," written as $\sim p$.

A **conditional** is an if-then statement.

$p \rightarrow q$

The **converse** is the statement formed by exchanging the hypothesis and the conclusion.

$q \rightarrow p$

The **inverse** is the statement formed by negating the hypothesis and the conclusion.

$\sim p \rightarrow \sim q$

The **contrapositive** is the statement formed by both exchanging and negating the hypothesis and the conclusion.

$\sim q \rightarrow \sim p$

If given a conditional with hypothesis p and conclusion q, the three related conditionals can be written based on this conditional using the definitions given here.

The given conditional's hypothesis and conclusion are switched to write the converse.

The given conditional's hypothesis and conclusion are both negated, but not switched, to write the inverse.

The given conditional's hypothesis and conclusion are switched and negated to write the contrapositive.

Write the converse, inverse, and contrapositive of the conditional statement. Find the truth value of each.

If an animal is a dog, then it has four legs.

p: An animal is a dog.
q: It has four legs.

Identify the hypothesis, p, and the conclusion, q.

Related Conditional	Symbol	Statement	Truth Value
Conditional	$p \rightarrow q$	If an animal is a dog, then it has four legs.	True
Converse	$q \rightarrow p$	If an animal has four legs, then the animal is a dog.	False
Inverse	$\sim p \rightarrow \sim q$	If an animal is **not** a dog, then it does **not** have four legs.	False
Contrapositive	$\sim q \rightarrow \sim p$	If an animal does **not** have four legs, then the animal is **not** a dog.	True

Counterexample for the **converse**:
A cat is an animal with 4 legs that is not a dog.

Counterexample for the **inverse**:
A cat is an animal that is not a dog yet it does have 4 legs.

In this example, the given conditional statement is "If an animal is a dog, then it has four legs." Identify the hypotheses and the conclusion of this given conditional and then use the definitions given above to write the conditional's converse, inverse, and contrapositive. Remember, negating a statement simply means to add "not" to the statement.

Notice that the truth value of the conditional and the contrapositive are both true and that the truth value of the converse and inverse are both false. Related conditional statements that have the same truth value are called logically equivalent statements. The Law of Contrapositive states that a conditional and its contrapositive will always be logically equivalent.

Using Deductive Reasoning to Verify Conjectures

Key Objectives

- Apply the Law of Detachment and the Law of Syllogism in logical reasoning.

Key Terms

- **Deductive reasoning** is the process of using logic to draw conclusions based on given facts, definitions, and properties.

Theorems, Postulates, Corollaries, and Properties

- **Law of Detachment** If $p \rightarrow q$ is a true statement and p is true, then q is true.
- **Law of Syllogism** If $p \rightarrow q$ and $q \rightarrow r$ are true statements, then $p \rightarrow r$ is a true statement.

Proving that a conjecture is false can be simple because only one counterexample is required. The process for proving that a conjecture is true is more involved. Deductive reasoning, reasoning based on given facts, definitions, and properties, must be used to prove that a conjecture is true. The Laws of Detachment and Syllogism are forms of deductive reasoning that will be explained in this lesson.

Example 1 Determining if a Conclusion is Based on Inductive or Deductive Reasoning

Inductive reasoning is logic based on specific examples. Using a counterexample to disprove a conjecture is an example of inductive reasoning. Deductive reasoning is logic based on given facts, definitions, and properties. Conclusions made based on known definitions, research, or data are examples of deductive reasoning.

Determine whether the conclusion is a result of inductive or deductive reasoning.

There is a myth that tapping the side of a carbonated beverage will prevent it from foaming over. However, if you were to observe a carbonated beverage that has been shaken, tapped, and then opened, you would see that this myth is false.

Since the conclusion is based on a pattern of observation, it is a result of inductive reasoning.

In this example, the conclusion is that the myth is false. This conclusion is based on observation of a specific example, opening a carbonated beverage after it is shaken and tapped, which is a counterexample to the given conditional. Therefore, the type of reasoning used is inductive reasoning.

Determine whether the conclusion is a result of inductive or deductive reasoning.

There is a myth that if a person goes swimming less than one hour after eating they will likely get a cramp and drown. However, studies have shown that there has never been a death from drowning or a near drowning attributed to eating before swimming. Therefore, the myth cannot be true.

The conclusion is based on logical reasoning from scientific research. It is a result of deductive reasoning.

In this example, the conclusion is that the myth is false, which was also the conclusion in the previous example. However, this conclusion is based on scientific research instead of on a specific example. Therefore, the type of reasoning used is deductive reasoning.

Example 2 Verifying Conjectures by Using the Law of Detachment

One case in which a conditional statement is true is when the hypothesis is true and the conclusion is true. If the hypothesis of a conditional statement is true and the conclusion is false, then the conditional statement cannot be true. Therefore, it can be deduced that when it is known that a conditional statement is true and it is known that the hypothesis is true, it must be the case that the conclusion is also true. The Law of Detachment summarizes this concept. According to the Law of Detachment, if $p \rightarrow q$ (a conditional) is a true statement and p (its hypothesis) is true, then q (its conclusion) is true.

Determine if the conjecture is valid by the Law of Detachment.

Given: If M is the midpoint of a line segment, then M bisects the segment. M is the midpoint of $\overline{AB}$.

Conjecture: $\overline{AM} \cong \overline{MB}$

Identify the hypothesis and conclusion in the given conditional.

If M is the midpoint of a line segment, then M bisects the segment. $p \to q$

The given statement M is the midpoint of $\overline{AB}$ matches the hypothesis of a true conditional. p is true.

By the Law of Detachment $\overline{AM} \cong \overline{MB}$. q is true.

The conjecture is valid.

In order for a conjecture to be valid by the Law of Detachment, a conditional and its hypothesis must be true and the conjecture must be equivalent to the conditional's conclusion.

In this example, the conditional is given to be true and the given statement matches the hypothesis of the given conditional. The conjecture is equivalent to the conditional's conclusion, so the conjecture holds by the Law of Detachment.

Determine if the conjecture is valid by the Law of Detachment.

Given: If you are absent more than 2 times, you will fail the class. Ed failed the class.

Conjecture: Ed was absent more than 2 times.

Identify the hypothesis and conclusion in the given conditional.

If you are absent more than 2 times, you will fail the class. $p \to q$

The given statement "Ed failed the class" matches the conclusion of a true conditional. q is true.

But this does not mean the hypothesis is true. p is not true.

The conjecture is not valid.

In this example, the given conditional is true, but the given statement matches the conclusion of the given conditional, not the conditional's hypothesis. So, the conjecture is not valid by the Law of Detachment.

Example 3 Verifying Conjectures by Using the Law of Syllogism

The Law of Detachment is one example of deductive reasoning. Another example is the Law of Syllogism, which states that if $p \to q$ and $q \to r$ are true statements, then $p \to r$ is a true statement. In other words, if two given conditionals are true, and the same statement is used for the conclusion of one conditional and for the hypothesis of the other conditional, then the conditional formed by the non-matching hypothesis and conclusion must also be true. Specifically, if the first conditional's hypothesis matches the second conditional's conclusion, then by the Law of Syllogism, the conditional formed from the hypothesis of the first conditional and the conclusion of the second conditional must be true.

Determine if the conjecture is valid by the Law of Syllogism.

Given: If $90° < m\angle A < 180°$, then $\angle A$ is obtuse. If $\angle A$ is obtuse, then it is not an acute angle.

Conjecture: If $90° < m\angle A < 180°$, then it is not an acute angle.

Let p, q, and r represent the following.

p: The measure of an angle is between 90° and 180°.

q: The angle is obtuse.

r: The angle is not an acute angle.

It is given that $p \to q$ and $q \to r$.

Thus $p \to r$ since $p \to q$ and $q \to r$ are true statements.

The conjecture is valid by the Law of Syllogism.

In order for a conjecture to be valid by the Law of Syllogism, two conditionals where the same statement is used in the hypothesis of one and in the conclusion of the other must be true, and the conjecture must match the conditional formed by the non-matching hypothesis and conclusion.

In this example, two conditionals are given to be true, and the conclusion of the first conditional is the same statement as the hypothesis of the second conditional, "$\angle A$ is obtuse." So, if the conjecture is the conditional formed by the hypothesis of the first conditional and the conclusion of the second, then the conjecture must hold by the Law of Syllogism. Notice that the conjecture's hypothesis is the hypothesis of the first conditional and the conjecture's conclusion is the conclusion of the second conditional. Therefore, the conjecture is valid by the Law of Syllogism.

Determine if the conjecture is valid by the Law of Syllogism.

Given: If a number is a multiple of 10, then it is a multiple of 5. If the last digit of a number is 0, then it is a multiple of 5.
Conjecture: If a number is a multiple of 10, then the last digit of the number is 0.

Let p, q, and r represent the following.

p: A number is a multiple of 10.
q: A number is a multiple of 5.
r: The last digit of a number is zero.

It is given that $p \rightarrow q$ and $r \rightarrow q$.

The Law of Syllogism cannot be used since q is the conclusion of both conditionals.

Even though the conjecture is true, the logic used to draw the conclusion is not valid.

The first given conditional's conclusion is "it is a multiple of 5." The second given conditional's hypothesis is "the last digit of a number is 0," which does not match the first conditional's conclusion.

Try reversing the conditionals so that "If the last digit of a number is 0, then it is a multiple of 5" is the first conditional and "If a number is a multiple of 10, then it is a multiple of 5" is the second conditional. Now, the conclusion of the first is "it is a multiple of 5" and the hypothesis of the second is "If a number is a multiple of 10." These statements are still not the same.

Therefore, since the conclusion of neither conditional matches the hypothesis of the other conditional, this conjecture cannot be valid by the Law of Syllogism.

So, even though this conjecture is true, it is not true by the Law of Syllogism.

Example 4 Applying the Laws of Deductive Reasoning

Draw a conclusion from the given information.

If Ed runs 10 miles on Monday, then he will not run the next day.
If Ed does not run on Tuesday, then he will run 10 miles on Wednesday.
Ed ran 10 miles on Monday.

Let p and q represent the following.

p: Ed ran 10 miles on Monday.
q: Ed did not run the next day.

It is given that p is a true statement.

Thus by the Law of Detachment q is true.

Let x and y represent the following.

x: Ed does not run on Tuesday.
y: Ed will run 10 miles on Wednesday.

It is given that x is a true statement, since Tuesday is the day after Monday.

Thus by the Law of Detachment, y is true.

Conclusion: Ed will run 10 miles on Wednesday.

In this example, two conditionals are given to be true and the statement "Ed ran 10 miles on Monday" is also given to be true. A letter can be assigned to each statement in the two given conditionals to organize the statements. Applying the Law of Detachment to the first conditional results in the conclusion that the statement "Ed did not run the next day" is true. Since Tuesday is the day after Monday, the statement "Ed did not run on Tuesday" can also be assumed to be true. When the Law of Detachment is applied to this statement and the second given conditional, the result is that "Ed will run 10 miles on Wednesday" can be concluded. Therefore, if the Law of Detachment is applied twice to the given information, the conclusion "Ed will run 10 miles on Wednesday" must be true.

Draw a conclusion from the given information.

If a triangle is equilateral, then it has three 60° angles. If a triangle has three 60° angles, then it is not an obtuse triangle. $\triangle ABC$ is equilateral.

Let p and q represent the following.

p: The triangle is equilateral.
q: The triangle has 3 60° angles.

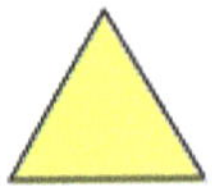

It is given that p is true.

Thus by the Law of Detachment q is true.

Let x and y represent the following.

x: The triangle has 3 60° angles.
y: The triangle is not obtuse.

It is given that x is true.

Thus by the Law of Detachment y is true.

Conclusion: $\triangle ABC$ is not an obtuse triangle.

Again, use the given information and the law of Detachment twice to reach a logical conclusion based on deductive reasoning.

Biconditional Statements and Definitions

Key Objectives

- Write and analyze biconditional statements.

Key Terms

- A **biconditional statement** is a statement that can be written in the form "p if and only if q." This means "if p then q" and "if q then p."
- A **definition** is a statement that describes an object and can be written as a true biconditional.
- A **polygon** is a closed plane figure formed by three or more segments, where each segment intersects exactly two other segments only at their endpoints and no two segments with a common endpoint are collinear.
- A **triangle** is a three-sided polygon.
- A **quadrilateral** is a four-sided polygon.

Example 1 Identifying the Conditionals within a Biconditional Statement

A **biconditional statement** is a statement that can be written in the form "p if and only if q."

$p \leftrightarrow q$ means $p \rightarrow q$ and $q \rightarrow p$

A biconditional statement can be thought of as a combination of two conditionals. The usual notation used for a biconditional statement is the double arrow. Sometimes a biconditional may be written using "iff" instead of the double arrow.

Remember, if a conditional relates two statements p and q where p is the conditional's hypothesis and q is the conditional's conclusion, then the converse of that conditional is the statement where q is the hypothesis and p is the conclusion. In other words, the converse of a conditional just switches the hypothesis and the conclusion.

A conditional has a p statement and a q statement. To write a conditional and its converse from a biconditional, identify the p statement and the q statement in the biconditional. Then, write the conditional as "if p, then q" and write the converse as "if q, then p."

Write the conditional statement and the converse within each biconditional.

An angle is a right angle if and only if the angle measures 90°.

An angle is a right angle if and only if the angle measures 90°.

The phrase "if and only if" indicates a biconditional statement.

Related Conditional	Symbols	Statement
Biconditional	$p \leftrightarrow q$	An angle is a right angle if and only if the angle measures 90°.
Conditional	$p \rightarrow q$	If an angle is a right angle, then the angle measures 90°.
Converse	$q \rightarrow p$	If an angle measures 90°, then the angle is a right angle.

In this example, the p statement in the biconditional is "an angle is a right angle" and the q statement in the biconditional is "the angle measures 90°."

Write the conditional statement and the converse within each biconditional.

A solution is an acid ↔ it has a pH less than 7.

A solution is an acid ↔ it has a pH less than 7.

The symbol ↔ is read as "if and only if."

Related Conditional	Symbols	Statement
Biconditional	$p \leftrightarrow q$	A solution is an acid if and only if it has a pH less than 7.
Conditional	$p \rightarrow q$	If a solution is an acid, then it has a pH less than 7.
Converse	$q \rightarrow p$	If a solution has a pH less than 7, then it is an acid.

In this example, p is defined to be "a solution is an acid" and q is defined to be "it has a pH less than 7."

Example 2 Writing a Biconditional Statement

To write the converse of a given conditional, just reverse the hypothesis and the conclusion. To write the biconditional from a given conditional, remove "if" and replace "then" with "if and only if."

For the conditional, write the converse and a biconditional statement.

If $10 = 7x - 4$, then $x = 2$.

Related Conditional	Symbols	Statement
Conditional	$p \to q$	If $10 = 7x - 4$, then $x = 2$.
Converse	$q \to p$	If $x = 2$, then $10 = 7x - 4$.
Biconditional	$p \leftrightarrow q$	$10 = 7x - 4$ if and only if $x = 2$.

If a triangle has three equal sides, then it is equilateral.

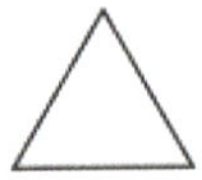

Related Conditional	Symbols	Statement
Conditional	$p \to q$	If a triangle has three equal sides, then it is equilateral.
Converse	$q \to p$	If a triangle is equilateral, then it has three equal sides.
Biconditional	$p \leftrightarrow q$	A triangle has three equal sides if and only if it is equilateral.

When writing a related conditional or biconditional from a given conditional, it is acceptable to replace pronouns such as "it" or "they."

The hypothesis of the conditional given in the first example is "$10 = 7x - 4$" and the conclusion is "$x = 2$." Switch the hypothesis and the conclusion to write the converse. Leave the hypothesis and the conclusion as they are in the original conditional to write the biconditional, but remove "if" and replace "then" with "if and only if."

In the conditional given in the second example, the hypothesis is "a triangle has three equal sides" and the conclusion is "it is equilateral."

Example 3 Analyzing the Truth Value of a Biconditional Statement

Remember, a conditional has a truth value of either true or false. A conditional can be proved to be false by providing a single counterexample to the conditional where the hypothesis is true, but the conclusion is false.

A biconditional statement is true only when the related conditional and its converse are true. If either the related conditional or its converse are false, then the biconditional statement is also false.

Determine if the biconditional is true. If false, give a counterexample.

A rectangle has length 10 and width 3 if and only if it has an area of 30.

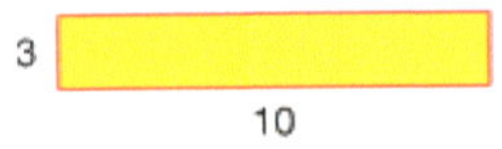

Related Conditional	Symbols	Statement	
Biconditional	$p \leftrightarrow q$	A rectangle has length 10 and width 3 if and only if it has an area of 30.	
Conditional	$p \to q$	If a rectangle has length 10 and width 3, then it has an area of 30.	The conditional is true.
Converse	$q \to p$	If the area of a rectangle is 30, then it has length 10 and width 3.	The converse is false.

Counterexample for the **converse**:

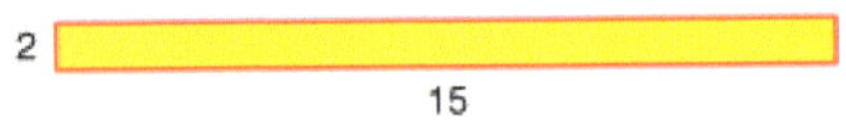

Area = 15(2) = 30

If the area of the rectangle is 30, then the length may be 2 and the width may be 15, which is not 10 and 3.

Because the converse is false, the biconditional is false.

The given conditional is true only if the related conditional and its converse are true. So, write the conditional and its converse using the hypothesis p and the conclusion q from the given biconditional. Then, determine the truth value of each. If a counterexample can be found for either the conditional or the converse, then the biconditional is false.

A counterexample can be found for the converse since a rectangle with area 30 does exist where its length and width are not 10 and 3, respectively. Specifically, a rectangle with length 15 and width 2 has area 30, but does not have length 10 and width 3. Therefore, since the converse is false, the biconditional is false.

Determine if the biconditional is true. If false, give a counterexample.

$\angle ABC$ is and acute angle $\leftrightarrow m\angle ABC < 90^\circ$.

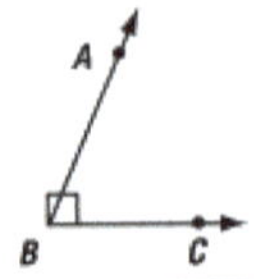

Related Conditional	Symbols	Statement	
Biconditional	$p \leftrightarrow q$	$\angle ABC$ is an acute angle if and only if $m\angle ABC < 90^\circ$.	
Conditional	$p \to q$	If $\angle ABC$ is an acute angle, then $m\angle ABC < 90^\circ$.	The conditional is true.
Converse	$q \to p$	If $m\angle ABC < 90^\circ$, then $\angle ABC$ is an acute angle.	The converse is true.

Since the conditional and the converse are true, the biconditional is true.

In this example, use the definition of acute angles to determine the truth value of the conditional and its converse. An acute angle is defined to be an angle with measure that is less than 90°.

Example 4 Writing Definitions as Biconditional Statements

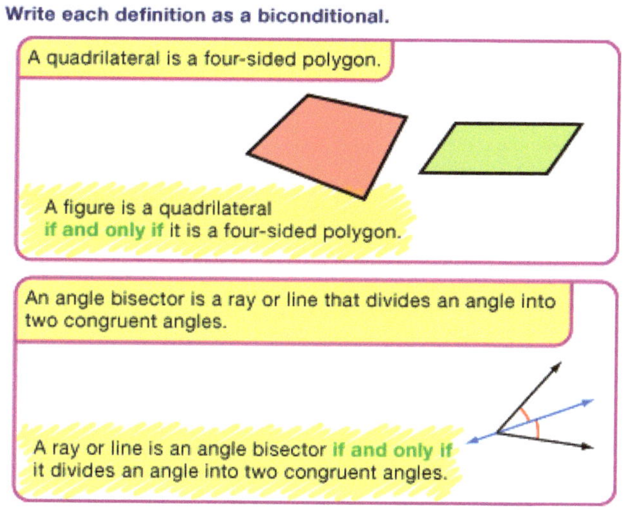

A definition is a statement that describes an object and can be written as a true biconditional.

In the first example, the definition can be written as a biconditional by simply replacing "is" with "if and only if" and rephrasing the definition slightly.

The definition in the second example can be written as a biconditional by replacing "that" with "if and only if" and rephrasing the definition.

Algebraic Proof

Key Objectives

- Review properties of equality and use them to write algebraic proofs.
- Identify properties of equality and congruence.

Key Terms

- A **proof** is an argument that uses logic, definitions, properties, and previously proven statements to show that a conclusion is true.

Theorems, Postulates, Corollaries, and Properties

- **Properties of Equality** There are eight Properties of Equality: Addition, Subtraction, Multiplication, Division, Reflexive, Symmetric, Transitive, and Substitution.
- **Distributive Property** $a(b + c) = ab + ac$ and $a(b - c) = ab - ac$
- **Properties of Congruence** There are three Properties of Congruence: Reflexive, Symmetric, and Transitive.

Properties of Equality	
Addition Property of Equality	If $a = b$, then $a + c = b + c$.
Subtraction Property of Equality	If $a = b$, then $a - c = b - c$.
Multiplication Property of Equality	If $a = b$, then $ac = bc$.
Division Property of Equality	If $a = b$ and $c \neq 0$, then $\frac{a}{c} = \frac{b}{c}$.
Reflexive Property of Equality	$a = a$.
Symmetric Property of Equality	If $a = b$, then $b = a$.
Transitive Property of Equality	If $a = b$ and $b = c$, then $a = c$.
Substitution Property of Equality	If $a = b$, then b can be substituted for a in any expression.

The eight Properties of Equality are listed above. The first four, Addition, Subtraction, Multiplication, and Division are commonly used when solving algebraic equations. Basically, these four properties show that an equation can be added, subtracted, multiplied, or divided by any number, as long as that same number and operation is applied to both sides of the equation. The Substitution Property of Equality is commonly used when evaluating algebraic expressions and when checking that the solution to an algebraic equation is correct.

Example 1 Solving an Equation in Algebra

To solve an algebraic equation, use the Properties of Equality and manipulate the equation so that the variable will be isolated on one side of the equation. Remember to apply any operation to both sides of the equation. Solving an equation and justifying the steps using properties or definitions is an example of an algebraic proof.

Solve the equation $-8 = 5n + 2$. Write a justification for each step.

$-8 = 5n + 2$	Given
$-2 \quad\quad -2$	Subtraction Property of Equality
$-10 = 5n$	Simplify.
$\frac{-10}{5} = \frac{5n}{5}$	Division Property of Equality
$-2 = n$	Simplify.
$n = -2$	Symmetric Property of Equality

The variable in this equation is n. In the original equation, 2 is added to $5n$. So, use the Subtraction Property of Equality and subtract 2 from both sides of the equation. After the equation is simplified, n is multiplied by 5. So, use the Division Property of Equality and divide both sides of the equation by 5. After the equation is simplified again, the result is $-2 = n$. Since the variable n is isolated on one side of the equation, the equation is solved. However, the Symmetric Property of Equality can be applied to write the equation with n on the left side, as equations are more commonly written.

Example 2 Technology Application

To simulate the motion of an object in a computer game, the designer uses the formula $sr = 3.6p$ to find the number of pixels the object must travel during each second of animation. In the formula, s is the desired speed of the object in kilometers per hour, r is the scale of pixels per meter, and p is the number of pixels traveled per second.

The graphics in a game are based on a scale of 8 pixels per meter. The designer wants to simulate a vehicle moving at 54 km/h. How many pixels must the vehicle travel each second? Solve the equation for p and justify each step.

$s \cdot r = 3.6p$	Given equation
$(54)(8) = 3.6p$	Substitution Property of Equality
$432 = 3.6p$	Simplify.
$\frac{432}{3.6} = \frac{3.6p}{3.6}$	Division Property of Equality
$120 = p$	Simplify.
$p = 120$	Symmetric Property of Equality

The vehicle must travel at 120 pixels per second!

The given equation in this example contains three variables, s, r, and p. Begin by using the Substitution Property of Equality to substitute given values for s and r. 54 is substituted into the equation for s because s is defined to be the speed of the object in km/h, and 54 km/h is the given speed of the object. The variable r is defined to be the scale of pixels per meter. So, since the given scale of pixels per meter is 8, 8 is substituted into the equation for r.

After the substitution is complete and the equation is simplified by multiplying, the only operation that needs to be undone in the resulting equation is multiplication by 3.6. So, by the Division Property of Equality, both sides of the equation can be divided by 3.6. This operation isolates the variable p on one side of the equation, so the equation is solved.

Example 3 Solving an Equation in Geometry

Write a justification for each step.

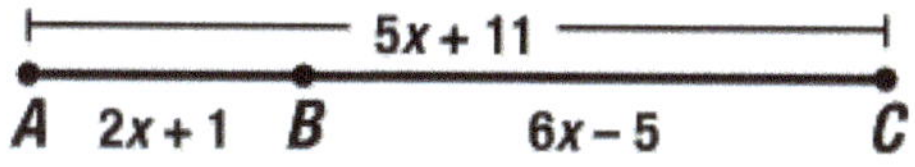

$AC = AB + BC$	Segment Addition Postulate
$5x + 11 = (2x + 1) + (6x - 5)$	Substitution Property of Equality
$5x + 11 = 8x - 4$	Simplify.
$-3x + 11 = -4$	Subtraction Property of Equality
$-3x = -15$	Subtraction Property of Equality
$x = 5$	Division Property of Equality

In this example, the figure and the solving steps are given, and the steps must be justified using postulates, definitions, properties, or information that is given. To determine the justification, consider the corresponding equation and the equation that precedes it. Determine how the equation has changed and then identify the property, postulate, or definition that allows such a change. For example, consider step 6, the final step. The corresponding equation is $x = 5$. The equation that precedes this is $-3x = -15$. In order to get from $-3x = -15$ to $x = 5$, the preceding equation must be divided by -3. The property that allows this change is the Division Property of Equality.

Example 4 Identifying Properties of Equality and Congruence

Reflexive Property of Congruence

A figure is congruent to itself and any part of a figure is congruent to itself.
(Reflex. Prop. of $\cong$)

$\triangle ABC \cong \triangle ABC$ $\quad \overline{AB} \cong \overline{AB}$

Symmetric Property of Congruence

If figure $A \cong$ figure B, then figure $B \cong$ figure A.
(Sym. Prop. of $\cong$)

If $\angle 1 \cong \angle 2$, then $\angle 2 \cong \angle 1$.

Transitive Property of Congruence

If figure $A \cong$ figure B and figure $B \cong$ figure C, then figure $A \cong$ figure C.
(Trans. Prop. of $\cong$)

If $\overline{PQ} \cong \overline{RS}$ and $\overline{RS} \cong \overline{TU}$, then $\overline{PQ} \cong \overline{TU}$.

Equality and congruence are similar, but equality is used to describe a relationship between numbers or things that represent numbers (such as variables) and congruence describes a relationship between figures. These three properties of congruence are similar to the Reflexive, Symmetric, and Transitive Properties of Equality.

Identify the property that justifies each statement.

$80° = 80°$

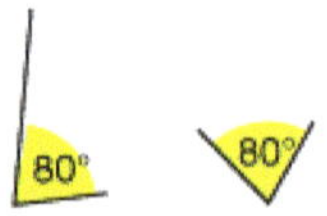

Reflexive Property of Equality

$\overline{AB} \cong \overline{CD}$, so $\overline{CD} \cong \overline{AB}$.

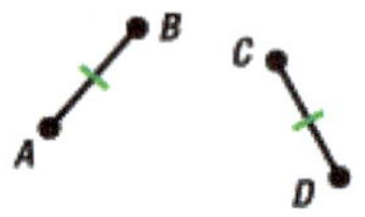

Symmetric Property of Congruence

$\angle G \cong \angle G$

Reflexive Property of Congruence

$\overline{XY} \cong \overline{VW}$ and $\overline{VW} \cong \overline{AB}$. So $\overline{XY} \cong \overline{AB}$.

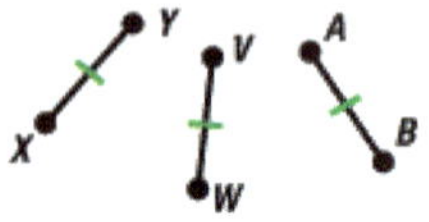

Transitive Property of Congruence

The statement $80° = 80°$ shows that a degree measurement is equal to itself. The Reflexive Property of Equality states that $a = a$, or that any number a is equal to itself. So, the Reflexive Property of Equality justifies the statement.

The second statement shows that one segment is congruent to another, so the second segment is congruent to the first. The Symmetric Property of Congruence states that if a figure is congruent to a second figure, then it must be true that the second figure is congruent to the first figure. So, the statement is justified by the Symmetric Property of Congruence.

The third statement shows that an angle is congruent to itself. This appears to be the same property as the first statement. However, notice that congruency is used here, not equality. Therefore, the Reflexive Property of Congruence justifies this statement.

In the third statement, the relationship between three congruent segments is described. Generally speaking, this statement shows that if segment 1 is congruent to segment 2 and segment 2 is congruent to segment 3, then segment 1 is congruent to segment 3. This is an example of the Transitive Property of Congruence.

Geometric Proof

Key Objectives

- Write two-column proofs.
- Prove geometric theorems by using deductive reasoning.

Key Terms

- A **theorem** is any statement that you can prove.
- A **two-column proof** is a type of proof where the statements and corresponding reasons are listed in two columns with the statements in the left column and the reasons in the right column.

Theorems, Postulates, Corollaries, and Properties

- **Linear Pair Theorem** If two angles form a linear pair, then they are supplementary.
- **Congruent Supplements Theorem** If two angles are supplementary to the same angle (or to two congruent angles), then the two angles are congruent.
- **Right Angle Congruence Theorem** All right angles are congruent.
- **Congruent Complements Theorem** If two angles are complementary to the same angle (or to two congruent angles), then the two angles are congruent.

A proof is an argument where logic, definitions, properties, and previously proven statements are used to demonstrate that a conjecture is always true. There are several types of proofs in geometry, but a two-column proof is most commonly seen.

The process of writing a proof is the same for all types of proofs. A proof is based on a conjecture. The conjecture contains a hypothesis (statements providing information that can be assumed to be true) and a conclusion (the statement that is to be proven true). In a proof, deductive reasoning is used to write a list of connected steps beginning with the hypothesis and ending with the conclusion.

Each step in a proof contains two parts: a statement and a justification, or reason. "Given" is the reason for statements in the hypothesis. The reason for all other statements in the proof must be either a theorem, postulate, property, or definition.

The given statements are typically listed at the start of a proof. After the given statements are made in a proof, each subsequent statement is first a conclusion based on a preceding step (or steps), and then that conclusion becomes a hypothesis for conclusions made in later steps. In other words, once a statement is justified, that statement can be used in the hypothesis for a later conclusion. The final statement in a proof is the original conjecture's conclusion. Once the conjecture's conclusion is stated and justified using at least one previous statement, the conjecture has been proven.

If a figure is not given with the conjecture, it is often helpful to draw a figure and use markings, such as tick marks, to show the given information. Mark only the given information on the figure; do not mark the information from the statement to be proven.

Example 1 Writing Justifications

In the example below, the proof's statements are complete and the justifications, or reasons, must be added.

The information given in the problem is "$\angle E$ and $\angle G$ are supplementary" and "$\angle E \cong \angle H$." So, those two statements are the hypothesis of the conjecture to be proven.

Notice that the final statement is "$\angle H$ and $\angle G$ are supplementary." Since this is the final statement of the proof, "$\angle H$ and $\angle G$ are supplementary" must be the conclusion of the conjecture to be proven.

Write a justification for each step, given that ∠E and ∠G are supplementary and ∠E ≅ ∠H.

1. ∠E and ∠G are supplementary. — Given information
2. m∠E + m∠G = 180° — Def. of supp. ∠s
3. ∠E ≅ ∠H — Given information
4. m∠E = m∠H — Def. of ≅ ∠s
5. m∠H + m∠G = 180° — *Steps 2, 4* — Subst. Prop. of =
6. ∠H and ∠G are supplementary. — Def. of supp. ∠s

The statements given in steps 1 and 3 are the conjecture's hypothesis, so the corresponding reason for each is "Given information."
Notice that the step 2 statement is a conclusion based on information from step 1. Specifically, step 1 is the hypothesis for the conclusion made in step 2: IF "∠E and ∠G are supplementary" (step 1), THEN "m∠E + m∠G = 180°" (step 2). The conclusion is true because the definition of supplementary angles states that the sum of two supplementary angles is 180°. So, the reason for the statement in step 2 is the definition of supplementary angles, which can be abbreviated as "Def. of supp. ∠s."
The statement in step 5 is a conclusion made using statements 2 and 4 as the hypothesis (noted to the left of the step 5 reason).

Example 2 Completing a Two-Column Proof

A theorem is any statement that can be proven. One example of a theorem is the Linear Pair Theorem, which states that if two angles form a linear pair, then those two angles are supplementary. This conjecture is proven in the example below. The conjecture is divided into its hypothesis, provided in the "Given" statement (and figure), and its conclusion, provided in the "Prove" statement. So, the information that can be assumed to be true for a proof is provided in a "Given" statement and/or in a given figure, and the statement to be proven is provided in the "Prove" statement.

The proof below is an example of a two-column proof. Notice that the statements are listed in the left column and the corresponding reasons for each statement, or justifications, are listed in the right column.

Fill in the blanks to complete a two-column proof of the Linear Pair Theorem.

Given:
∠1 and ∠2 form a linear pair.

Prove:
∠1 and ∠2 are supplementary.

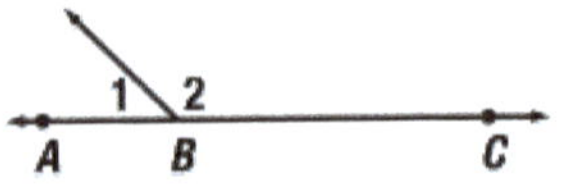

Statements:	Reasons:
1. ∠1 and ∠2 form a linear pair.	1. Given information
2. $\overrightarrow{BA}$ and $\overrightarrow{BC}$ form a line.	2. Def. of lin. pair
3. m∠ABC = 180°	3. Def. of straight ∠
4. m∠1 + m∠2 = m∠ABC	4. ∠ Add. Post.
5. m∠1 + m∠2 = 180°	5. Subst. *Steps 3, 4*
6. ∠1 and ∠2 are supplementary.	6. Def. of supp. ∠s

In this example, most of the proof is complete, and a few parts must be added (the second part of the statement in step 3 and the reasons for steps 1, 2, 5, and 6). The statement listed in step 1 is the "Given" statement. Therefore, the reason is "Given information." To determine the reason for step 2, consider the statement in step 1 as the hypothesis and the statement in step 2 as the conclusion in a conditional. What reason makes the conclusion true? The definition of linear pair states that noncommon sides of the two angles in a linear pair are opposite rays that form a line. Therefore, the reason for step 2 is the definition of linear pair, abbreviated as "Def. of lin. pair."

Note that "Subst.", which is the abbreviation for the Substitution Property of Equality, is listed as the reason in step 5.

Example 3 Writing a Two-Column Proof from a Plan

Another example of a theorem is the Right Angle Congruence Theorem, which states that all right angles are congruent. This theorem is proven in the example below.

Use the given plan to write a two-column proof of the Right Angle Congruence Theorem.

Given: $\angle C$ and $\angle D$ are right angles.

Prove: $\angle C \cong \angle D$

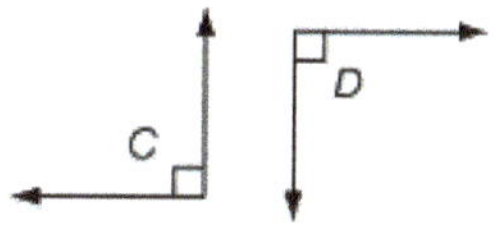

Use the definition of a right angle to write the measure of each angle. Then use the Transitive Property and the definition of congruent angles.

Statements:	Reasons:
1. $\angle C$ and $\angle D$ are right angles.	1. Given
2. $m\angle C = 90°$ and $m\angle D = 90°$	2. Def. of rt. $\angle$
3. $m\angle C = m\angle D$	3. Trans. Prop. of =
4. $\angle C \cong \angle D$	4. Def. of $\cong$ $\angle$s

The only given information in the conjecture proven here is that $\angle C$ and $\angle D$ are right angles. The conjecture's conclusion (the statement to be proven) is that $\angle C \cong \angle D$.

Begin by making a plan for the two-column proof. Think of the proof as a puzzle where the first piece is the given information and each subsequent piece depends on the previous piece (or pieces). So, use the given information to deduce new information. And then assume that the new information is true and deduce new information from that assumption.

Flowchart and Paragraph Proofs

Key Objectives

- Write flowchart and paragraph proofs.
- Prove geometric theorems by using deductive reasoning.

Key Terms

- A **flowchart proof** uses boxes and arrows to show the structure of a proof.
- In a **paragraph proof**, the statements and reasons of the proof are presented as sentences in a paragraph.

Theorems, Postulates, Corollaries, and Properties

- **Common Segments Theorem** Given collinear points A, B, C, and D arranged on a segment such that A and D are the endpoints, B is between A and C, and C is between B and D. If $\overline{AB} \cong \overline{CD}$, then $\overline{AC} \cong \overline{BD}$.
- **Vertical Angles Theorem** Vertical angles are congruent.
- **Theorem** If two congruent angles are supplementary, then each angle is a right angle.

Example 1 Writing a Flowchart Proof

A flowchart proof uses boxes and arrows to show the structure of a proof. The steps in a flowchart proof move from left to right or from top to bottom, shown by arrows connecting each box. The boxes used in a flowchart proof contain the statements and the reason for each statement is listed under the corresponding box.

Use the given two-column proof to write a flowchart proof.

Given: $\angle 1 \cong \angle 3$

Prove: $m\angle 2 = m\angle 4$

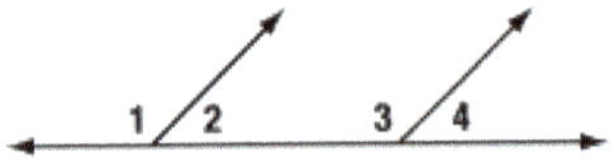

Statements:	Reasons:
1. $\angle 1 \cong \angle 3$	1. Given
2. $\angle 1$ and $\angle 2$ are supplementary. $\angle 3$ and $\angle 4$ are supplementary.	2. Lin. Pair Thm.
3. $\angle 2 \cong \angle 4$	3. $\cong$ Supps. Thm.
4. $m\angle 2 = m\angle 4$	4. Def. of $\cong \angle$s

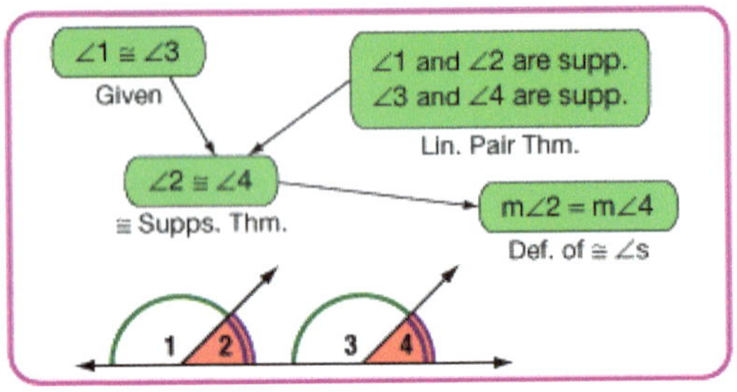

In this example, the conjecture, figure, and two-column proof are given and a flowchart proof must be written, based on the two-column proof.

The statement and reason from step 1 in the two-column proof should be listed in and under either the top box or the box farthest to the left. Here, the flowchart proof is using a top/down layout, so the given statement is shown in the top box with the corresponding reason underneath the box.

The statements and reasons from steps 2, 3, and 4 are each listed in and under a box. The arrows are used to show the progression of the steps.

Notice that the information from steps 1 and 3 are shown in the figure at the bottom.

Example 2 Reading a Paragraph Proof

In a paragraph proof, the statements and reasons of the proof are presented as sentences in a paragraph. The sentences are written in a logical order, typically beginning with a sentence (or sentences) summarizing the information given in the "Given" statement(s) and in the figure, and ending with a sentence that summarizes the conjecture's conclusion.

Use the given paragraph proof to write a two-column proof.

Given: $\angle 1 \cong \angle 3$

Prove: $\angle 2 \cong \angle 4$

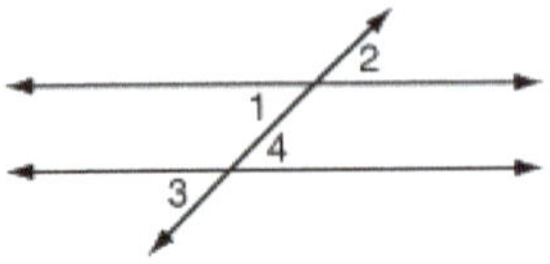

Paragraph Proof

By the Vertical Angles Theorem, $\angle 1 \cong \angle 2$ and $\angle 3 \cong \angle 4$. It is given that $\angle 1 \cong \angle 3$. By the Transitive Property of Congruence, $\angle 3 \cong \angle 2$, and thus $\angle 2 \cong \angle 4$.

Statements:	Reasons:
1. $\angle 1 \cong \angle 3$	1. Given
2. $\angle 1 \cong \angle 2$ and $\angle 3 \cong \angle 4$	2. Vert. $\angle$s Thm.
3. $\angle 3 \cong \angle 2$	3. Trans. Prop. of $\cong$
4. $\angle 2 \cong \angle 4$	4. Trans. Prop. of $\cong$

In this example, the conjecture, figure, and paragraph proof are given and then a two-column proof is written based on the paragraph proof.

The order of the steps in the two-column proof does not have to exactly match the order of the corresponding sentences from the paragraph proof, but the order should be similar if not the same.

Example 3 Writing a Paragraph Proof

The Common Segments Theorem states that given collinear points A, B, C, and D arranged on a segment such that A and D are the endpoints, B is between A and C, and C is between B and D. If $\overline{AB} \cong \overline{CD}$, then $\overline{AC} \cong \overline{BD}$. This theorem is proven in the example below.

Use the given two-column proof to write a paragraph proof.

Given: $\overline{PQ} \cong \overline{RS}$

Prove: $\overline{PR} \cong \overline{QS}$

Two-Column Proof

Statements:	Reasons:
1. $\overline{PQ} \cong \overline{RS}$	1. Given
2. $PQ = RS$	2. Def. of ≅ segs.
3. $QR = QR$	3. Reflex. Prop. of =
4. $PQ + QR = QR + RS$	4. Add. Prop. of =
5. $PQ + QR = PR$ and $QR + RS = QS$	5. Seg. Add. Post
6. $PR = QS$	6. Substitution
7. $\overline{PR} \cong \overline{QS}$	7. Def. of ≅ segs.

Paragraph Proof

Notice that $\overline{PQ} \cong \overline{RS}$. By the definition of congruent segments, $PQ = RS$. By the Reflexive Property of Equality, $QR = QR$. Applying the Additive Property of Equality, $PQ + QR = QR + RS$. By the Segment Addition Postulate, $PQ + QR = PR$, and similarly, $QR + RS = QS$. Thus $PR = QS$ by substitution. Since the segment lengths are equal, $\overline{PR} \cong \overline{QS}$ by the definition of congruent segments.

In this example, the conjecture, figure, and two-column proof are given and then a paragraph proof is written based on the two-column proof.

Each sentence in the paragraph proof is simply a combination of a statement and its corresponding reason from the two-column proof.

The paragraph proof begins with the information given in the conjecture and ends with the conjecture's conclusion.

3

Planes, Lines, and Angles

Key Objectives

- Identify parallel, perpendicular, and skew lines.
- Identify the angles formed by two lines and a transversal.

Key Terms

- **Parallel lines** (||) are coplanar and do not intersect.
- **Perpendicular lines** ($\perp$) intersect at 90° angles.
- **Skew lines** are not coplanar. Skew lines are not parallel and do not intersect.
- **Parallel planes** are planes that do not intersect.
- A **transversal** is a line that intersects two coplanar lines at two different points.
- **Corresponding angles** lie on the same side of the transversal and on the same side of the other two lines.
- **Alternate interior angles** are nonadjacent angles that lie on opposite sides of the transversal and between the other two lines.
- **Alternate exterior angles** lie on opposite sides of the transversal and outside the other two lines.
- **Same-side interior angles** lie on the same side of the transversal and between the other two lines.

Example 1 Classifying Lines and Planes

Three classifications of pairs of lines are identified in this example: parallel, perpendicular, and skew lines. Additionally, a pair of planes are also classified as parallel planes.

Classify each of the following.

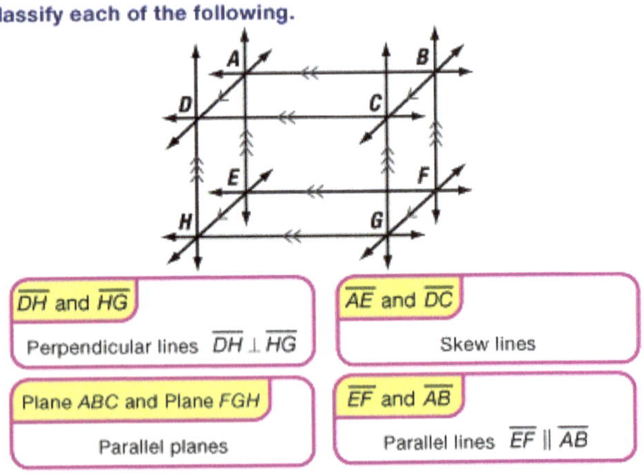

Perpendicular lines intersect at 90°. There are many pairs of perpendicular lines in the given figure, including $\overline{DH}$ and $\overline{HG}$, which can be written symbolically as $\overline{DH} \perp \overline{HG}$. Additionally, $\overline{AB} \perp \overline{BF}$, $\overline{HG} \perp \overline{FG}$, and $\overline{BC} \perp \overline{CD}$.

Skew lines are neither coplanar nor parallel, nor do they intersect. There are many pairs of skew lines in the given figure including $\overline{AE}$ and $\overline{DC}$, as well as $\overline{HE}$ and $\overline{CG}$.

Parallel planes do not intersect. Plane *ABC* is || to plane *FGH*.

Lines are parallel if they are coplanar and do not intersect. The line segments $\overline{EF}$ and $\overline{AB}$ are two coplanar lines that do not intersect, so $\overline{EF} \parallel \overline{AB}$. There are many other pairs of parallel lines in the given figure including $\overline{EF} \parallel \overline{HG}$, $\overline{DH} \parallel \overline{AE}$, and $\overline{BC} \parallel \overline{AD}$, to name a few.

Example 2 Classifying Pairs of Angles

In the figure given below, line *p* is a transversal since it is a line that intersects two coplanar lines, lines *m* and *n*, at two different points. Eight angles are formed by the intersections of lines *p*, *m* and *n*. These angles are numbered 1 through 8. Pairs of these angles are classified in this example as either corresponding angles, alternative interior angles, alternate exterior angles, or same-side interior angles.

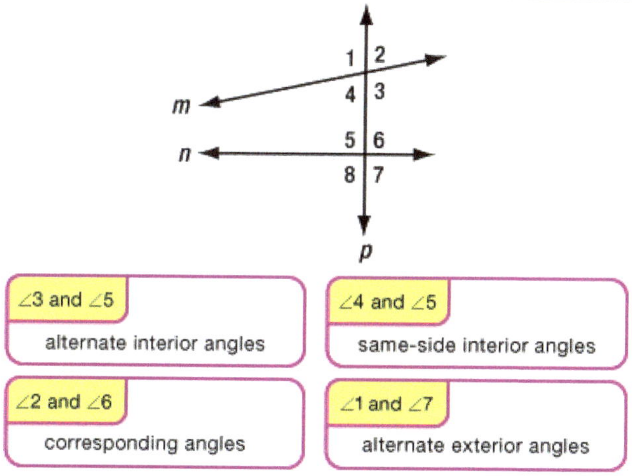

Notice that $\angle 3$ and $\angle 5$ lie on opposite sides of the transversal, *p*. They are also in the space between the two other lines, *m* and *n*, so $\angle 3$ and $\angle 5$ are in the interior of *m* and *n*. Therefore, $\angle 3$ and $\angle 5$ are alternate interior angles.

$\angle 4$ and $\angle 5$ lie on the same side of the transversal, *p*. They are also in the interior of *m* and *n*. Therefore, $\angle 4$ and $\angle 5$ are same-side interior angles.

Notice that $\angle 2$ and $\angle 6$ are in the same relative position within each group of four angles. In other words, these angles lie on the same side of *p* and on the same side of *m* and *n* ($\angle 2$ is above *m* and $\angle 6$ is above *n*). Therefore, $\angle 2$ and $\angle 6$ are corresponding angles. Another pair of corresponding angles is $\angle 4$ and $\angle 8$.

$\angle 1$ and $\angle 7$ lie on opposite sides of the transversal, *p*, and outside the other two lines, *m* and *n*, so $\angle 1$ and $\angle 7$ are said to be in the exterior of lines *m* and *n*. So, $\angle 1$ and $\angle 7$ are alternate exterior angles.

Angles, Parallel Lines, and Transversals

Key Objectives

- Prove and use theorems about the angles formed by parallel lines and a transversal.

Theorems, Postulates, Corollaries, and Properties

- **Corresponding Angles Postulate** If two parallel lines are cut by a transversal, then the pairs of corresponding angles are congruent.
- **Alternate Interior Angles Theorem** If two parallel lines are cut by a transversal, then the pairs of alternate interior angles are congruent.
- **Alternate Exterior Angles Theorem** If two parallel lines are cut by a transversal, then the two pairs of alternate exterior angles are congruent.
- **Same-Side Interior Angles Theorem** If two parallel lines are cut by a transversal, then the two pairs of same-side interior angles are supplementary.

Example 1 Using Congruent Angle Theorems and Postulates

Corresponding Angles Postulate

If two parallel lines are cut by a transversal, then the pairs of corresponding angles are congruent.
(Corr. ∠s Post.)

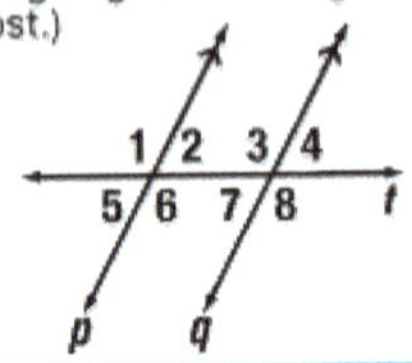

$\angle 1 \cong \angle 3$
$\angle 2 \cong \angle 4$
$\angle 5 \cong \angle 7$
$\angle 6 \cong \angle 8$

By the Corresponding Angles Postulate, if two parallel lines are cut by a transversal, then the pairs of corresponding angles are congruent.

Corresponding angles are on the same side of the transversal and on the same sides of the other lines.

Alternate Interior Angles Theorem

If two parallel lines are cut by a transversal, then the pairs of alternate interior angles are congruent.
(Alt. Int. ∠s Thm.)

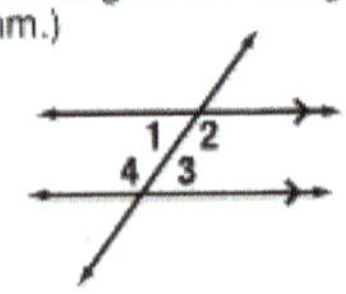

$\Rightarrow$ $\angle 1 \cong \angle 3$
$\angle 2 \cong \angle 4$

By the Alternate Interior Angles Theorem, if two parallel lines are cut by a transversal, then the pairs of alternate interior angles are congruent.

Alternate interior angles are on opposite sides of the transversal and on the inside of the other two lines. There are two pairs of alternate interior angles in the figure given here. $\angle 1$ and $\angle 3$ are alternate interior angles, as well as $\angle 2$ and $\angle 4$.

Alternate Exterior Angles Theorem

If two parallel lines are cut by a transversal, then the pairs of alternate exterior angles are congruent.
(Alt. Ext. ∠s Thm.)

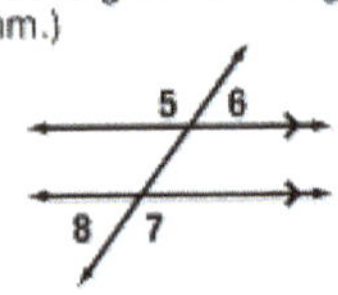

$\Rightarrow$ $\angle 5 \cong \angle 7$
$\angle 6 \cong \angle 8$

By the Alternate Exterior Angles Theorem, if two parallel lines are cut by a transversal, then the pairs of alternate exterior angles are congruent.

Alternate exterior angles are on opposite sides of the transversal and on the outside of the other two lines.

Find m∠*PQT*.

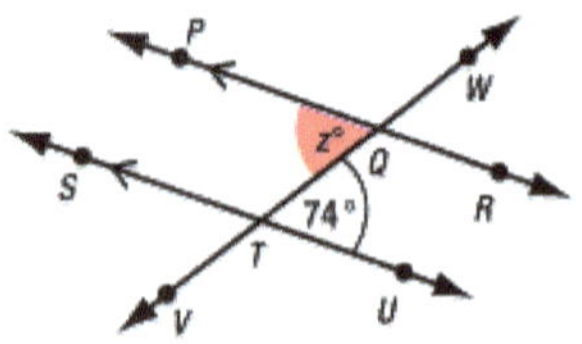

By the Alternate Interior Angles Theorem, $\angle PQT \cong \angle UTW$.
Therefore, $m\angle PQT = m\angle UTW$.
Since $m\angle UTW = 74°$, $m\angle PQT = 74°$.

$m\angle PQT = 74°$

In this example, the measure of $\angle UTW$ is given to be 74° and the measure of $\angle PQT$ is to be found. Notice that $\angle UTW$ and $\angle PQT$ are alternate interior angles. Furthermore, notice that line *PR* is given to be parallel to line *SU*. Since line $PR \parallel$ line *SU*, the Alternate Interior Angles Theorem can be applied. The Alternate Interior Angles Theorem states that if two parallel lines are cut by a transversal, then the alternate interior angles are congruent. Therefore, by the Alternate Interior Angles Theorem, $m\angle UTW = m\angle PQT$. So, since $m\angle UTW = 74°$, $m\angle PQT = 74°$ as well.

Find m∠*STQ*.

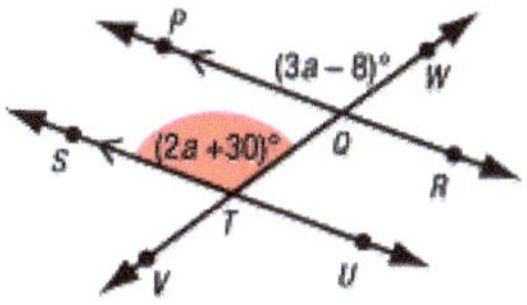

∠*STQ* and ∠*PQW* are corresponding angles.
By the Corresponding Angles Postulate,
∠*STQ* ≅ ∠*PQW*.
Therefore, m∠*STQ* = m∠*PQW*.

Set the expression for m∠*STQ* equal to the expression for m∠*PQW* and solve for *a*.

$$m\angle STQ = m\angle PQW$$
$$2a + 30 = 3a - 8$$
$$38 = a$$

Substitute *a* = 38 into the expression for m∠*STQ* and simplify.

$$m\angle STQ = (2(38) + 30)^\circ$$
$$= (76 + 30)^\circ$$
$$= 106^\circ$$

m∠*STQ* = 106°

Notice that the lines *PR* and *SU* are given to be parallel, that expressions for the measures of ∠*STQ* and ∠*PQW* are given, and that m∠*STQ* is to be found. First, classify the pair ∠*STQ* and ∠*PQW*. Since ∠*STQ* and ∠*PQW* are in the same relative position in each group of 4 angles, ∠*STQ* and ∠*PQW* are corresponding angles. So, since the lines are parallel and ∠*STQ* and ∠*PQW* are corresponding angles, the Corresponding Angles Postulate can be applied, resulting in the fact that m∠*STQ* = m∠*PQW*.

Since m∠*STQ* = m∠*PQW*, it follows that $2a + 30 = 3a - 8$. Solve this equation to find the value of *a*.

Once *a* is known, substitute that value into the expression for m∠*STQ*, $2a + 30$, and simplify.

Example 2 Using the Same-Side Interior Angles Theorem

Same-Side Interior Angles Theorem

If two parallel lines are cut by a transversal, then the two pairs of same-side interior angles are supplementary.
(Same-Side Int. ∠s Thm.)

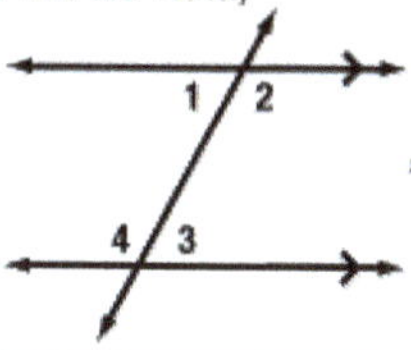

⇒ m∠1 + m∠4 = 180°
m∠2 + m∠3 = 180°

If two parallel lines are cut by a transversal, then the two pairs of same-side interior angles are supplementary.

Same-side interior angles are on the same side of the transversal and on the inside of the other two lines. There are two pairs of same-side interior angles.

Find m∠*UTQ*.

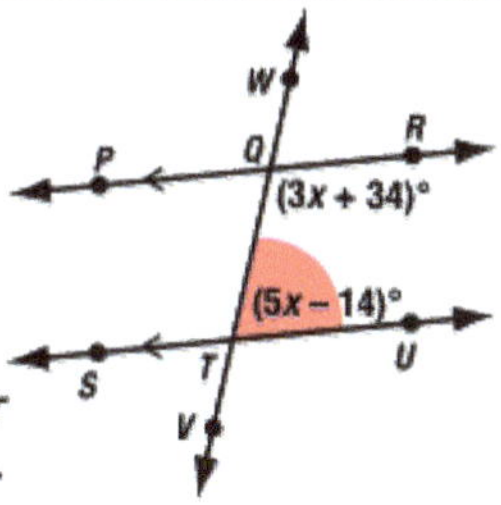

∠*RQT* and ∠*UTQ* are same-side interior angles.
By the Same-Side Interior Angles Theorem,
∠*RQT* and ∠*UTQ* are supplementary.
Therefore, m∠*RQT* + m∠*UTQ* = 180°.

Set the sum of the expression for m∠*RQT* and m∠*UTQ* equal to 180° and solve for *x*.

$$m\angle RQT + m\angle UTQ = 180^\circ$$
$$(3x + 34) + (5x - 14) = 180$$
$$8x + 20 = 180$$
$$8x = 160$$
$$x = 20$$

Substitute *x* = 20 into the expression for m∠*UTQ* and simplify.

$$m\angle UTQ = (5(20) - 14)^\circ$$
$$= (100 - 14)^\circ$$
$$= 86^\circ$$

m∠*UTQ* = 86°

The process for finding m∠*UTQ* is very similar to the process used to find m∠*STQ* in Example 2. Again, first classify the pair ∠*UTQ* and ∠*RQT*. Since the two angles are on the same side of the transversal and inside of the other lines, ∠*UTQ* and ∠*RQT* are same-side interior angles. Therefore, since lines *PR* and *SU* are parallel, ∠*UTQ* and ∠*RQT* are supplementary by the Same-Side Interior Angles Theorem. Use this fact to write an equation, then simplify and solve.

Once *x* is known, substitute that value into the expression for m∠*UTQ*, $5x - 14$, and simplify to find m∠*UTQ*.

Find m∠*ABC*.

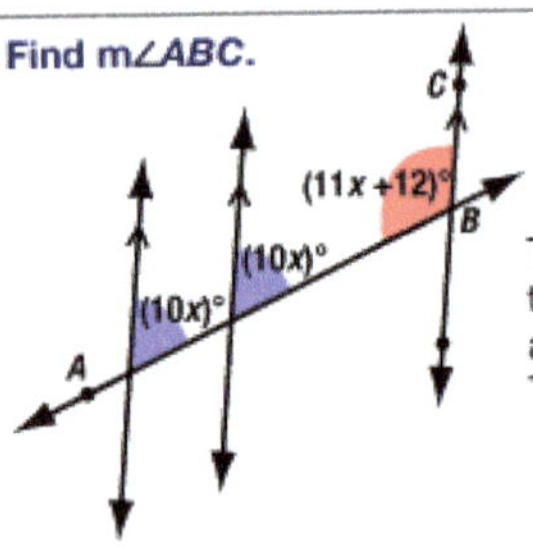

The angle represented by $(10x)°$ and the angle represented by $(11x + 12)°$ are same-side interior angles. Therefore, these angles are supplementary.

Set the sum of $(10x)°$ and $(11x + 12)°$ equal to 180° and solve for x.

$$(10x) + (11x + 12) = 180$$
$$21x + 12 = 180$$
$$21x = 168$$
$$x = 8$$

Substitute $x = 8$ into the expression for m∠*ABC* and simplify.

$$m\angle ABC = (11(8) + 12)°$$
$$= (88 + 12)°$$
$$= 100°$$

$m\angle ABC = 100°$

In this example the $m\angle ABC$ is found by using the Same-Side Interior Angles Theorem.

Since the three lines are parallel and the angles represented by $10x$ and $11x + 12$ are same-side interior angles, these two angles are supplementary. It follows that $(10x) + (11x + 12) = 180$.

Simplify and solve the equation to find the value of x. Then, substitute that value for x into the expression for $m\angle ABC$, $11x + 12$, to find $m\angle ABC$.

Proving that Lines are Parallel

Key Objectives

- Use the angles formed by a transversal to prove two lines are parallel.

Theorems, Postulates, Corollaries, and Properties

- **Converse of the Corresponding Angles Postulate** If two coplanar lines are cut by a transversal so that a pair of corresponding angles are congruent, then the two lines are parallel.
- **Parallel Postulate** Through a point P not on line ℓ, there is exactly one line parallel to ℓ.
- **Converse of the Alternate Interior Angles Theorem** If two coplanar lines are cut by a transversal so that a pair of alternate interior angles are congruent, then the two lines are parallel.
- **Converse of the Alternate Exterior Angles Theorem** If two coplanar lines are cut by a transversal so that a pair of alternate exterior angles are congruent, then the two lines are parallel.
- **Converse of the Same-Side Interior Angles Theorem** If two coplanar lines are cut by a transversal so that a pair of same-side interior angles are supplementary, then the two lines are parallel.

Example 1 Showing that Lines are Parallel

Converse of the Corresponding Angles Postulate

If two coplanar lines are cut by a transversal so that a pair of corresponding angles are congruent, then the two lines are parallel.

(Conv. of Corr. ∠s Post.) $\angle 1 \cong \angle 2$

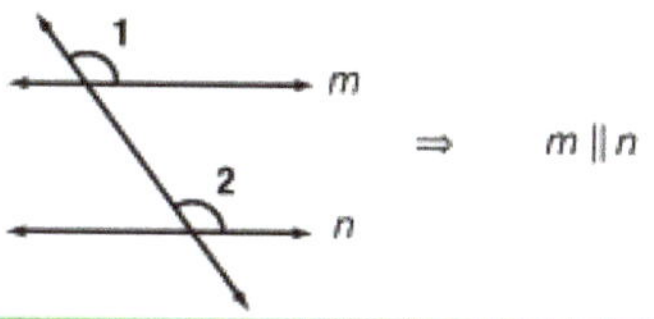

By the Converse of the Corresponding Angles Postulate, if two coplanar lines are cut by a transversal so that a pair of corresponding angles are congruent, then the two lines are parallel.

Converse of the Alternate Exterior Angles Theorem

If two coplanar lines are cut by a transversal so that a pair of alternate exterior angles are congruent, then the two lines are parallel.

(Conv. of Alt. Ext. ∠s Thm.) $\angle 3 \cong \angle 4$

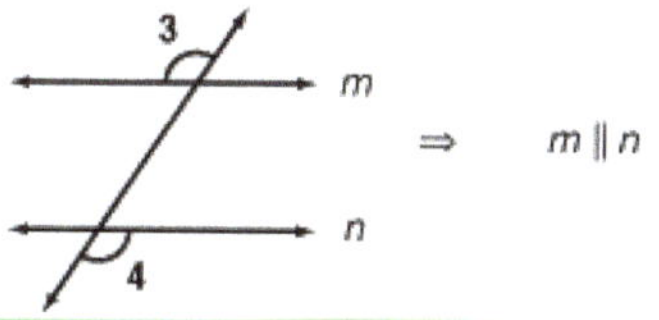

By the converse of the Alternate Exterior Angles Theorem, if two coplanar lines are cut by a transversal so that a pair of alternate exterior angles are congruent, then the two lines are parallel.

Converse of the Alternate Interior Angles Theorem

If two coplanar lines are cut by a transversal so that a pair of alternate interior angles are congruent, then the two lines are parallel.

(Conv. of Alt. Int. ∠s Thm.) $\angle 5 \cong \angle 6$

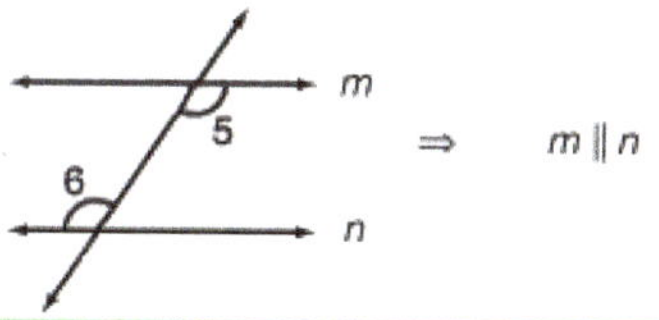

By the converse of the Alternate Interior Angles Theorem, if two coplanar lines are cut by a transversal so that a pair of alternate interior angles are congruent, then the two lines are parallel.

Use the given information to prove that $p \parallel q$.

Given: $m\angle 2 = (9z - 2)°$, $m\angle 4 = (7z + 14)°$ and $m\angle 6 = (11z - 18)°$

Prove: $p \parallel q$

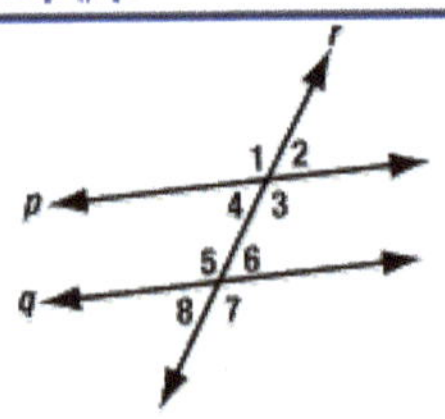

Solve for z.

$m\angle 2 = m\angle 4$ *Vertical ∠s*

$9z - 2 = 7z + 14$ *Substitution.*

$z = 8$ *Solve for z.*

Substitute z = 8 into the expressions for m∠4 and m∠6.

$m\angle 4 = (7(8) + 14)° = (56 + 14)° = 70°$

$m\angle 6 = (11(8) - 18)° = (88 - 18)° = 70°$

Prove that $p \parallel q$.

Since $m\angle 4 = m\angle 6$, $\angle 4 \cong \angle 6$.

$\angle 4$ and $\angle 6$ are congruent alternate interior angles.

Therefore, $p \parallel q$ by the Converse of the Alternate Interior Angles Theorem.

In this example, expressions for the measures of angles 2, 4, and 6 are given. Use these expressions to prove that $p \parallel q$. Notice that $\angle 2$ and $\angle 4$ are vertical angles. Therefore, the $m\angle 2 = m\angle 4$. Use this fact to write an equation and solve for z.

Notice that $\angle 4$ and $\angle 6$ are alternate interior angles. So, if $\angle 4$ and $\angle 6$ are congruent, then it can be concluded that $p \parallel q$ by the Converse of the Alternate Interior Angles Theorem. Substitute the value of z into the expressions for $m\angle 4$ and $m\angle 6$. Since $m\angle 4 = 70°$ and $m\angle 6 = 70°$, $\angle 4 \cong \angle 6$.

Therefore, since these alternate interior angles are congruent, $p \parallel q$ by the Converse of the Alternate Interior Angles Theorem.

Example 2 Using the Converse of the Same Side Interior Angles Theorem

Converse of the Same-Side Interior Angles Theorem

If two coplanar lines are cut by a transversal so that a pair of same-side interior angles are supplementary, then the two lines are parallel.

(Conv. Same-Side Int. ∠s Thm.)

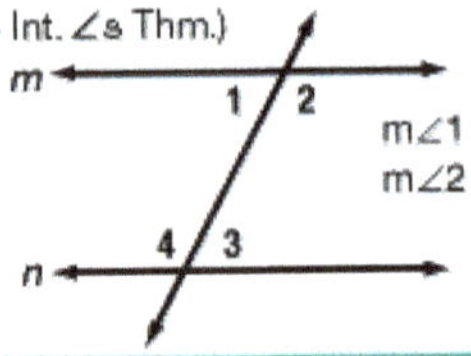

$m\angle 1 + m\angle 4 = 180° \Rightarrow m \parallel n$

$m\angle 2 + m\angle 3 = 180° \Rightarrow m \parallel n$

By the converse of the Same-Side Interior Angles Theorem, if two coplanar lines are cut by a transversal so that a pair of same-side interior angles are supplementary, then the two lines are parallel.

Use the given information to show that $m \parallel n$.

Given: $m\angle 8 = (6x + 5)°$, $m\angle 3 = (7x - 20)°$ and $x = 15$

Prove: $m \parallel n$

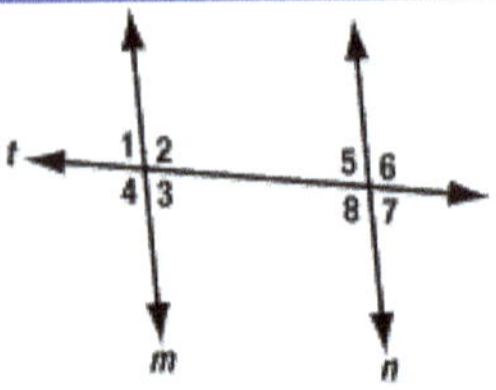

Substitute x = 15 into the expressions for m∠3 and m∠8.

$m\angle 8 = (6(15) + 5)° = (90 + 5)° = 95°$

$m\angle 3 = (7(15) - 20)° = (105 - 20)° = 85°$

Use these angle measures to prove that $m \parallel n$.

Since $m\angle 3 + m\angle 8 = 95° + 85° = 180°$, $\angle 3$ and $\angle 8$ are supplementary angles.

By definition, $\angle 3$ and $\angle 8$ are same-side interior angles.

Therefore, since $\angle 3$ and $\angle 8$ are same-side interior angles and they are supplementary, $m \parallel n$ by the Converse of the Same-Side Interior Angles Theorem.

In this example, two lines, m and n, are proven to be parallel by using the Converse of the Same-Side Interior Angles Theorem.

Expressions for the measures of angles 3 and 8 are given. Notice that $\angle 3$ and $\angle 8$ are same-side interior angles. If it can be shown that $\angle 3$ and $\angle 8$ are supplementary, then it can be concluded that $m \parallel n$ by the Converse of the Same-Side Interior Angles Theorem.

Remember, the sum of two supplementary angles is 180°.

Example 3 Proving that Lines are Parallel

Use the given information to show that $a \parallel b$.

Given:
$\angle 4 \cong \angle 8$ and $\angle 7 \cong \angle 16$

Prove: $a \parallel b$

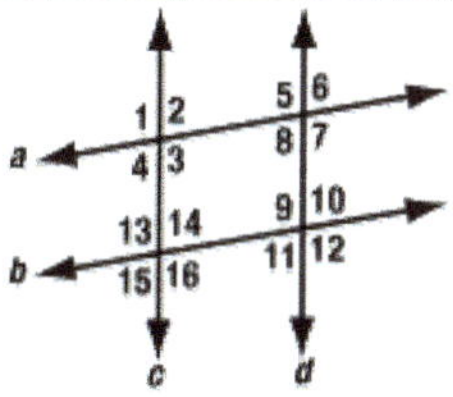

Statements	Reasons
1. $\angle 4 \cong \angle 8$	1. Given
2. $\angle 4$ and $\angle 8$ are corresp. $\angle$s	2. Def. corresp. $\angle$s
3. $c \parallel d$	3. Conv. Corresp. $\angle$s Post.
4. $\angle 7 \cong \angle 16$	4. Given
5. $\angle 16 \cong \angle 9$	5. Alt. Int. $\angle$s Thm.
6. $\angle 7 \cong \angle 9$	6. Trans. Prop. of $\cong$
7. $\angle 7$ and $\angle 9$ are alt. int. $\angle$s	7. Def. alt. int. $\angle$s
8. $a \parallel b$	8. Conv. Alt. Int. $\angle$s Thm.

The figure given in this example includes four lines instead of three, as was the case in the previous examples. To prove that $a \parallel b$, consider the given information: $\angle 4 \cong \angle 8$ and $\angle 7 \cong \angle 16$. Notice that $\angle 4$ and $\angle 8$ are corresponding angles formed by transversal line a. Therefore, it can be concluded that $c \parallel d$ by the Converse of the Corresponding Angles Postulate.

Since $c \parallel d$, it can be concluded that any corresponding, alternate interior, or alternate exterior angles formed by either transversal a or by transversal b are congruent.

There are many different ways to complete the proof. Instead of using the Alternative Interior Angles Theorem to conclude that $\angle 16 \cong \angle 9$, the Corresponding Angles Theorem could be used to conclude that $\angle 16 \cong \angle 12$. And since $\angle 16 \cong \angle 12$ and $\angle 7 \cong \angle 16$ (Given), by the Transitive Property of Congruence, $\angle 7 \cong \angle 12$. It follows that $a \parallel b$ by the Converse of the Corresponding Angles Postulate.

Properties of Perpendicular Lines

Key Objectives

- Prove and apply theorems about perpendicular lines.

Key Terms

- The **perpendicular bisector** of a segment is a line perpendicular to a segment at the segment's midpoint.
- The **distance from a point to a line** is the length of the perpendicular segment from the point to the line.

Theorems, Postulates, Corollaries, and Properties

- **Linear Pair of Congruent Angles Postulate** If two intersecting lines form a linear pair of congruent angles, then the lines are perpendicular.
- **Perpendicular Transversal Theorem** In a plane, if a transversal is perpendicular to one of two parallel lines, then it is perpendicular to the other line.
- **Lines Perpendicular to the Same Line Theorem** If two coplanar lines are perpendicular to the same line, then the two lines are parallel to each other.

Example 1 Distance from a Point to a Line

Write and solve an inequality for *x*.

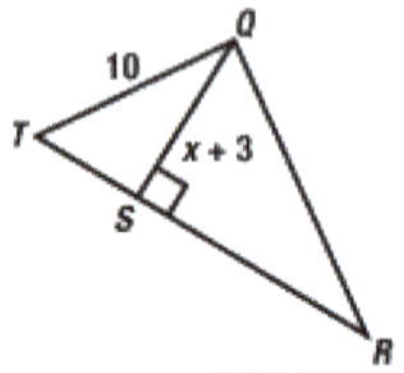

Use the definition of distance from a point to a line to write an inequality comparing *QS* and *QT*.

$\overline{QS}$ is a line segment from Q to $\overleftrightarrow{TR}$.
$\overline{QT}$ is a line segment from Q to $\overleftrightarrow{TR}$.
Since $\overline{QS} \perp \overleftrightarrow{TR}$, QS must be the shortest line segment from Q to $\overleftrightarrow{TR}$.
Therefore, $QS < QT$.

Substitute the expression for *QS*, *QS* = *x* + 3, and the length of *QT*, *QT* = 10, into the inequality and solve for *x*.

$QS < QT$
$x + 3 < 10$ *Substitute.*
$x < 7$ *Solve for x.*

The distance from a point to a line is used in this example to write and solve an inequality.

Notice that three line segments from the point Q to $\overleftrightarrow{TR}$ are shown in this figure, $\overline{QT}$, $\overline{QS}$, and $\overline{QR}$. The line segment that is also perpendicular to $\overleftrightarrow{TR}$, $\overline{QS}$, is the distance from Q to $\overleftrightarrow{TR}$. Therefore, QS must be the shortest distance from Q to $\overleftrightarrow{TR}$. It follows that $QS < QT$.

Substitute the given expression for QS, $x + 3$ and the given length for QT, 10, into the inequality, $QS < QT$, and solve for x.

Example 2 Proving Properties of Lines

Perpendicular Transversal Theorem
In a plane, if a transversal is perpendicular to one of two parallel lines, then it is perpendicular to the other line.
(⊥ Transv. Thm.)

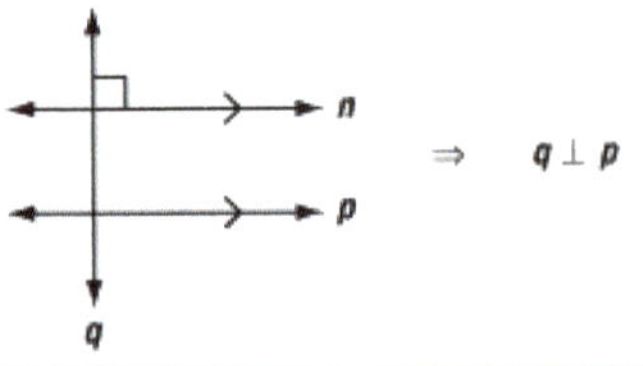

By the Perpendicular Transversal Theorem, in a plane, if a transversal is perpendicular to one of two parallel lines, then it is perpendicular to the other line.

Lines Perpendicular to the Same Line Theorem
If two coplanar lines are perpendicular to the same line, then the two lines are parallel to each other.
(2 lines ⊥ to same line → 2 lines ||.)

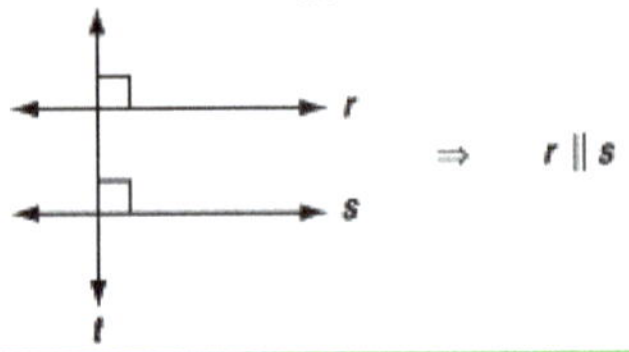

By the Lines Perpendicular to the Same Line Theorem, if two coplanar lines are perpendicular to the same line, then the two lines are parallel to each other.

Linear Pair of Congruent Angles Postulate

If two lines intersect and form a linear pair such that the angles are congruent, then the lines are perpendicular.

(Lin. Pair. ≅ ∠s Post.)

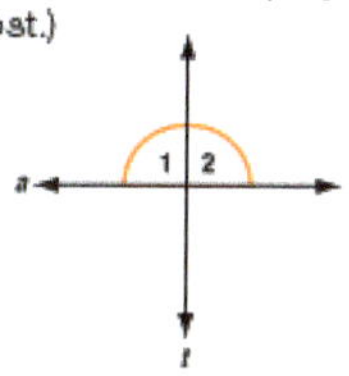

$\angle 1 \cong \angle 2 \Rightarrow a \perp t$

By the Linear Pair of Congruent Angles Postulate, if two lines intersect and form a linear pair such that the angles are congruent, then the lines are perpendicular.

Write a two-column proof.

Given: $\angle 1 \cong \angle 2$, $a \parallel b$ and $c \parallel d$

Prove: $b \perp d$

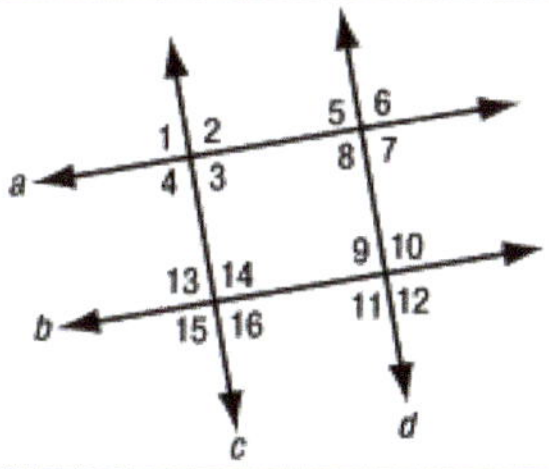

Statements	Reasons
1. $\angle 1 \cong \angle 2$	1. Given
2. $\angle 1$ and $\angle 2$ form a linear pair	2. Def. linear pair
3. $a \perp c$	3. Lin. Pair of ≅ ∠s Post.
4. $a \parallel b$	4. Given
5. $b \perp c$	5. ⊥ Transversal Thm.
6. $c \parallel d$	6. Given
7. $b \perp d$	7. ⊥ Transversal Thm.

Use the given information to prove that $b \perp d$.

Notice that $\angle 1$ and $\angle 2$ form a linear pair. So, since $\angle 1$ and $\angle 2$ are congruent (Given) and form a linear pair (definition of linear pair), $a \perp c$ by the Linear Pair of Congruent Angles Postulate.

It is also given that $a \parallel b$. So, since $a \perp c$ and $a \parallel b$, b must be perpendicular to c as well by the Perpendicular Transversal Theorem.

Finally, since $b \perp c$ and $c \parallel d$ (Given), b must be perpendicular to d as well by the Perpendicular Transversal Theorem.

Finding the Slope Given Two Points

Key Objectives

- Learn to find slope by using the slope formula.

Key Terms

- The **rise** is the difference in the y-values of two points on a line.
- The **run** is the difference in the x-values of two points on a line.
- The **slope** of a line is a measure of its steepness and is the ratio of rise to run.

Slope Formula

$$m = \frac{y_2 - y_1}{x_2 - x_1}$$

The formula given above can be used to calculate the slope of a line when the coordinates of two points on the line are known. The letter m is used to represent slope. The coordinates of two points on the line are represented by y_2, y_1, x_2, and x_1, where y_2 is the y-coordinate of the second point, x_2 is the x-coordinate of the second point, and so on.

Example 1

To find the slope of a line, substitute the coordinate values from two points on that line into the slope formula and simplify. It does not matter which point is designated as (x_1, y_1) and which point is designated as (x_2, y_2).

Finding Slope by Using the Slope Formula

Find the slope of the line that contains (5, –3) and (–1, 2).

(5, –3) and (–1, 2)
(x_1, y_1) (x_2, y_2)

$m = \frac{y_2 - y_1}{x_2 - x_1}$	*Use the slope formula.*
$= \frac{2-(-3)}{-1-5}$	*Substitute (5, –3) for (x_1, y_1) and (–1, 2) for (x_2, y_2).*
$= \frac{5}{-6}$	*Simplify.*
$= -\frac{5}{6}$	The slope of the line that contains (5, –3) and (–1, 2) is $-\frac{5}{6}$.

In this example, (5, –3) is designated as (x_1, y_1) and (–1, 2) is designated as (x_2, y_2). However, the slope may also be found by designating (–1, 2) as (x_1, y_1) and (5, –3) as (x_2, y_2). To demonstrate this, find the slope again, using this point designation.

$$m = \frac{-3-2}{5-(-1)} = \frac{-5}{6} = -\frac{5}{6}$$

The slope is again –5/6.
The order in which the points are chosen is irrelevant because the line is the same no matter which point is chosen as (x_1, y_1).

NOTE: Do *not* designate the x-coordinate from the first point as x_1 and then the y-coordinate from the *second* point as y_1.

Example 2

Again, to find the slope of a line, substitute the coordinate values from two points on that line into the slope formula and simplify. When the graph of a line is given, the slope formula can still be used to find the slope of the line. Simply identify two points on the line and then substitute their coordinates into the formula.

Finding Slope from Graphs and Tables

The graph shows a linear relationship. Find the slope.

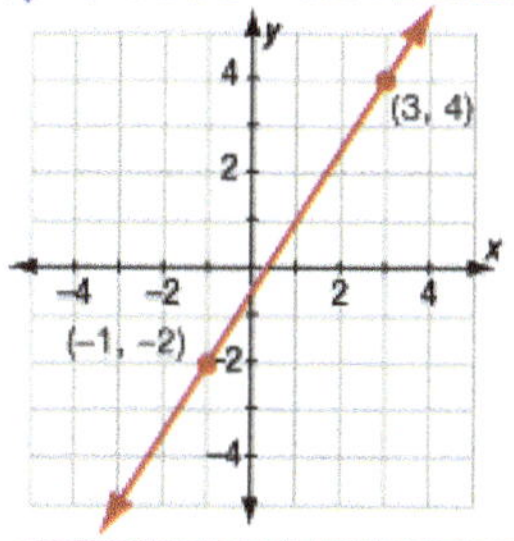

Let (−1, −2) be (x_1, y_1) and (3, 4) be (x_2, y_2).

$m = \frac{y_2 - y_1}{x_2 - x_1}$	*Use the slope formula.*
$= \frac{4 - (-2)}{3 - (-1)}$	*Substitute (−1, −2) for (x_1, y_1) and (3, 4) for (x_2, y_2).*
$= \frac{6}{4}$	*Simplify.*
$= \frac{3}{2}$	

Consider this exercise. Begin by choosing any two points on the line. Here, the points (–1, –2) and (3, 4) are identified. Substitute the coordinates into the slope formula. The slope is found to be 3/2. Note that different points could have been used to find the slope.

Find the slope of this line again, but now use the points (1, 1) and (3, 4).

$$m = \frac{4-1}{3-1} = \frac{3}{2}$$

As expected, the same slope is found.

Notice that the process was simpler this time because there was no need to simplify the fraction.

Remember, a table is a list of *x*- and *y*-values that form coordinate pairs. In other words, a table is a list of points.

The table shows a linear relationship. Find the slope.

x	0	2	3	7
y	4	4	4	4

Step 1 Choose any two points from the table. Let (0, 4) be (x_1, y_1) and (2, 4) be (x_2, y_2).

Step 2 Use the slope formula.

$m = \frac{y_2 - y_1}{x_2 - x_1}$	*Use the slope formula.*
$= \frac{4 - 4}{2 - 0}$	*Substitute (0, 4) for (x_1, y_1) and (2, 4) for (x_2, y_2).*
$= \frac{0}{2}$	*Simplify.*
$= 0$	

In this exercise, it is given that the table describes a linear relationship. This means that the points on the table will form a straight line.

Once again, the slope formula can be used.

Choose any two points from the table, substitute their coordinates into the formula, and simplify.

Since this is a linear relationship, it does not matter which two points are chosen from the table.

Example 3

In Example 3, the slope of a line is found by using the slope formula when a graph of a line is given, and then the meaning of the slope is explained in the context of the situation.

Consumer Application

The graph shows the value of a computer after *x* months. Find the slope of the line and explain what the slope represents.

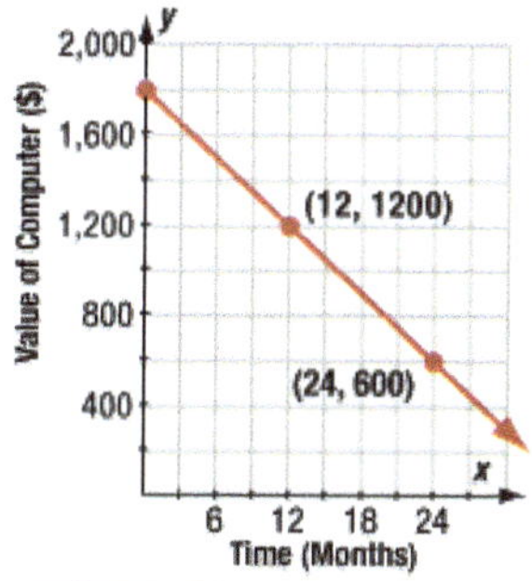

Step 1 Use the slope formula.

$$m = \frac{y_2 - y_1}{x_2 - x_1}$$
$$= \frac{600 - 1200}{24 - 12}$$
$$= \frac{-600}{12} = \frac{-100}{2} = -50$$

Step 2 Tell what the slope represents.

In this situation, *y* represents dollars and *x* represents time.

$$\text{slope} = \frac{\text{change in cost}}{\text{change in time}} \text{ in units of } \frac{\text{dollars}}{\text{month}}$$

The computer decreases in value $50 each month.

Consider this example. Find the slope by substituting the coordinates from the points (12, 1200) and (24, 600) into the slope formula.

Remember, slope is the ratio of the change in *y* to the change in *x*. Here, the variable *y* represents the value of the computer in dollars, so "change in *y*" can be translated as "change in value". The variable *x* represents the time (in months) that has passed since the computer was purchased, so "change in *x*" can be translated as "change in time". The slope was found to be –50, or –50/1. The negative indicates a decrease, 50 is the change in *value*, and 1 is the change in *time*. Therefore, a slope of –50 means that the value of the computer decreases by $50 each month.

Slope-Intercept Form

Key Objectives

- Write a linear equation in slope-intercept form.
- Graph a line using slope-intercept form.

Key Terms

- The **slope-intercept form** of a line is $y = mx + b$, where m is the slope and b is the y-intercept.

Example 1 Graphing by Using Slope and *y*-Intercept

Graph the line given the slope and *y*-intercept.

slope = $\frac{2}{3}$, y-intercept = −1

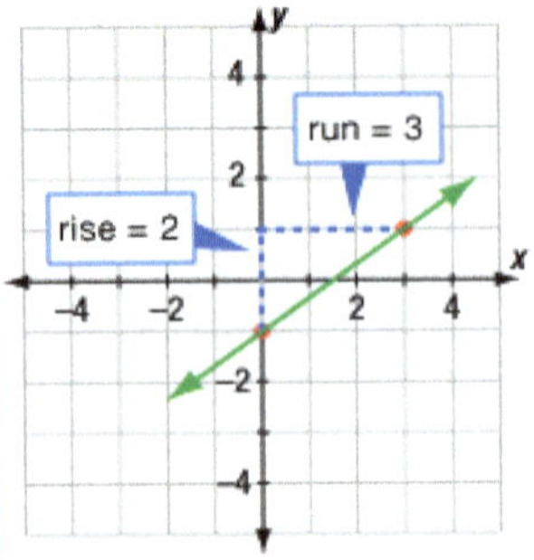

Step 1 The y-intercept is −1, so the line contains (0, −1). Plot (0, −1).

Step 2 Count 2 units up and 3 units right from (0, −1) and plot another point.

Step 3 Draw the line through the two points.

A line is graphed in this example using the slope and y-intercept. The slope and y-intercept of the line are given.

Step 1
For the y-intercept, $x = 0$ and $y = -1$. Plot the point (0, −1).

Step 2
The slope of the line is 2/3. From the first point, count up 2 units and then to the right 3 units to plot the second point, (3, 1).

Step 3
Draw a line through the points to complete the graph.

Graph the line given the slope and *y*-intercept.

slope = −3, y-intercept = 1

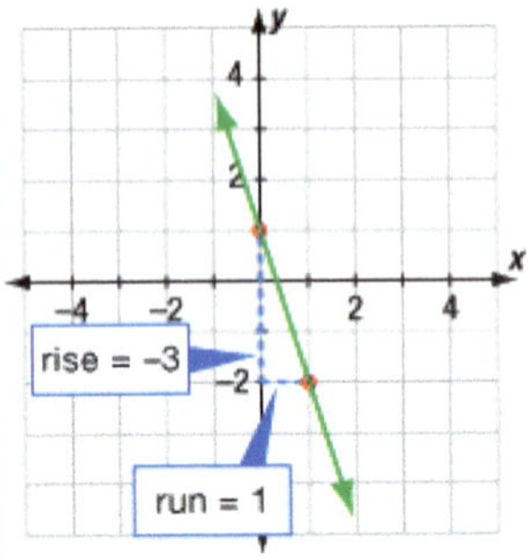

Step 1 The y-intercept is 1, so the line contains (0, 1). Plot (0, 1).

Step 2 Count 3 units down and and 1 unit right from (0, 1) and plot another point.

Step 3 Draw the line through the two points.

A line is graphed in this example using the slope and y-intercept. The slope and y-intercept of the line are given.

Step 1
For the y-intercept, $x = 0$ and $y = 1$. Plot the point (0, 1).

Step 2
The slope of the line is −3. From the first point, count down 3 units and then to the right 1 unit to plot the second point, (1, −2).

Step 3
Draw a line through the points to complete the graph.

Example 2 Writing Linear Equations in Slope-Intercept Form

Write the equation that describes the line in slope-intercept form.

slope = $\frac{3}{4}$, y-intercept = 5

$y = mx + b$ *Use the slope-intercept form of a linear equation.*

$y = \frac{3}{4}x + 5$ *Substitute the given values for m and b.*

An equation for a line is written in slope-intercept form in this example. The slope and y-intercept of the line are given.

Substitute the slope and y-intercept for m and b in the slope-intercept equation. The equation of the line is

$$y = \frac{3}{4}x + 5.$$

Write the equation that describes the line in slope-intercept form.

slope = –1, y-intercept = $-\frac{3}{5}$

$y = mx + b$	*Write the slope-intercept form.*
$y = (-1)x + \left(-\frac{3}{5}\right)$	*Substitute the given values for m and b.*
$y = -x - \frac{3}{5}$	*Simplify.*

An equation for a line is written in slope-intercept form in this example. The slope and y-intercept of the line are given.

Substitute the slope and y-intercept for m and b in the slope-intercept equation. The equation of the line is

$y = -x - \frac{3}{5}$.

Write the equation that describes the line in slope-intercept form.

slope = –2, y-intercept = 0

$y = mx + b$	*Write the slope-intercept form.*
$y = -2x + 0$	*Substitute the given values for m and b.*
$y = -2x$	*Simplify.*

To write the equation that describes the line with slope –2 and y-intercept 0, substitute the slope and y-intercept for m and b in the slope-intercept equation, $y = mx + b$. The equation of the line is $y = -2x$. Notice that since the y-intercept is 0, it does not appear as a factor in the equation.

Write the equation that describes the line in slope-intercept form.

slope = 0, y-intercept = –2

$y = mx + b$	*Write in slope-intercept form.*
$y = 0x + (-2)$	*Substitute the given values for m and b.*
$y = -2$	*Simplify.*

Substitute the slope and y-intercept for m and b in the slope-intercept equation. The equation of the line is $y = -2$. Notice that since the slope is 0, there is no x-term in the equation of the line.

A line with 0 slope is a horizontal line. The equation of a horizontal line is always $y = b$, where b is the y-intercept.

Similarly, a line with undefined slope is a vertical line. The equation of a vertical line is always $x = a$, where a is the x-intercept.

Write the equation that describes the line in slope-intercept form.

slope = 2, (1, –3) is on the line

Step 1 Find the y-intercept.

$y = mx + b$	*Write the slope-intercept form.*
$y = 2x + b$	
$-3 = 2(1) + b$	*Substitute 2 for m, 1 for x, and –3 for y.*
$-3 = 2 + b$	*Simplify.*
$-2 \quad -2$	*Solve for b. Subtract 2 from each side.*
$-5 = b$	

Step 2 Write the equation.

$y = mx + b$	*Write the slope-intercept form.*
$y = 2x - 5$	*Substitute 2 for m and –5 for b.*

An equation for a line is written in slope-intercept form in this example. The slope and a point on the line are given.

Step 1
Substitute the slope for m in the slope-intercept equation. Substitute the coordinates of the given point on the line for x and y and solve for b. The value of b, the y-intercept, is –5.

Step 2
Substitute the value for the slope and the y-intercept into the slope-intercept form. The equation of the line is $y = 2x - 5$.

Example 3 Using Slope-Intercept Form to Graph

Write the equation in slope-intercept form. Then graph the line described by the equation.

$y = -3x + 2$

$y = mx + b$

$y = -3x + 2$

slope: $m = -3$ y-intercept: $b = 2$

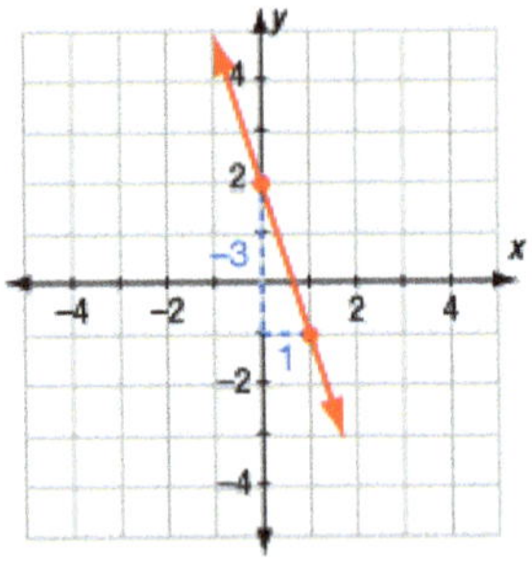

Step 1 Plot (0, 2).

Step 2 Count 3 units down and 1 unit right and plot another point.

Step 3 Draw the line connecting the two points.

An equation is written in slope-intercept form and graphed in this example. The equation of the line is given.

The equation as given is in slope-intercept form with $m = -3$ and $b = 2$.

Step 1
Plot the y-intercept $(0, b)$, or (0, 2).

Step2
Using the slope value of −3, count down 3 units from the y-intercept and then 1 unit to the right. Plot the second point.

Step 3
Draw a line through the points to complete the graph.

Write the equation in slope-intercept form. Then graph the line described by the equation.

$x - 3y = 12$

Step 1 Write the equation in slope-intercept form by solving for y.

$$x - 3y = 12$$
$$-x \qquad -x$$
Subtract x from both sides.

$$\frac{-3y}{-3} = \frac{-x + 12}{-3}$$
Divide both sides by −3.

$$y = \frac{1}{3}x + (-4)$$
$$\frac{-x}{-3} = \frac{1}{3}x$$

$$y = \frac{1}{3}x - 4$$

Step 2 Graph the line.

$y = \frac{1}{3}x + -4$ is in the form $y = mx + b$.

slope: $m = \frac{1}{3}$

y-intercept: $b = -4$

- Plot (0, −4).
- Count 1 unit up and 3 units right and plot another point.
- Draw the line connecting the two points.

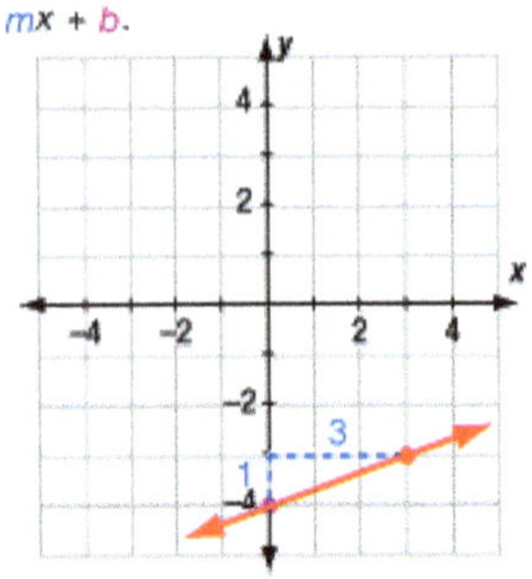

An equation is written in slope-intercept form and graphed in this example. The equation of the line is given.

Step 1
The equation is not in slope-intercept form. Rearrange the equation to isolate y.

Step2
Graph the line using the slope of 1/3 and the y-intercept of −4. Plot the y-intercept $(0, b)$, or (0, −4). Using the slope value of 1/3, count up 1 unit from the y-intercept and then 3 units to the right. Plot the second point. Draw a line through the points to complete the graph.

Example 4 Consumer Application

To rent a tuxedo, a formal wear store charges $50 plus $40 per day. The total cost as a function of the number of days rented is shown in the graph. Write an equation that represents the total cost as a function of the number of days. Identify the slope and y-intercept and describe their meanings. Find the cost of the tuxedo if it is rented for 3 days.

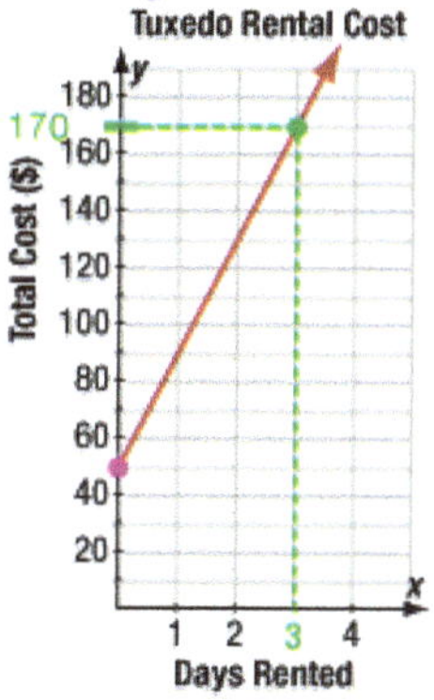

total cost	is	$40	for each day	plus	$50
y	=	$40	$\cdot x$	+	$50

$y = 40x + 50$

$m = 40$ — The rate of change of the cost, $40/day.

$b = 50$ — The one time charge, $50.

$$\begin{aligned} y &= 40x + 50 \\ &= 40(3) + 50 \quad \text{Substitute 3 for } x. \\ &= 120 + 50 \\ &= 170 \end{aligned}$$

The cost of the tuxedo for 3 days is $170.

An equation is written that describes the cost of a tuxedo rental in this application example. The equation is used to determine the cost of a 3-day rental. The rental rate and one-time charge are given.

To write the equation, refer to the given information. The cost, y, is equal to $40 per day, where x represents the number of days, plus the one-time charge of $50. The equation is $y = 40x + 50$. The graph can also be used to identify the y-intercept and the slope of the line to write the equation.

The slope of the line is 40, which is the daily-rate portion of the cost. The y-intercept is 50, which is the one-time charge, or the amount the cost starts at as soon as the rental begins.

To find the cost of a 3-day rental, substitute 3 for x in the equation and solve for y, the cost. The cost is $170.

Point-Slope Form

Key Objectives

- Graph a line and write a linear equation using point-slope form.
- Write a linear equation given two points.

Key Terms

- The **point-slope form** of a line is $y - y_1 = m(x - x_1)$, where m is the slope and (x_1, y_1) is a point on the line.

Example 1 Using Slope and a Point to Graph

Graph the line with the given slope that contains the given point.

slope = 2; (−1, 1)

Step 1
Plot (−1, 1).

Step 2
Use the slope to move from (−1, 1) to another point.

$$\text{slope} = \frac{\text{change in } y}{\text{change in } x} = 2 = \frac{2}{1}$$

Move 2 units up and 1 unit right and plot another point.

Step 3
Draw the line connecting the two points.

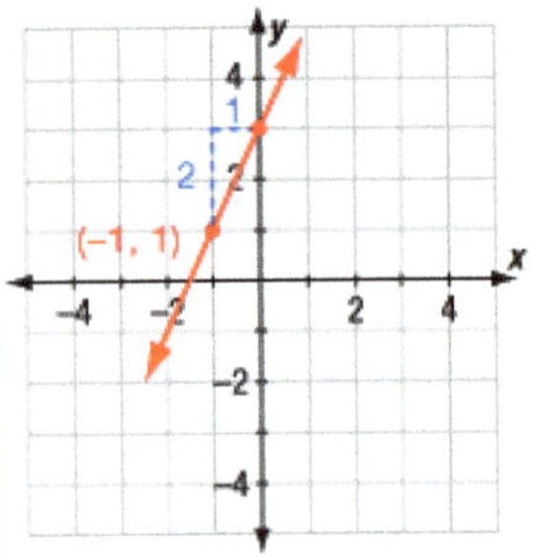

A line that passes through a given point and has a given slope is graphed in this example.

Step 1
Begin by plotting the given point, (−1, 1).

Step 2
Use the slope of the line, 2, to locate another point. Count from the first point up 2 units and then to the right 1 unit. Plot a second point.

Step 3
Draw a line through the points to complete the graph.

Graph the line with the given slope that contains the given point.

slope = $-\frac{2}{3}$; (−3, 4)

Step 1
Plot (−3, 4).

Step 2
Use the slope to move from (−3, 4) to another point.

$$\text{slope} = \frac{\text{change in } y}{\text{change in } x} = -\frac{2}{3} = \frac{-2}{3}$$

Move 2 units down and 3 units right and plot another point.

Step 3
Draw the line connecting the two points.

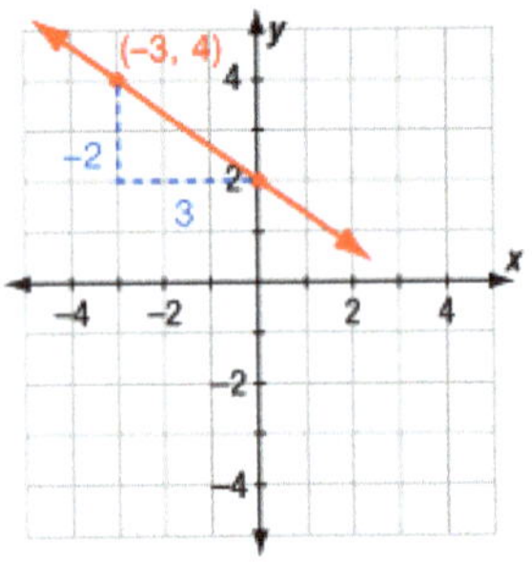

A line that passes through a given point and has a given slope is graphed in this example.

Step 1
Begin by plotting the given point, (−3, 4).

Step 2
Use the slope of the line, −2/3, to locate another point. Count from the first point down 2 units and then to the right 3 units. Plot a second point.

Step 3
Draw a line through the points to complete the graph.

Graph the line with the given slope that contains the given point.

slope = 0; (2, −3)

Step 1
Plot (2, −3).

Step 2
Use the slope to move from (2, −3) to another point.

$$\text{slope} = \frac{\text{change in } y}{\text{change in } x} = 0 = \frac{0}{1}$$

A line with slope 0 is horizontal.

Step 3
Draw the line connecting the two points.

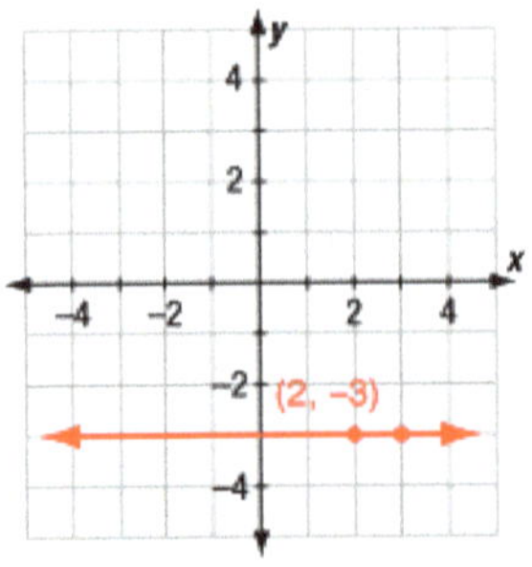

A line that passes through a given point and has a given slope is graphed in this example.

Step 1
Begin by plotting the given point, (2, −3).

Step 2
Use the slope of the line, 0, to locate another point. A slope of zero means there is no rise, or vertical displacement. A line with slope 0 is horizontal. Count from the first point to the right 1 unit. Plot a second point.

Step 3
Draw a line through the points to complete the graph.

Example 2 Writing Linear Equations in Point-Slope Form

Write an equation in point-slope form for the line with the given slope that contains the given point.

slope = $\frac{4}{3}$; (−1, 0)

$y - y_1 = m(x - x_1)$ *Use point-slope form.*

$y - 0 = \frac{4}{3}(x - (-1))$ *Substitute the slope and the x and y values from the point for m, x_1 and y_1.*

$y - 0 = \frac{4}{3}(x + 1)$ *Rewrite subtraction of negative numbers as addition.*

or

$y = \frac{4}{3}(x + 1)$

An equation is written in point-slope form in this example. The slope and one point on the line are given.

Begin by writing out the point-slope form for an equation of a line. Substitute the given slope value for m. Substitute the given point coordinates for x_1 and y_1. Simplify the equation.

The point-slope form of the equation is $y = \frac{4}{3}(x+1)$.

Write an equation in point-slope form for the line with the given slope that contains the given point.

slope = −3; (5, 4)

$y - y_1 = m(x - x_1)$ *Use point-slope form.*

$y - 4 = -3(x - 5)$ *Substitute the slope and the x and y values from the point for m, x_1 and y_1.*

An equation is written in point-slope form in this example. The slope and one point on the line are given.

Begin by writing out the point-slope form for an equation of a line. Substitute the given slope value for m. Substitute the given point coordinates for x_1 and y_1. Simplify the equation. The point-slope form of the equation is $y - 4 = -3(x - 5)$.

Write an equation in point-slope form for the line with the given slope that contains the given point.

slope = 0; (−5, −1)

$y - y_1 = m(x - x_1)$ *Use point-slope form.*

$y - (-1) = 0(x - (-5))$ *Substitute the slope and the x and y values from the point for m, x_1 and y_1.*

$y + 1 = 0$ *Rewrite subtraction of negative numbers as addition.*

An equation is written in point-slope form in this example. The slope and one point on the line are given.

Begin by writing out the point-slope form for an equation of a line. Substitute the given slope value for m. Substitute the given point coordinates for x_1 and y_1. Simplify the equation. The point-slope form of the equation is $y + 1 = 0$. Notice that with a slope of zero the line does not depend on x.

Example 3 Writing Linear Equations in Slope-Intercept Form

Write an equation in slope-intercept form for the line with slope −2 that contains (−3, −1).

slope = −2; (−3, −1)

Step 1 Write the equation in point-slope form.

$y - y_1 = m(x - x_1)$ *Use point-slope form.*

$y - (-1) = -2(x - (-3))$ *Substitute the slope and the x and y values from the point for m, x_1 and y_1.*

$y + 1 = -2(x + 3)$ *Rewrite subtraction of negative numbers as addition.*

Step 2 Write the equation in slope-intercept form by solving for y.

$y + 1 = -2(x + 3)$ *Use the Distributive Property.*

$y + 1 = -2x - 6$ *Distribute −2 on the right side.*

$\underline{-1} \quad \underline{-1}$ *Subtract 1 from both sides.*

$y = -2x - 7$

The equation $y = -2x - 7$ is in slope-intercept form.

An equation is written in slope-intercept form in this example. The slope and one point on the line are given.

Begin by writing out the point-slope form for an equation of a line. Substitute the given slope value for m. Substitute the given point coordinates for x_1 and y_1. The point-slope form of the equation is $y + 1 = -2(x + 3)$.

Simplify and isolate y on one side of the equation to write the equation in slope-intercept form. The slope-intercept form of the equation is $y = -2x - 7$.

Example 4 Using Two Points to Write an Equation

Write an equation in slope-intercept form for the line through the two points. (1, 4) and (3, 14) **Step 1** Find the slope. $m = \frac{y_2 - y_1}{x_2 - x_1} = \frac{14 - 4}{3 - 1} = \frac{10}{2} = 5$ **Step 2** Substitute the slope and one of the points into the point-slope form. $y - y_1 = m(x - x_1)$ *Use the point-slope form.* $y - 4 = 5(x - 1)$ *Substitute the slope and the x and y values from the point for m, x_1 and y_1.* **Step 3** Write the equation in slope-intercept form. $y - 4 = 5(x - 1)$ $y - 4 = 5x - 5$ *Distribute 5 on the right side.* $+4 \quad +4$ *Add 4 to both sides.* $y = 5x - 1$	An equation is written in slope-intercept form in this example. Two points on the line are given. **Step 1** Begin by finding the slope for the line by finding the difference in y coordinates of the two points over the difference in the x coordinates. **Step 2** Substitute the slope as m in the point-slope form. Substitute one of the given point coordinates for x_1 and y_1. The point-slope form of the equation is $y - 4 = 5(x - 1)$. **Step 3** Simplify and isolate y on one side of the equation to write the equation in slope-intercept form. The slope-intercept form of the equation is $y = 5x - 1$.
Write an equation in slope-intercept form for the line through the two points. (4, −5) and (0, 3) **Step 1** Find the slope. $m = \frac{y_2 - y_1}{x_2 - x_1} = \frac{3 - (-5)}{0 - 4} = \frac{8}{-4} = -2$ **Step 2** Substitute the slope and one of the points into the point-slope form. $y - y_1 = m(x - x_1)$ *Use the point-slope form.* $y - 3 = -2(x - 0)$ *Substitute the slope and the x and y values from the point for m, x_1 and y_1.* **Step 3** Write the equation in slope-intercept form. $y - 3 = -2(x - 0)$ $y - 3 = -2x - 0$ *Distribute −2 on the right side.* $+3 \quad +3$ *Add 3 to both sides.* $y = -2x + 3$	An equation is written in slope-intercept form in this example. Two points on the line are given. **Step 1** Begin by finding the slope for the line by finding the difference in y coordinates of the two points over the difference in the x coordinates. **Step 2** Substitute the slope as m in the point-slope form. Substitute one of the given point coordinates for x_1 and y_1. The point-slope form of the equation is $y - 3 = -2(x - 0)$. **Step 3** Simplify and isolate y on one side of the equation to write the equation in slope-intercept form. The slope-intercept form of the equation is $y = -2x + 3$.

Example 5 Problem-Solving Application

The cost to place an ad in a newspaper for one week is a linear function of the number of lines in the ad. The costs for 2, 4, and 7 lines are shown. Write an equation in slope-intercept form that represents the function.
Then find the cost of an ad that is 15 lines long.

Lines	2	4	7
Cost ($)	7.00	10.00	14.50

Step 1 Choose any two ordered pairs from the table to find the slope.

$$m = \frac{y_2 - y_1}{x_2 - x_1} = \frac{10 - 7}{4 - 2} = \frac{3}{2}$$ *Use (4, 10.00) and (2, 7.00).*

Step 2 Substitute the slope and any ordered pair from the table into the point-slope form.

$$y - y_1 = m(x - x_1)$$
$$y - 7 = \frac{3}{2}(x - 2)$$ *Use (2, 7.00).*

Step 3 Write the equation in slope-intercept form by solving for y.

$$y - 7 = \frac{3}{2}(x - 2)$$
$$y - 7 = \frac{3}{2}x - 3$$ *Distribute $\frac{3}{2}$ on the right side.*
$$+7 \qquad +7$$ *Add 7 to both sides.*
$$y = \frac{3}{2}x + 4$$

Step 4 Find the cost of an ad containing 15 lines by substituting 15 for x.

$$y = \frac{3}{2}x + 4$$
$$y = \frac{3}{2}(15) + 4 = 26.50$$ The cost of an ad containing 15 lines is $26.50.

An equation is written in slope-intercept form to describe the cost of a newspaper ad in this application example. The equation is used to find the cost of a 15-line ad. The costs for ads of three different line lengths are given.

Step 1
Begin by finding the slope for the line by finding the difference in y coordinates (cost) of two of the points over the difference in the x coordinates (lines).

Step 2
Substitute the slope as m in the point-slope form. Substitute one of the the given point coordinates for x_1 and y_1. The point-slope form of the equation is $y - 7 = \frac{3}{2}(x - 2)$.

Step 3
To write the equation in slope-intercept form, simplify and isolate y on one side of the equation. The slope-intercept form of the equation is $y = \frac{3}{2}x + 4$.

Step 4
To find the cost of a 15-line ad, substitute 15 for x and solve for the cost, y. The cost of the ad is $26.50.

Slopes of Parallel and Perpendicular Lines

Key Objectives

- Identify and graph parallel and perpendicular lines.
- Write equations to describe lines parallel or perpendicular to a given line.

Key Terms

- **Parallel lines** are lines in the same plane that have no points in common.
- **Perpendicular lines** are lines that intersect to form right angles (90°).

Theorems, Postulates, Corollaries, and Properties

- **Parallel Lines Theorem** In a coordinate plane, two nonvertical lines are parallel if and only if they have the same slope. Any two vertical lines are parallel.
- **Perpendicular Lines Theorem** In a coordinate plane, two nonvertical lines are perpendicular if and only if the product of their slopes is −1. Vertical and horizontal lines are perpendicular.

Example 1 Identifying Parallel Lines

Identify which lines are parallel.

Equation	Slope
$y = \frac{1}{2}x - 5$	$\frac{1}{2}$
$y = -5$	0
$y = 1$	0
$y = \frac{1}{2}x + 1$	$\frac{1}{2}$

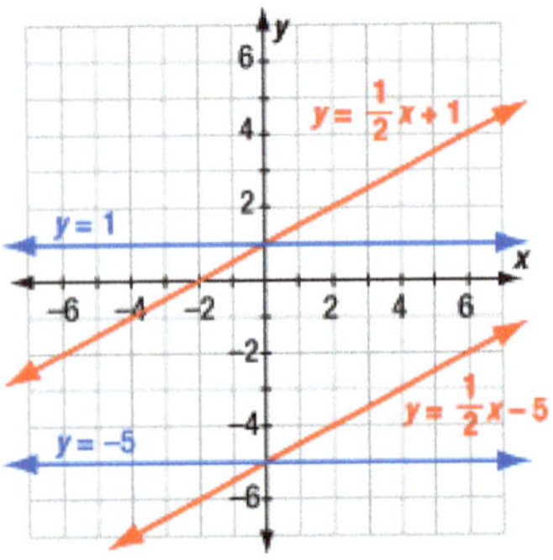

The lines $y = -5$ and $y = 1$ both have slope 0. These lines are parallel.

The lines $y = \frac{1}{2}x - 5$ and $y = \frac{1}{2}x + 1$ both have slope $\frac{1}{2}$. These lines are parallel.

Parallel lines are identified in this example. The equations for the lines are given.

To determine whether lines are parallel, find whether their slopes are equal. The slopes of two of the lines are both 1/2. These two lines are parallel. The slopes of the other two lines are 0. These two lines are parallel.

Identify which lines are parallel.

Equation	Slope
$y = -\frac{3}{2}x + 4$	$-\frac{3}{2}$
$y = -x + 2$	-1
$3x + 2y = 8$	$-\frac{3}{2}$
$y - 5 = -(x + 7)$	-1

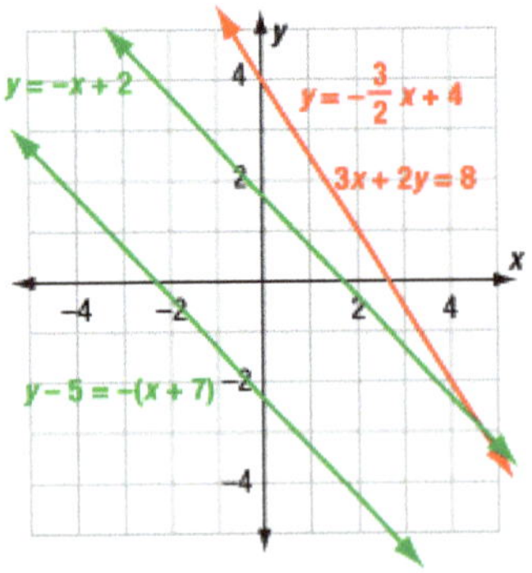

$y = -x + 2$ and $y - 5 = -(x + 7)$ represent parallel lines. They each have a slope of −1.

$y = -\frac{3}{2}x + 4$ and $3x + 2y = 8$ have the same slope, but they are not parallel lines. They are the same line.

Parallel lines are identified in this example. The equations for the lines are given.

To determine whether lines are parallel, find whether their slopes are equal. The slopes of two of the equations are −3/2. However, the equations are for the same line. The slopes of the other two equations are −1. These two lines are parallel.

Example 2 Geometry Application

Show that *ABCD* is a parallelogram.

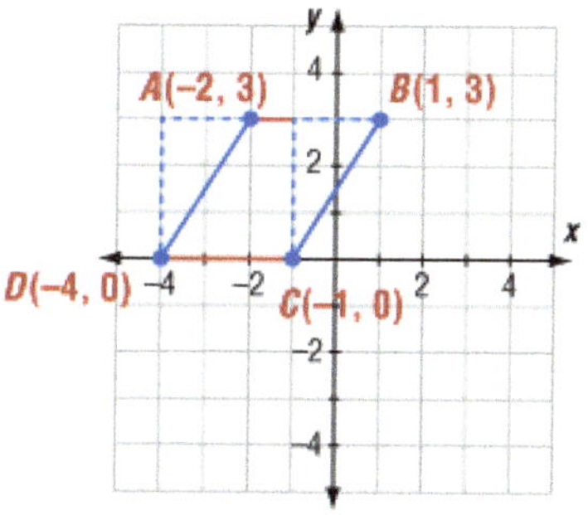

Use the ordered pairs and the slope formula to find the slopes of $\overline{AD}$ and $\overline{BC}$.

slope of $\overline{BC} = \frac{3-0}{1-(-1)} = \frac{3}{2}$

slope of $\overline{AD} = \frac{3-0}{-2-(-4)} = \frac{3}{2}$

$\overline{AB}$ is parallel to $\overline{CD}$ because they are both horizontal.
$\overline{AD}$ is parallel to $\overline{BC}$ because they have the same slope.
Therefore, *ABCD* is a parallelogram because both pairs of opposite sides are parallel.

A figure in the coordinate plane is shown to be a parallelogram in this example. The coordinates of the vertices of the figure are given.

To show that the figure is a parallelogram, show that the pairs of opposite sides are parallel. Find the slopes of $\overline{AD}$ and $\overline{BC}$ by finding the difference of the y coordinates of their endpoints over the difference of the x coordinates of their endpoints. The slopes of $\overline{AD}$ and $\overline{BC}$ both equal 3/2. These lines are parallel. The slopes of $\overline{AB}$ and $\overline{CD}$ are both equal to 0. These lines are parallel.

ABCD is a parallelogram because opposite sides are parallel.

Example 3 Identifying Perpendicular Lines

Identify which lines are perpendicular.

Equation	Slope
$y = -1$	0
$y = -\frac{1}{5}x + 1$	$-\frac{1}{5}$
$y - 7 = 5(x - 1)$	5
$x = 4$	undefined

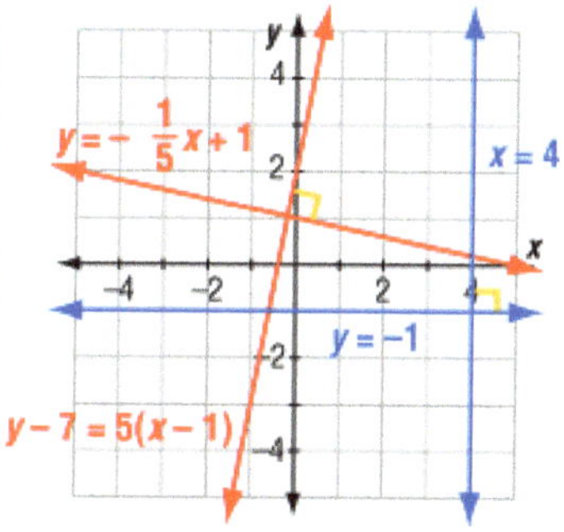

$y = -\frac{1}{5}x + 1$ and $y - 7 = 5(x - 1)$ are perpendicular because the product of their slopes is −1.

$y = -1$ and $x = 4$ are perpendicular because they intersect to form a 90° angle.

Perpendicular lines are identified in this example. The equations for the lines are given.

To determine whether lines are perpendicular, find whether the product of their slopes equals −1. The slopes of two of the equations are 5 and −1/5, whose product equals −1. These lines are perpendicular.

The slopes of the other two equations are 0 and undefined (the line is vertical). These lines are a special case, but because one is horizontal and one is vertical, they intersect to form a 90° angle and are perpendicular.

Example 4 Geometry Application

Show that *PQR* is a right triangle.

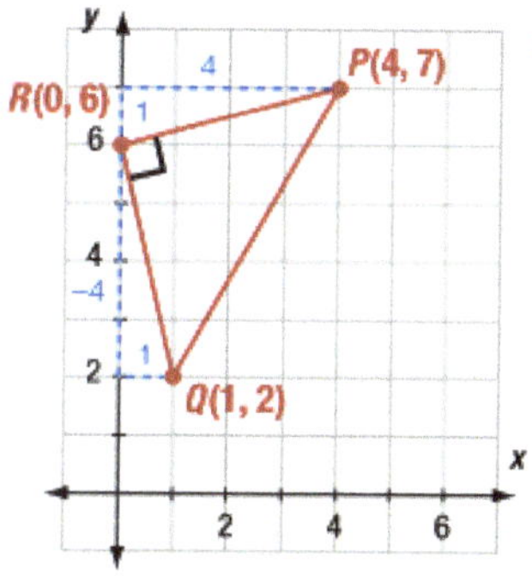

slope of $\overline{PR} = \frac{7-6}{4-0} = \frac{1}{4}$

slope of $\overline{RQ} = \frac{2-6}{1-0} = -\frac{4}{1}$

$\frac{1}{4} \cdot \frac{-4}{1}$ *Multiply the slopes.*

−1

If *PQR* is a right triangle, $\overline{PR}$ will will be perpendicular to $\overline{RQ}$.

$\overline{PR}$ is perpendicular to $\overline{RQ}$ because $\frac{1}{4} \cdot \frac{-4}{1} = -1$.

Therefore, *PQR* is a right triangle because it contains a right angle.

A figure in the coordinate plane is shown to be a right triangle in this example. The coordinates of the vertices of the figure are given.

To show the figure is a right triangle, show that two of the sides meet to form a right angle. Find the slopes of $\overline{PR}$ and $\overline{RQ}$ by finding the difference of the y coordinates of their endpoints over the difference of the x coordinates of their endpoints. The slope of $\overline{PR}$ is 1/4 and the slope of $\overline{RQ}$ is −4. The product of the slopes is −1. These line segments are perpendicular.

$\triangle PQR$ is a right triangle because two sides are perpendicular.

Example 5 Writing Equations of Parallel and Perpendicular Lines

Write an equation in slope-intercept form for the line that passes through (5, −6) and is parallel to the line described by $y = -x + 3$.

Step 1 Find the slope of the line.

$y = -x + 3$ *The equation is in slope-intercept form. The slope is −1.*

The parallel line also has a slope of −1.

Step 2 Write the equation in point-slope form.

$$y - y_1 = m(x - x_1)$$
$$y - (-6) = -1(x - 5)$$
$$y + 6 = -1(x - 5)$$

Step 3 Write the equation in slope-intercept form.

$$y + 6 = -1(x - 5)$$
$y + 6 = -x + 5$ *Distribute −1.*

$-6 \quad -6$ *Subtract 6 from both sides.*

$$y = -x - 1$$

An equation is written for a line parallel to a given line that passes through a given point in this example.

Step 1
Find the slope of the given line. Since the equation for the line is in slope-intercept form, the slope is −1. The parallel line will also have a slope of −1.

Step 2
Write the equation of the parallel line in point-slope form by substituting the slope as m and the coordinates of the given point as x_1 and y_1.

Step 3
Simplify and isolate y to write the equation in slope-intercept form. The equation is $y = -x - 1$.

Write an equation in slope-intercept form for the line that passes through (8, 5) and is perpendicular to the line described by $y = -4x + 7$.

Step 1 Find the slope of the line.

$y = -4x + 7$ *The equation is in slope-intercept form. The slope is −4.*

The perpendicular line has a slope of $\frac{1}{4}$ because $-4 \cdot \frac{1}{4} = -1$.

Step 2 Write the equation in point-slope form.

$$y - y_1 = m(x - x_1)$$
$$y - 5 = \frac{1}{4}(x - 8)$$

Step 3 Write the equation in slope-intercept form.

$$y - 5 = \frac{1}{4}(x - 8)$$
$y - 5 = \frac{1}{4}x - 2$ *Distribute $\frac{1}{4}$.*

$+5 \quad +5$ *Add 5 to both sides.*

$$y = \frac{1}{4}x + 3$$

An equation is written for a line perpendicular to a given line that passes through a given point in this example.

Step 1
Find the slope of the given line. Since the equation for the line is in slope-intercept form, the slope is −4. The perpendicular line will have a slope of 1/4, since the product of the two slopes equals −1.

Step 2
Write the equation of the perpendicular line in point-slope form by substituting the slope as m and the coordinates of the given point as x_1 and y_1.

Step 3
Simplify and isolate y to write the equation in slope-intercept form. The equation is $y = \frac{1}{4}x + 3$.

4

4 Triangle Congruence

Classifying Triangles

Key Objectives

- Classify triangles by their angle measures and side lengths.
- Use triangle classifications to find angle measures and side lengths.

Key Terms

- An **acute triangle** has three acute angles.
- An **equiangular triangle** has three congruent angles.
- A **right triangle** has one right angle.
- An **obtuse triangle** has one obtuse angle.
- An **equilateral triangle** has three congruent sides.
- An **isosceles triangle** has at least two congruent sides.
- A **scalene triangle** has no congruent sides.

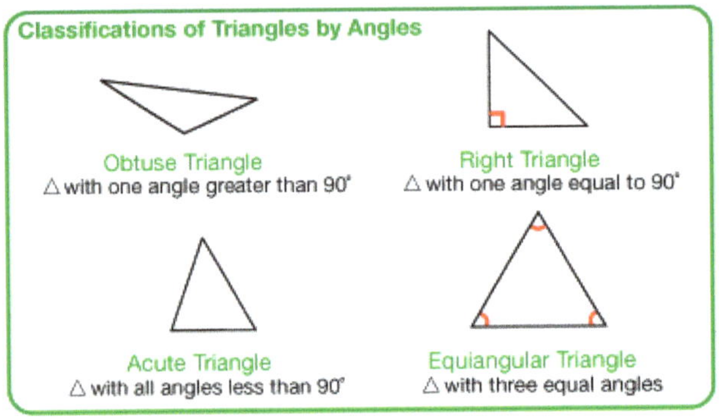

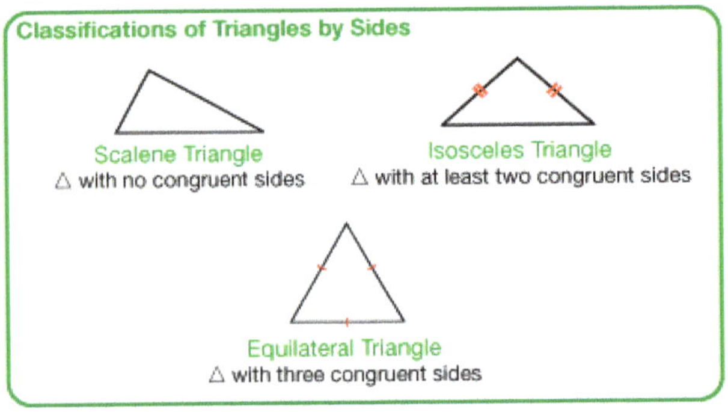

A triangle is a closed plane figure with three sides and three vertices. Each of a triangle's sides is a line segment and each vertex is the endpoint of two of these segments. Since a triangle always has three sides and three vertices, a triangle also has three interior angles. All triangles can be classified by the measure of its interior angles or by the lengths of its sides.

Example 1 Classifying Triangles by Angle Measures

A triangle can be classified by its angle measures. When a triangle is classified by its angle measures, there are four possibilities: acute (all angles are acute), equiangular (all angles are congruent), right (one right angle), and obtuse (one obtuse angle).

Classify each triangle by its angle measures.

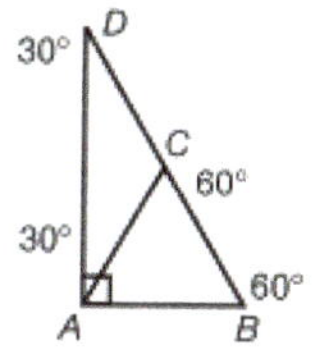

The figure given here is a large triangle, $\triangle ABD$, that is divided into two smaller triangles, $\triangle ACD$ and $\triangle ACB$, by $\overline{AC}$. Notice that the figure contains a right angle mark at $\angle DAB$. Therefore, $m\angle DAB = 90°$.

$\triangle ACB$

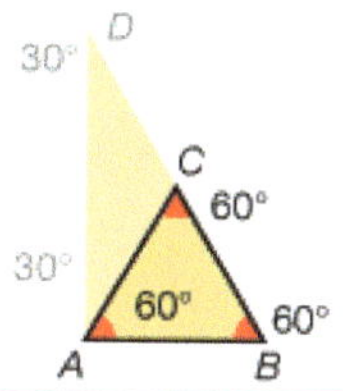

$m\angle DAB = 90°$
and
$m\angle DAC = 30°$

$m\angle CAB = 90° - 30° = 60°$

$m\angle ACB = 60°$ $m\angle ABC = 60°$ $m\angle CAB = 60°$

$\triangle ACB$ is an equiangular triangle.

The first triangle to classify is $\triangle ACB$. Since two of its angles are given to be 60°, the measure of only one angle, $m\angle CAB$, must be found in order to classify $\triangle ACB$ by its angles. By the Angle Addition Postulate,
$m\angle DAC + m\angle CAB = m\angle DAB$, or
$m\angle CAB = m\angle DAB - m\angle DAC$. It is given in the figure that $m\angle DAB = 90°$ and $m\angle DAC = 30°$. So,
$m\angle CAB = 90° - 30° = 60°$. Therefore, the measure of each angle in $\triangle ACB$ is 60°, and so all three angles are congruent. Thus, this is an equiangular triangle. Note that because the measures of all three angles in $\triangle ACB$ are less than 90°, $\triangle ACB$ is also an acute triangle.

$\triangle ACD$

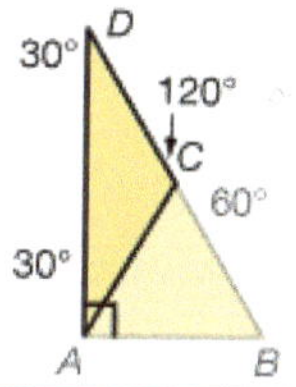

$m\angle ACD = 180° - 60° = 120°$

$m\angle ADC = 30°$ $m\angle DAC = 30°$ $m\angle ACD = 120°$

$\triangle ACD$ is an obtuse triangle.

Classify $\triangle ACD$. Two of its angle measures are known: $m\angle ADC = 30°$ and $m\angle DAC = 30°$. So, find $m\angle ACD$. Since $\angle ACD$ and $\angle ACB$ form a straight angle,
$m\angle ACD + m\angle ACB = 180°$. Therefore,
$m\angle ACD = 180° - 60° = 120°$, which is an obtuse angle. So, this is an obtuse triangle.

Example 2 Classifying Triangles by Side Lengths

A triangle can be classified by its side lengths. When a triangle is classified by its side lengths, there are three possibilities: scalene (no congruent sides), isosceles (at least two congruent sides), and equilateral (three congruent sides).

Classify each triangle by its side lengths.

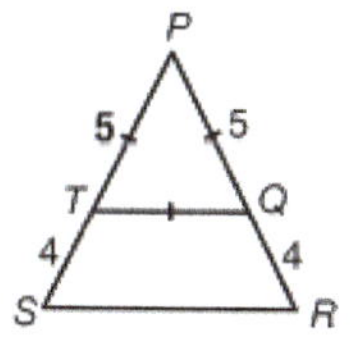

There are two triangles in the figure given here, $\triangle SPR$ and $\triangle TPQ$. Each triangle will be classified by its side lengths.

$\triangle TPQ$

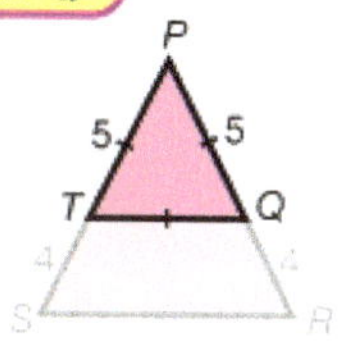

The three sides of $\triangle TPQ$ are congruent.

$\triangle TPQ$ is an equilateral triangle.

First, classify $\triangle TPQ$. Notice that the triangle contains single tick marks on $\overline{PT}$, $\overline{PQ}$, and $\overline{TQ}$. These tick marks indicate that those three sides are congruent. So, all three sides of $\triangle TPQ$ are congruent. A triangle with three congruent sides is classified as equilateral. Therefore, $\triangle TPQ$ is an equilateral triangle.

$\triangle SPR$

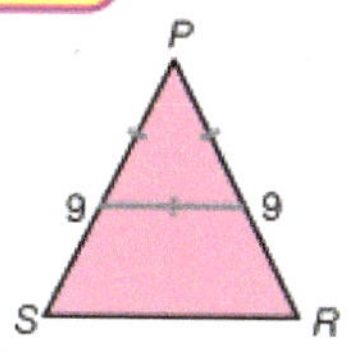

Two sides of $\triangle SPR$ are congruent.

$\triangle SPR$ is isosceles.

Next, classify $\triangle SPR$. In the given figure, $\overline{PS}$ is divided into two parts, $\overline{PT}$ and $\overline{TS}$, where $PT = 5$ and $TS = 4$. So, by the Segment Addition Postulate, $PS = PT + TS = 5 + 4 = 9$. The length of $\overline{PR}$ can be found in a similar way, and $PR = 9$ as well. So, since $PS = 9$ and $PR = 9$, the two sides of $\triangle SPR$ are congruent. A triangle with at least two congruent sides is classified as isosceles. Therefore, $\triangle SPR$ is an isosceles triangle.

Example 3 Using Triangle Classification

Find the side lengths of $\triangle ABC$.

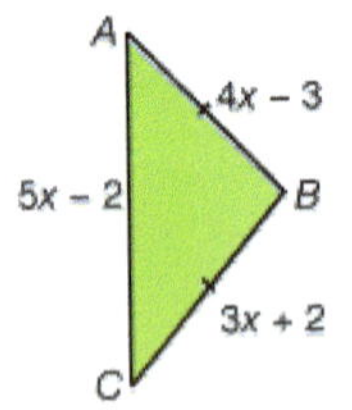

Set the expressions from the congruent sides equal to each other and solve for x.

$\overline{AB} \cong \overline{BC}$	*Given*
$AB = BC$	*Def. of ≅ segs.*
$4x - 3 = 3x + 2$	*Substitute.*
$x - 3 = 2$	
$x = 5$	

Substitute 5 into each expression for x to find the length of each side.

$AB = 4(5) - 3 = 20 - 3 = 17$

$BC = 3(5) + 2 = 15 + 2 = 17$

$AC = 5(5) - 2 = 25 - 2 = 23$

This figure has tick marks indicating congruency on $\overline{AB}$ and $\overline{BC}$. When two segments are congruent, their lengths are equal. Therefore, $AB = BC$ by the definition of congruent segments.

In the figure, expressions are given for AB and BC.
$AB = 4x - 3$
$BC = 3x + 2$.
So, by the Substitution Property of Equality, $4x - 3 = 3x + 2$.
Solve this equation for x.

Once the value of x is found, substitute that value into the expression for the length of each side of $\triangle ABC$ and simplify. The result is the length of each side.

Example 4 Music Application

A manufacturer produces musical triangles by bending pieces of steel into the shape of an equilateral triangle. Each side of a triangle is 3 inches long. How many triangles can be formed from an 80 inch piece of steel?

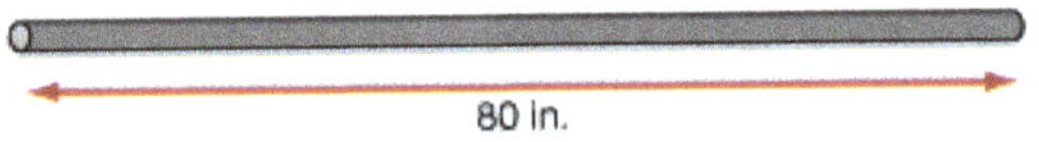

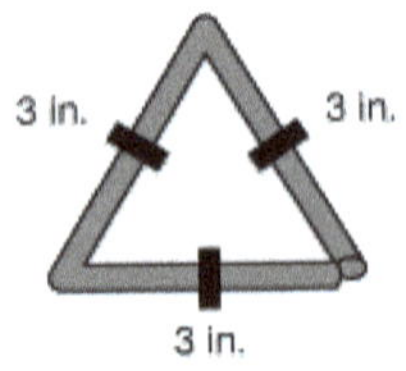

3 in. + 3 in. + 3 in. = 9 in. for 1 triangle.

80 in. ÷ 9 in. = $8\frac{8}{9}$ in.

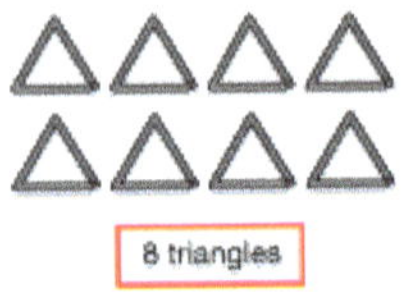

In this example, a musical triangle is described as an equilateral triangle. Therefore, the musical triangle must have three congruent sides, or three sides with equal length.

It is given that each side of the musical triangle is 3 inches long. Since a triangle has 3 sides, the length of steel used to make one musical triangle is 3 inches + 3 inches + 3 inches, or 9 inches. The full length of steel is 80 inches. So, divide 80 by 9 to find the maximum number of musical triangles that can be formed from the 80 inches of steel.

The quotient of 80 and 9 is not a whole number. However, the number of musical triangles formed should be a whole number. Even though 8 and 8/9 is closer to 9 than it is to 8, the answer is 8 musical triangles because the extra 8/9 inches of steel cannot be used to form an additional musical triangle.

Angle Relationships in Triangles

Key Objectives

- Find the measure of interior and exterior angles of triangles.
- Apply theorems about the interior and exterior angles of triangles.

Key Terms

- An **auxiliary line** is a line that is added to a figure to aid in a proof.
- A **corollary** is a theorem whose proof follows directly from another theorem.
- The **interior** of a figure is the set of all points inside the figure.
- The **exterior** of a figure is the set of all points outside the figure.
- An **interior angle** is formed by two sides of a polygon.
- An **exterior angle** is formed by one side of a polygon and the extension of an adjacent side.
- Each exterior angle has two remote interior angles. A **remote interior angle** is an interior angle that is not adjacent to the exterior angle.

Theorems, Postulates, Corollaries, and Properties

- **Triangle Sum Theorem** The sum of the angle measures of a triangle is 180°.
- **Corollary** The acute angles of a right triangle are complementary.
- **Corollary** The measure of each angle of an equiangular triangle is 60°.
- **Exterior Angle Theorem** The measure of an exterior angle of a triangle is equal to the sum of the measures of its remote interior angles.
- **Third Angles Theorem** If two angles of one triangle are congruent to two angles of another triangle, then the third pair of angles are congruent.

Example 1 Triangle Sum Theorem

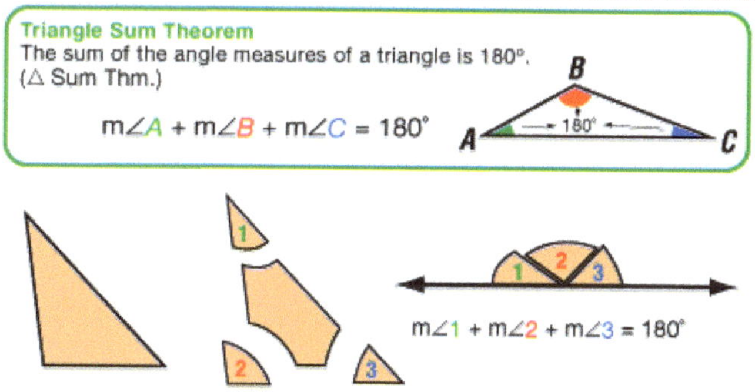

The Triangle Sum Theorem states that the sum of the interior angles of any triangle is 180°. This conclusion can be visualized by manipulating a triangle's angles. Begin with a triangle of any type or size and cut the angles (corners) off so that each piece of the triangle contains one vertex. Each piece represents one of the triangle's interior angles. Then, place the three angles so that they meet at their vertices. The three angles always form a straight angle, no matter what type or size of triangle is used. The measure of a straight angle is always 180°. Therefore, the sum of the angle measures of a triangle is always 180°.

Use the diagram to find m∠*CBD* and m∠*EAB*.

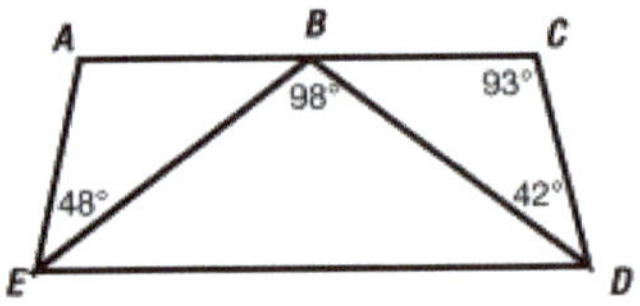

To find m∠*CBD*, use the Triangle Sum Theorem with △*BCD*.

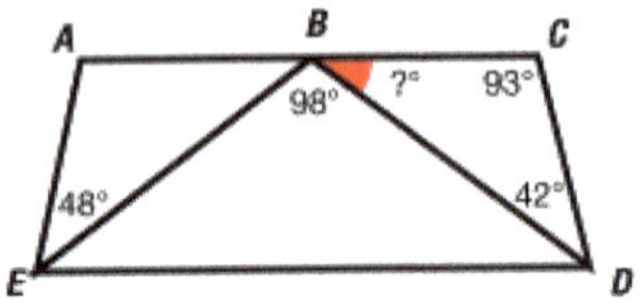

$$?^\circ + 93^\circ + 42^\circ = 180^\circ$$
$$?^\circ + 135^\circ = 180^\circ$$
$$?^\circ = 180^\circ - 135^\circ$$
$$?^\circ = 45^\circ$$

To find m∠*EAB*, first use the definition of a straight angle to find m∠*ABE*, then use the Triangle Sum Theorem with △*ABE*.

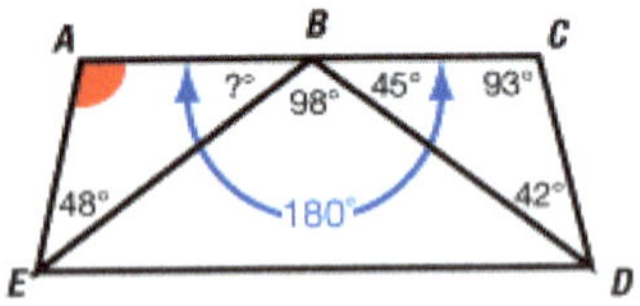

$$?^\circ + 98^\circ + 45^\circ = 180^\circ$$
$$?^\circ + 143^\circ = 180^\circ$$
$$?^\circ = 180^\circ - 143^\circ$$
$$?^\circ = 37^\circ$$

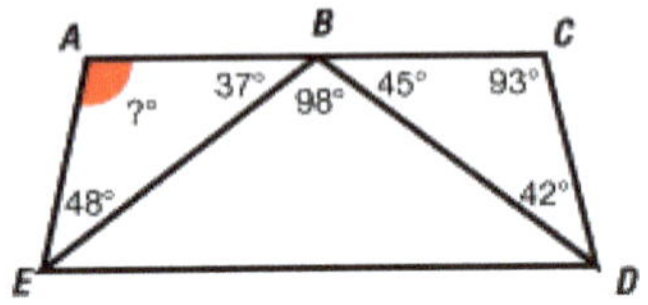

$$?^\circ + 48^\circ + 37^\circ = 180^\circ$$
$$?^\circ + 85^\circ = 180^\circ$$
$$?^\circ = 180^\circ - 85^\circ$$
$$?^\circ = 95^\circ$$

The figure given here is a four-sided polygon that is divided into three triangles: △*ABE*, △*BDE*, and △*BCD*. The measures of four angles are given.

First, find m∠*CBD*. Notice that ∠*CBD* is an angle in △*BCD* and the measures of the other two angles in △*BCD* are given, m∠*C* = 93° and m∠*BDC* = 42°. Use the Triangle Sum Theorem (sum of a triangle's three angles is 180°) to write an equation and solve the equation to find m∠*CBD*.

Now, find m∠*EAB*. Notice that ∠*EAB* is an angle in △*ABE*. However, the measure of only one other angle in △*ABE* is given, m∠*AEB* = 48°. So, m∠*ABE* must be found before the Triangle Sum Theorem can be used. Since ∠*ABE* is one of three angles that form a straight angle and the measures of the other two angles are known, use the definition of a straight angle (a straight angle measures 180°) to write an equation.

m∠*ABE* + m∠*EBD* + m∠*CBD* = 180°

Now substitute the known measures of ∠*EBD* and ∠*CBD* into the equation and solve to find m∠*ABE*.

Once m∠*ABE* is found, the Triangle Sum Theorem can be used with △*ABE* to find m∠*EAB*.

Example 2 Finding Angle Measures in Right Angles

Corollary

The acute angles of a right triangle are complementary.

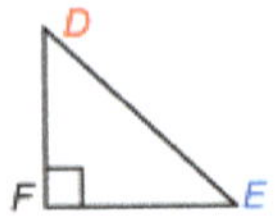

∠*D* and ∠*E* are complementary.
m∠*D* + m∠*E* = 90°

Corollary

The measure of each angle of an equiangular triangle is 60°.

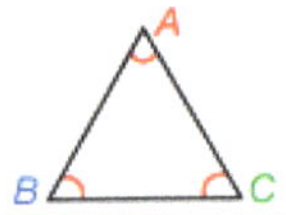

m∠*A* = m∠*B* = m∠*C* = 60°

The first corollary states that in a right triangle, the two acute angles must be complementary (their sum is 90°). This corollary follows from the Triangle Sum Theorem. By the Triangle Sum Theorem, the sum of the angle measures of a triangle is always 180°. In a right triangle, one of those three angles is 90°. So, the sum of the other two angles must be 180° – 90°, or 90°, which means those two angles are complementary.

The second corollary states that each angle measures 60° in an equiangular triangle. Again, by the Triangle Sum Theorem, the sum of the angle measures of a triangle is always 180°. In an equiangular triangle all three angles are congruent. Therefore, the measure of each angle is 180°/3, or 60°.

One of the acute angles of a right triangle measures 15.3°. What is the measure of the other acute angle?

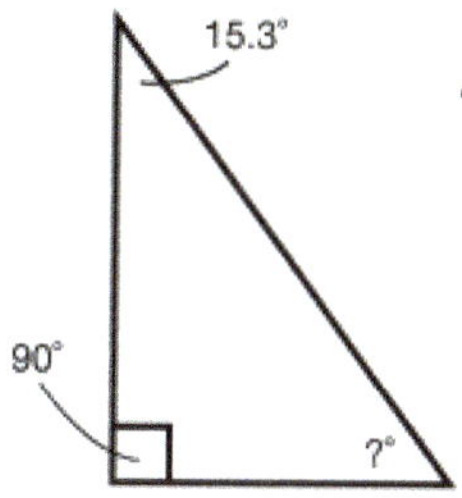

$?^\circ + 15.3^\circ + 90^\circ = 180^\circ$

$?^\circ + 105.3^\circ = 180^\circ$

$? = 180^\circ - 105.3^\circ$

$? = 74.7^\circ$

Check Your Work

$74.7^\circ + 15.3^\circ + 90^\circ = 180^\circ$ ✓

A right triangle is given in this example. One acute angle's measure is given to be 15.3°. The measure of the second acute angle can be found in two ways. Both methods are explained below, but Method 1 is used in the example to the left.

Method 1 Use the Triangle Sum Theorem to find the measure of the unknown angle. By the Triangle Sum Theorem, the sum of the measure of the unknown angle (represented here by ?°), 15.3°, and 90° must be equal to 180°. Use this fact to write an equation and then solve for ?°.

Method 2 Use the fact that the sum of the acute angles of a right triangle is 90° to find the measure of the unknown angle. So, the sum of ?° and 15.3° must be 90°.

$?^\circ + 15.3^\circ = 90^\circ$

It follows that $?^\circ = 90^\circ - 15.3^\circ = 74.7^\circ$. So, the measure of the other acute angle is 74.7°.

Example 3 Applying the Exterior Angle Theorem

The angles of a triangle are classified in three ways in the theorem below: exterior angles, interior angles, and remote interior angles. A triangle's interior angles are just the three angles inside the triangle. In the figure below, ∠1, ∠2, and ∠3 are interior angles. An exterior angle is formed when one of the triangle's sides is extended beyond the vertex. The sides of the exterior angle are the extension and the adjacent side of the triangle. Here, ∠4 is an exterior angle. Each exterior angle has two remote interior angles, where each remote interior angle is an interior angle that is not adjacent to the exterior angle. Here, ∠1 and ∠2 are the remote interior angles to ∠4.

Exterior Angle Theorem

The measure of an exterior angle of a triangle is equal to the sum of the measures of its remote interior angles.
(Ext. ∠ Thm.)

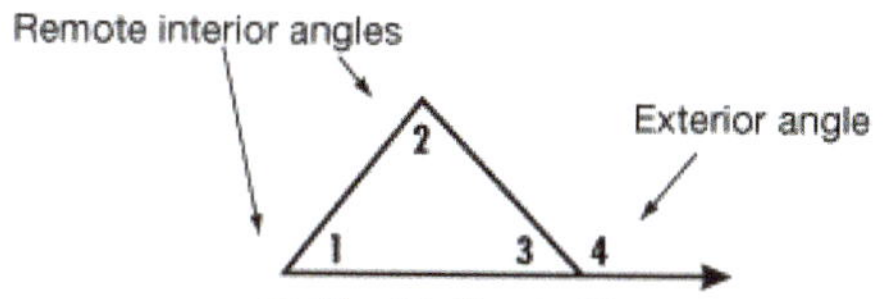

m∠4 = m∠1 + m∠2

The Exterior Angle Theorem relates the measure of an exterior angle to the measures of its remote interior angles. If any exterior angle is drawn on a triangle, its measure is equal to the sum of the measures of its remote interior angles.

Find m∠*H*.

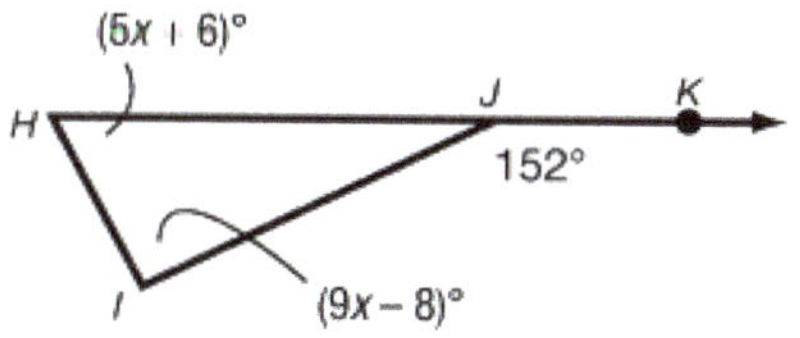

Use the Exterior Angle Theorem to write an equation. Solve the equation for *x*.

m∠*IJK* = m∠*H* + m∠*I*	*Ext. ∠ Thm.*
$152 = (5x + 6)^\circ + (9x - 8)^\circ$	*Substitute 152 for m∠IJK, 5x + 6 for m∠H and 9x – 8 for m∠I.*
$152 = 14x - 2$	*Combine like terms.*
$154 = 14x$	*Add 2 to both sides.*
$x = 11$	*Divide both sides by 14.*

Use $x = 11$ to find m∠*H*.

m∠*H* $= (5x + 6)^\circ = (5(11) + 6)^\circ = 61^\circ$

In this figure, ∠*IJK* is an exterior angle because its sides are formed by one side of the triangle and an extension of the adjacent side. The remote interior angles to ∠*IJK* in △*HIJ* are ∠*H* and ∠*I*. So, by the Exterior Angle Theorem, m∠*IJK* must be equal to the sum of m∠*H* and m∠*I*. Use this fact to write an equation. Substitute the expressions given in the figure for m∠*H* and m∠*I* and the given measure of ∠*IJK* into the equation and solve for *x*. Then, substitute the value of *x* into the expression given for m∠*H* and simplify to find m∠*H*.

Example 4 Applying the Third Angles Theorem

Third Angles Theorem

If two angles of one triangle are congruent to two angles of another triangle, then the third pair of angles are congruent. (Third ∠s Thm.)

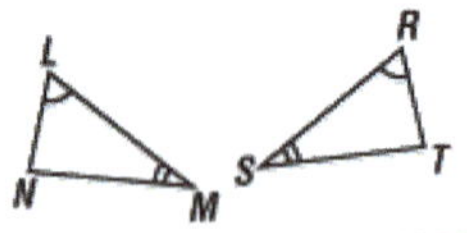

$\angle N \cong \angle T$

The Third Angles Theorem states that if there are two triangles such that two angles in one triangle are congruent to two angles in the other triangle, then the third angles in the two triangles must be congruent.

Find m∠X and m∠M.

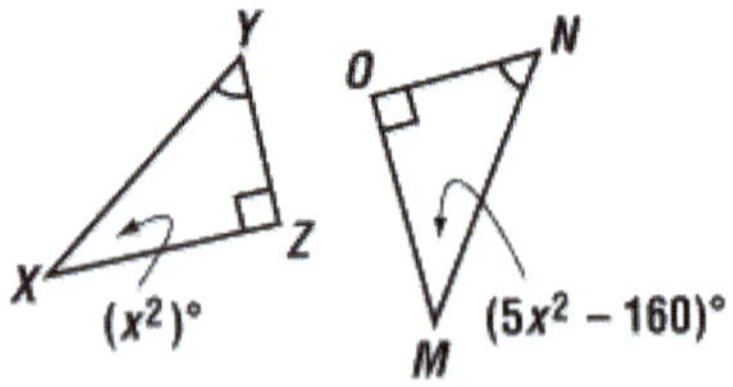

Use the Third Angles Theorem to write an equation.
Solve the equation for x^2.

$\angle X \cong \angle M$	*Third ∠s Thm.*
$m\angle X = m\angle M$	*Def. of ≅ ∠s*
$x^2 = 5x^2 - 160$	*Substitute x^2 for m∠X and substitute $5x^2 - 160$ for m∠M.*
$160 = 4x^2$	*Subtract x^2 and add 160 to both sides.*
$40 = x^2$	*Divide each side by 4.*

Use $x^2 = 40$ to find m∠M.

$m\angle M = (5x^2 - 160)^\circ = (5(40) - 160)^\circ = 40^\circ$

In this example, it is given that ∠Z and ∠O are right angles, so they are congruent. It is also given that ∠Y is congruent to ∠N. Therefore, two angles in △XYZ are congruent to two angles in △MNO. So, by the Third Angles Theorem, ∠X (the third angle of △XYZ) must be congruent to ∠M (the third angle of △MNO).

Use the fact that $\angle X \cong \angle M$ to write an equation. Substitute the expressions given in the figure for m∠X and m∠M into the equation and solve for x^2. Notice that x^2 is used in each expression, not just x, so solving for x^2 is sufficient.

Once the value of x^2 is known, m∠X is also known since $m\angle X = x^2$. So, $m\angle X = 40^\circ$.

Substitute 40° for x^2 in the expression for m∠M and simplify to find m∠M.

Congruent Triangles

Key Objectives

- Use properties of congruent triangles.
- Prove triangles congruent by using the definition of congruence.

Key Terms

- **Corresponding angles** of polygons are the angles in the same relative position in two different polygons that have the same number of angles.
- **Corresponding sides** of polygons are the sides in the same relative position in two different polygons that have the same number of sides.
- Two polygons are **congruent polygons** if and only if their corresponding angles and sides are congruent.

Theorems, Postulates, Corollaries, and Properties

- **Corresponding Angles of Congruent Polygons Property** The corresponding angles of congruent polygons are congruent.
- **Corresponding Sides of Congruent Polygons Property** The corresponding sides of congruent polygons are congruent.

Example 1 Naming Congruent Corresponding Parts

In a congruence statement, such as $\triangle ABC \cong \triangle DEF$, each polygon's vertices must be listed in the order of the corresponding parts. So, a congruence statement indicates not only that the two polygons are congruent, but also which parts of those two polygons are corresponding.

Identify all pairs of congruent corresponding parts.

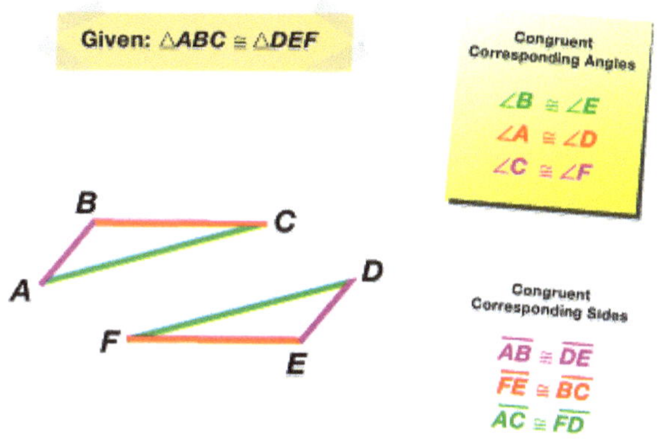

In this example, it is given that the two triangles are congruent. Therefore, by the properties of congruent polygons, the corresponding parts (sides and angles) of $\triangle ABC$ and $\triangle DEF$ must also be congruent. The corresponding parts are the sides or angles that are in the same relative position.

Note that the corresponding parts of the two triangles could be identified using only the triangle's names given in the congruence statement.

Example 2 Using Corresponding Parts of Congruent Triangles

Given: $\triangle ABD \cong \triangle CBD$

Find the value of x.

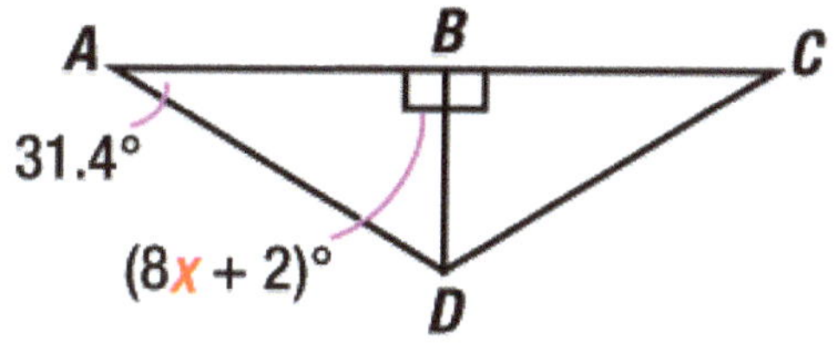

$m\angle CBD = 90^\circ$	Given
$m\angle ABD = 90^\circ$	Supp. ∠s Thm.
$90^\circ = (8x + 2)^\circ$	Substitute value for $m\angle ABD$.
$88 = 8x$	Subtract 2 from both sides.
$x = 11$	Divide both sides by 8.

In this example, it is given that $\triangle ABD$ is congruent to $\triangle CBD$. It is also given in the figure that $\angle CBD$ is a right angle. So, $m\angle CBD = 90^\circ$. It follows that $m\angle ABD = 90^\circ$ by the Supplementary Angles Theorem.

Alternatively, it can be deduced that $m\angle ABD = 90^\circ$ because $\angle ABD$ and $\angle CBD$ are corresponding angles in congruent triangles and $m\angle CBD = 90^\circ$.

The expression given in the figure for $m\angle ABD$ is $(8x + 2)^\circ$. Substitute this expression for $m\angle ABD$ into the equation $m\angle ABD = 90^\circ$ and solve for x.

Given: $\triangle ABD \cong \triangle CBD$

Find $m\angle BDC$.

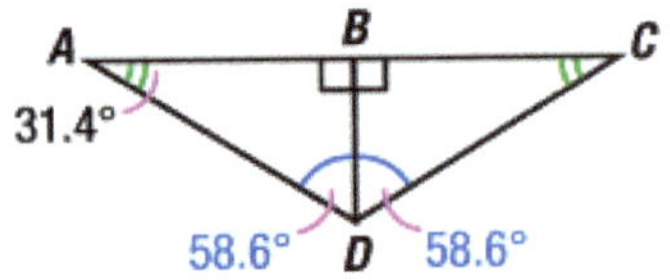

$\angle ABD \cong \angle CBD$	Corr. $\angle$s of $\cong$ $\triangle$s are $\cong$.
$\angle BAD \cong \angle CBD$	Corr. $\angle$s of $\cong$ $\triangle$s are $\cong$.
$\angle BDA \cong \angle BDC$	Corr. $\angle$s of $\cong$ $\triangle$s are $\cong$.
$m\angle BDA = m\angle BDC$	Def. of $\cong$ $\angle$s
$m\angle BDA + m\angle ABD + m\angle A = 180°$	$\triangle$ Sum Thm.
$m\angle BDA + 90° + 31.4° = 180°$	Sub. values for $m\angle ABD$ and $m\angle A$.
$m\angle BDA + 121.4° = 180°$	Simplify.
$m\angle BDA = 58.6°$	Subtract 121.4 from both sides.
$\angle BDA \cong \angle BDC$	Corr. $\angle$s of $\cong$ $\triangle$s are $\cong$.
$m\angle BDA = m\angle BDC$	Def. of $\cong$ $\angle$s
$m\angle BDC = 58.6°$	Trans. Prop. of =

The figure from the previous example is used again in this example. It is given that $\triangle ABD \cong \triangle CBD$, $\angle CBD$ is a right angle, and $m\angle A = 31.4°$.

Make a plan for finding $m\angle BDC$. Notice that the measures of two angles in $\triangle ABD$ are known. Therefore, the Triangle Sum Theorem can be used to find the measure of the third angle, $\angle BDA$. The two triangles are congruent, so their corresponding angles must also be congruent. The angle that corresponds with $\angle BDA$ is $\angle BDC$. So, since corresponding angles of congruent triangles are congruent, $\angle BDA \cong \angle BDC$ and by the definition of congruence, $m\angle BDA = m\angle BDC$.

Example 3 Proving Triangles Congruent

Given:
$\angle ACD$ and $\angle ACB$ are right $\angle$s.
C is the midpoint of $\overline{BD}$.
$\overline{AD} \cong \overline{AB}$
$\overline{AC}$ bisects $\angle A$.

Prove:
$\triangle ACD \cong \triangle ACB$

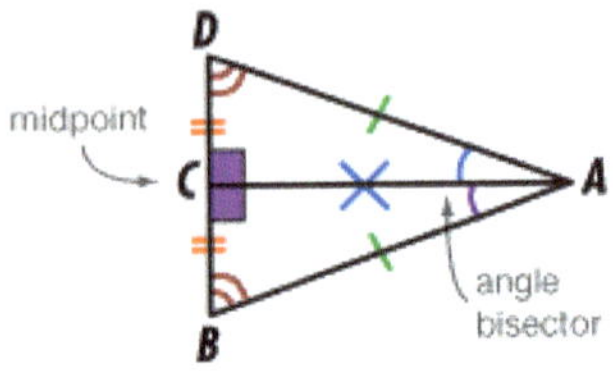

sides: $\overline{AD} \cong \overline{AB}$, $\overline{AC} \cong \overline{CA}$, $\overline{CD} \cong \overline{CB}$	$\angle$s: $\angle ACD \cong \angle ACB$, $\angle CAD \cong \angle CAB$, $\angle D \cong \angle B$

Statements:	Reasons:
1. $\angle ACD$ and $\angle ACB$ are right $\angle$s.	1. Given
2. C is the midpoint of $\overline{BD}$.	2. Given
3. $\overline{AD} \cong \overline{AB}$	3. Given
4. $\overline{AC}$ bisects $\angle A$.	4. Given
5. $\overline{AC} \cong \overline{CA}$	5. Reflex. Prop. of $\cong$
6. $\overline{CD} \cong \overline{CB}$	6. Def. of midpoint
7. $\angle ACD \cong \angle ACB$	7. Right $\angle$ $\cong$ Thm.
8. $\angle CAD \cong \angle CAB$	8. Def. of $\angle$ bisector
9. $\angle D \cong \angle B$	9. Third $\angle$s Thm.
10. $\triangle ACD \cong \triangle ACB$	10. Def. of $\cong$ $\triangle$s

Start this proof by listing all of the given information in the statements and by noting all of the given information on the figure. Now consider the conclusion to be proven: $\triangle ACD \cong \triangle ACB$. One way to show that two triangles are congruent is by showing that all of their corresponding parts are congruent. Showing that all of the corresponding parts of two triangles are congruent means showing that three pairs of angles are congruent and that three pairs of sides are congruent.

There are two important facts to remember for this proof. The first is that a segment's midpoint divides a segment into two congruent parts (definition of midpoint). The second is that if two triangles share a side, that side is congruent to itself (the Reflexive Property of Congruence).

Example 4 Proving Overlapping Triangles Congruent

Given:
$\overline{AD} \perp \overline{DC}$, $\overline{BC} \perp \overline{DC}$,
$\angle DAC \cong \angle CBD$
$\overline{AD} \cong \overline{BC}$, $\overline{BD} \cong \overline{AC}$

Prove:
$\triangle ACD \cong \triangle BDC$.

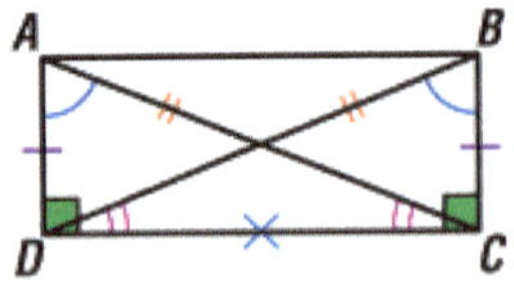

Congruent Corresponding Sides

$\overline{AD} \cong \overline{BC}$ $\overline{BD} \cong \overline{AC}$ $\overline{CD} \cong \overline{CD}$

Congruent Corresponding Angles

$\angle DAC \cong \angle CBD$ $\angle ADC \cong \angle BCD$ $\angle ACD \cong \angle BDC$

Statement:	Reason:
1. $\overline{AD} \cong \overline{BC}$	1. Given
2. $\overline{BD} \cong \overline{AC}$	2. Given
3. $\overline{CD} \cong \overline{CD}$	3. Reflex. Prop. of $\cong$
4. $\angle DAC \cong \angle CBD$	4. Given
5. $\angle ADC \cong \angle BCD$	5. Right $\angle$ $\cong$ Thm.
6. $\angle ACD \cong \angle BDC$	6. Third $\angle$'s Thm.
7. $\triangle ACD \cong \triangle BDC$	7. Def. of $\cong$ $\triangle$'s

In this example, as in the previous example, two triangles are proven to be congruent. The same general plan is used: show that the corresponding sides are congruent and that the corresponding angles are congruent.

It is given that two pairs of segments are perpendicular. If two segments are perpendicular, then a right angle is formed by those two segments.

The Third Angle Theorem is often used when proving that two triangles are congruent. The Third Angle Theorem states that if two angles in one triangle are congruent to two angles in a second triangle, then the third angles must also be congruent.

Triangle Congruence: SSS and SAS

Key Objectives

- Apply SSS and SAS to construct triangles and to solve problems.
- Prove triangles congruent by using SSS and SAS.

Key Terms

- The property of **triangle rigidity** states that if the side lengths of a triangle are given, the triangle can have only one shape.
- An **included angle** is an angle formed by two adjacent sides of a polygon.

Theorems, Postulates, Corollaries, and Properties

- **Side-Side-Side (SSS) Congruence Postulate** If three sides of one triangle are congruent to three sides of another triangle, then the triangles are congruent.
- **Side-Angle-Side (SAS) Congruence Postulate** If two sides and the included angle of one triangle are congruent to two sides and the included angle of another triangle, then the triangles are congruent.

Example 1 Using SSS to Prove Triangle Congruence

Side-Side-Side (SSS) Congruence

If three sides of one triangle are congruent to three sides of another triangle, then the triangles are congruent.

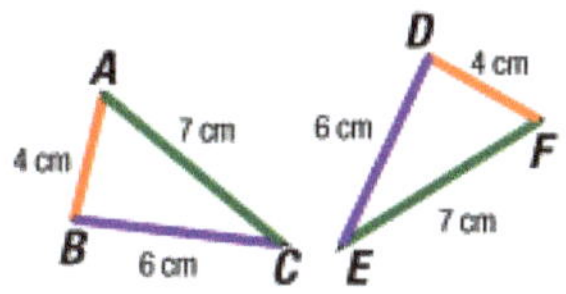

$\triangle ABC \cong \triangle FDE$

By the property of triangle rigidity, if the sides of a triangle are known, then that triangle can have only one shape. It follows that if the sides of another triangle are also known, and those sides are all congruent to the first triangle's sides, then the two triangles must have the same shape. Any two figures with the same size and shape are congruent. Therefore, if the sides of one triangle are congruent to the sides of another triangle, then those two triangles must be congruent.

Use SSS to explain why $\triangle MNP \cong \triangle OPN$.

The property that every line segment is congruent to itself is called the **Reflexive Property of Congruence.**

$\triangle MNP \cong \triangle OPN$ by SSS

$\overline{MP} \cong \overline{ON}$
$\overline{MN} \cong \overline{OP}$
$\overline{NP} \cong \overline{NP}$

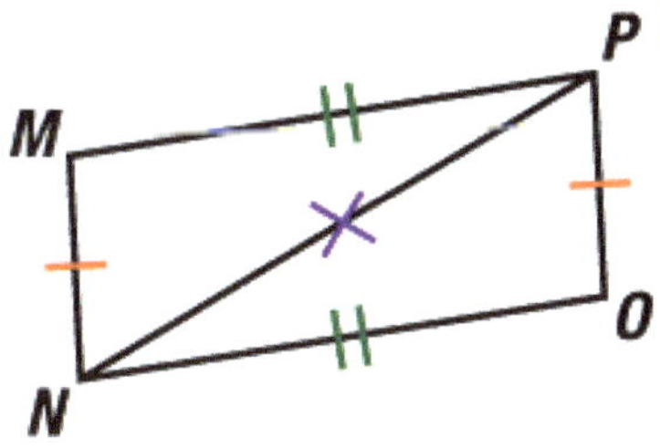

To show that two triangles are congruent by SSS, just show that the three pairs of corresponding sides are congruent.

In this example, the given figure is a quadrilateral (four-sided polygon) with a segment $\overline{NP}$ that divides it into two triangles. In the figure it is given that each pair of opposite sides of the quadrilateral are congruent, $\overline{MN} \cong \overline{OP}$ and $\overline{MP} \cong \overline{ON}$. So, two sides of $\triangle MNP$ are congruent to two sides of $\triangle OPN$. The third side of each triangle is formed by the diagonal segment, $\overline{NP}$. $\overline{NP}$ is a side for each triangle and it is congruent to itself by the Reflexive Property of Congruence. So, since the three sides of $\triangle MNP$ are congruent to the corresponding three sides of $\triangle OPN$, $\triangle MNP \cong \triangle OPN$ by SSS.

Example 2 Using SAS to Prove Triangle Congruence

Side-Angle-Side (SAS) Congruence

If two sides and the included angle of one triangle are congruent to two sides and the included angle of another triangle, then the triangles are congruent.

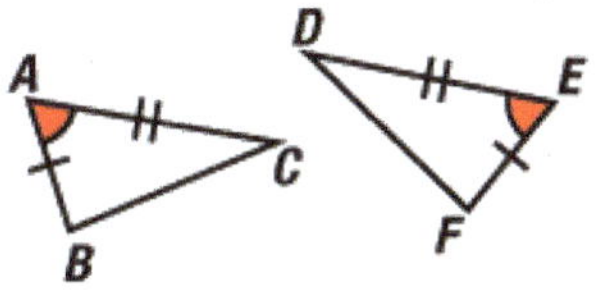

$\triangle ABC \cong \triangle EFD$

By the SAS Congruence Postulate, two triangles are congruent when two sides and the included angle of one triangle are congruent to two sides and the included angle of another triangle. Note that the included angle is the angle formed by two adjacent sides.

Use SAS to explain why $\triangle ACB \cong \triangle ACD$.

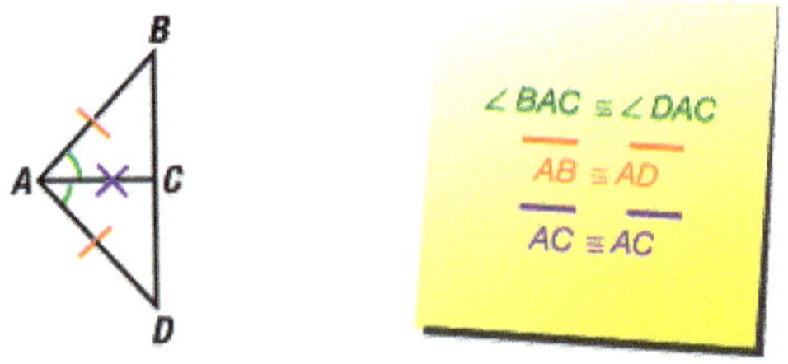

$\triangle ACB \cong \triangle ACD$ by SAS

To show that two triangles are congruent by SAS, show that two sides and their included angle in the first triangle are congruent to two sides and their included angle in the second triangle.

In this example, one pair of angles and one pair of sides are given to be congruent in $\triangle ACB$ and $\triangle ACD$. So, to use SAS to show that these two triangles are congruent, show that a second pair of sides are congruent. However, the pair of sides must be the sides that make $\angle BAC$ and $\angle DAC$ the included angle.

Example 3 Verifying Triangle Congruence

Show that the triangles are congruent for the given value of the variable.

Claim: $\triangle UVW \cong \triangle YXZ$ when $x = 5$

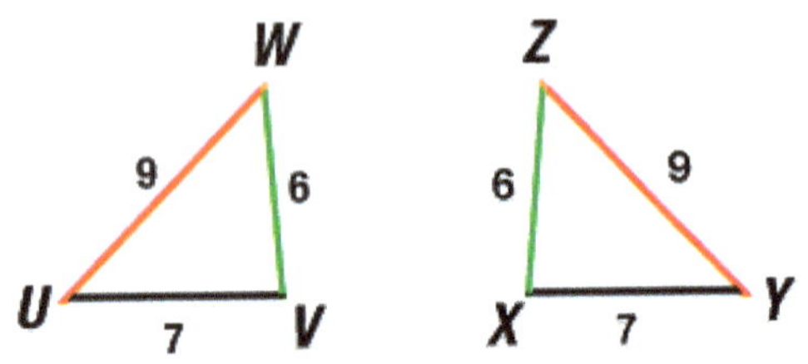

$\overline{VW} \cong \overline{XZ}$ $\overline{UW} \cong \overline{YZ}$ $\overline{UV} \cong \overline{XY}$

$\triangle UVW \cong \triangle YXZ$ by **SSS Congruence**

The lengths of the sides of $\triangle YXZ$ are unknown in this example. But expressions that represent the length of each side of $\triangle YXZ$ are given.

$XZ = x + 1$
$YZ = x + 4$
$XY = 2x - 3$

Substitute 5 for x into each expression and simplify.

$XZ = 5 + 1 = 6$
$YZ = 5 + 4 = 9$
$XY = 2(5) - 3 = 7$

Therefore, $VW = XZ$, $UW = YZ$, and $UV = XY$. So, by the definition of congruence, each of these pairs of sides is congruent. Since the three sides of $\triangle UVW$ are congruent to the three sides of $\triangle YXZ$, the two triangles are congruent by SSS.

Show that the triangles are congruent for the given value of the variable.

Claim: $\triangle ABC \cong \triangle DEF$ when $x = 4$

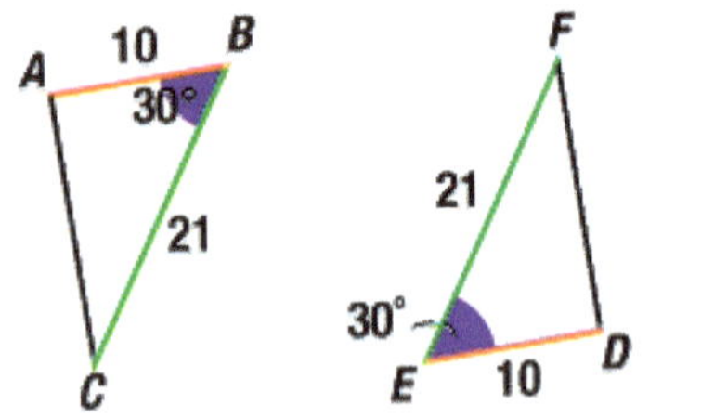

$\overline{BC} \cong \overline{EF}$ $\overline{AB} \cong \overline{DE}$ $\angle B \cong \angle E$

$\triangle ABC \cong \triangle DEF$ by **SAS Congruence**

In this example, $m\angle E$, EF, and DE are represented by expressions.

$m\angle E = (7x + 2)°$
$EF = 5x + 1$
$DE = 3x - 2$

To show that the triangles are congruent for the given value of the variable, $x = 4$, first substitute 4 into each expression and simplify.

$m\angle E = (7(4) + 2)° = 30°$
$EF = 5(4) + 1 = 21$
$DE = 3(4) - 2 = 10$

Therefore, $m\angle B = m\angle E$, $BC = EF$, and $AB = DE$. So, by the definition of congruence, each of these pairs of sides is congruent and the included angles are congruent. Since two sides and the included angle of $\triangle ABC$ are congruent to two sides and the included angle of $\triangle DEF$, the two triangles are congruent by SAS.

Example 4 Proving Triangles Congruent

Given:
$\ell \parallel m$, $\overline{AB} \cong \overline{CD}$

Prove:
$\triangle ADB \cong \triangle ADC$

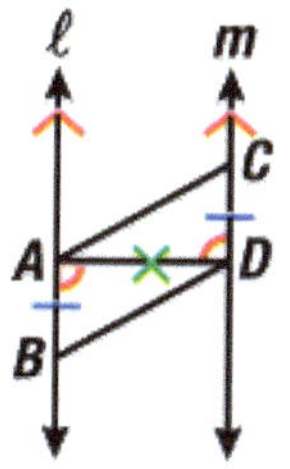

Statements:		Reasons:
1. $\overline{AB} \cong \overline{CD}$	side	1. Given
2. $\overline{AD} \cong \overline{AD}$	side	2. Reflex. Prop. of ≅
3. $\ell \parallel m$		3. Given
4. $\angle DAB \cong \angle ADC$	included angle	4. Alt. Int. ∠ Thm.
5. $\triangle ADB \cong \triangle ADC$		5. SAS Steps 1, 2, 4

It is given that the two lines are parallel in this example. So, consider the theorems related to parallel lines, such as the Alternate Interior Angles Theorem. The Alternate Interior Angles Theorem states that if two lines are parallel and cut by a transversal, then the pairs of alternate interior angles are congruent. In this figure, $\angle DAB$ and $\angle ADC$ are a pair of alternate interior angles. Therefore, $\angle DAB \cong \angle ADC$.

Triangle Congruence: ASA, AAS, and HL

Key Objectives

- Apply ASA, AAS, and HL to construct triangles and to solve problems.
- Prove triangles congruent by using ASA, AAS, and HL.

Key Terms

- An **included side** is the common side of two consecutive angles in a polygon.

Theorems, Postulates, Corollaries, and Properties

- **Angle-Side-Angle (ASA) Congruence Postulate** If two angles and the included side of one triangle are congruent to two angles and the included side of another triangle, then the triangles are congruent.
- **Angle-Angle-Side (AAS) Congruence Theorem** If two angles and a nonincluded side of one triangle are congruent to the corresponding angles and nonincluded side of another triangle, then the triangles are congruent.
- **Hypotenuse-Leg (HL) Congruence Theorem** If the hypotenuse and a leg of a right triangle are congruent to the hypotenuse and a leg of another right triangle, then the triangles are congruent.

Example 1 Problem Solving with ASA

Angle-Side-Angle (ASA) Congruence

If two angles and the included side of one triangle are congruent to two angles and the included side of another triangle, then the triangles are congruent.

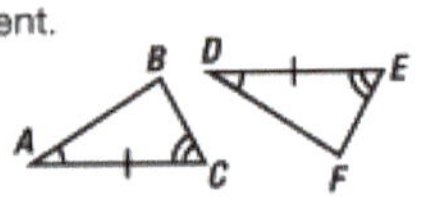

$\triangle ABC \cong \triangle DFE$

The common side of two consecutive angles in a polygon is the included side. The ASA Congruence Postulate allows for two triangles to be proven congruent when two angles and the included side in one triangle are congruent to two angles and the included side in the other triangle.

Example 2 Applying ASA Congruence

Determine if you can use ASA to prove triangles congruent. Explain.

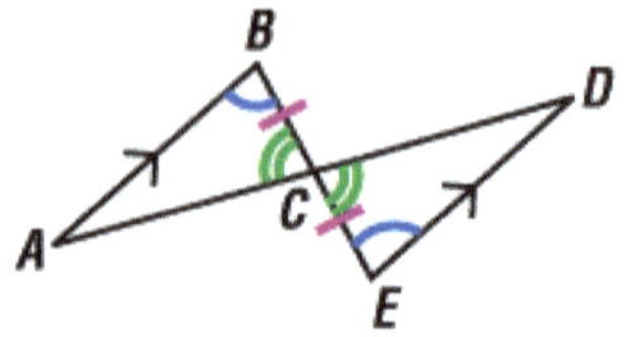

It is given that $\overline{BC} \cong \overline{EC}$ and $\overline{AB} \parallel \overline{ED}$.
Since $\overline{BE}$ transverses $\overline{AB}$ and $\overline{ED}$, by the Alternate Interior Angles Theorem, $\angle ABC \cong \angle DEC$.
By the Vertical Angles Theorem, $\angle BCA \cong \angle ECD$.
$\overline{BC} \cong \overline{EC}$ as given.
Therefore, $\triangle ABC \cong \triangle DEC$ by ASA.

The ASA Congruence Postulate can be used to prove two triangles congruent when two angles and their included side in one triangle are congruent to two angles and their included side in the other triangle. In this example it is given that the triangles have one pair of congruent sides. So, to prove that these triangles are congruent by ASA, show that the angles on each end of the congruent sides are congruent.

Example 3 Using AAS to Prove Triangles Congruent

Angle-Angle-Side (AAS) Congruence

If two angles and a nonincluded side of one triangle are congruent to the corresponding angles and nonincluded side of another triangle, then the triangles are congruent.

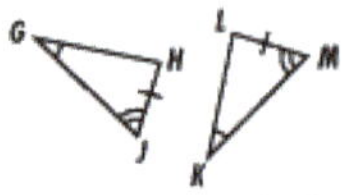

$\triangle GHJ \cong \triangle KLM$

The AAS Congruence Theorem uses two angles and a nonincluded side to show two triangles congruent. For any two angles in a triangle there are two nonincluded sides. Either nonincluded side can be used with the AAS Postulate. However, the nonincluded sides in each triangle must be corresponding sides.

Given:
$\angle Q \cong \angle S$
$\angle QRP$ and $\angle SRP$ are right angles.

Prove: $\triangle PQR \cong \triangle PSR$

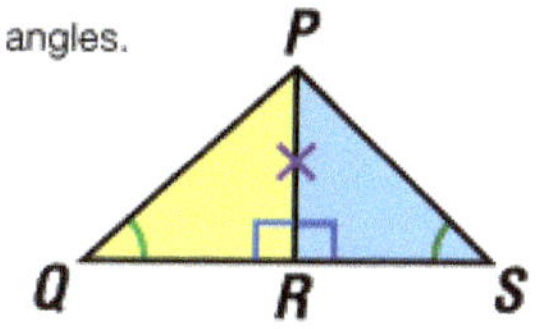

Statements:
1. $\angle Q \cong \angle S$ angle
2. $\angle QRP \cong \angle SRP$ angle
3. $\overline{PR} \cong \overline{PR}$ side
4. $\triangle PQR \cong \triangle PSR$

Reasons:
1. Given
2. Rt. $\angle \cong$ Thm.
3. Reflex. Prop. of $\cong$
4. AAS

It is given that $\angle Q$ is congruent to $\angle S$ in this example. It is also given that $\angle QRP$ and $\angle SRP$ are right angles. So, they are congruent by the Right Angle Congruence Theorem. Since two angles of one triangle are congruent to two angles of another triangle, ASA or AAS might be used to prove these two triangle congruent. Notice that the two triangles share a common side. This side is the nonincluded side to the congruent angles. Therefore, AAS can be used to prove these two triangles congruent.

Example 4 Applying HL Congruence

Hypotenuse-Leg (HL) Congruence

If the hypotenuse and a leg of a right triangle are congruent to the hypotenuse and a leg of another right triangle, then the triangles are congruent.

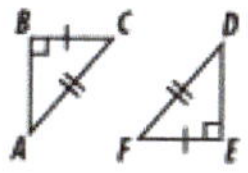

$\triangle ABC \cong \triangle DEF$

The HL Congruence Theorem can be used to show that two right triangles are congruent. Remember, the hypotenuse of a right triangle is the side that is opposite of the right angle. The two other sides are called the legs.

Determine if you can use the HL Congruence Theorem to prove $\triangle PQR \cong \triangle STU$. If not, tell what else you need to know.

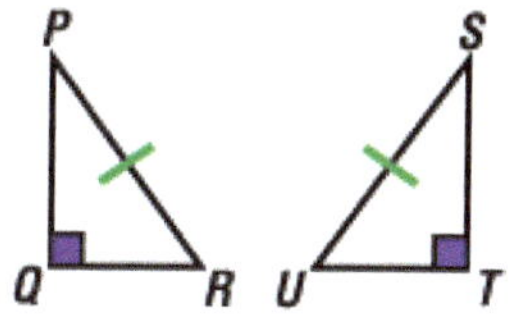

This conclusion cannot be proved by HL. According to the diagram, $\triangle PQR$ and $\triangle STU$ are right triangles, and the hypotenuses, $\overline{PR} \cong \overline{SU}$ are congruent.

You do not know that either leg of $\triangle PQR$ is congruent to either leg of $\triangle STU$.

In this example it is given that the two triangles are right triangles and that their hypotenuses are congruent. In order to prove two right triangles congruent by the HL Theorem, the hypotenuse and a leg of one right triangle must be congruent to the hypotenuse and a leg of the other right triangle. However, in this example the congruence of either pair of legs cannot be deduced. Therefore, the HL Theorem cannot be used to prove that these triangles are congruent.

Determine if you can use the HL Congruence Theorem to prove $\triangle ACB \cong \triangle ACD$. If not, tell what else you need to know.

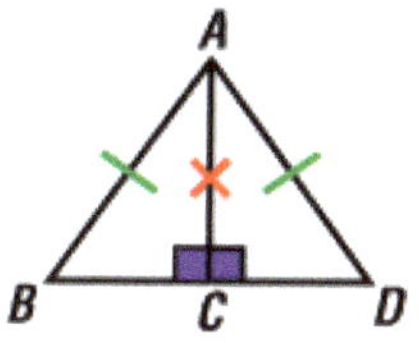

$\overline{AB} \cong \overline{AD}$ is given.
$m\angle ACB = 90°$ by the Definition of a Right Triangle.
By the Supplementary Angles Theorem, $m\angle ACD = 90°$. So $\triangle ACD$ and $\triangle ACB$ are right triangles with $\cong$ hypotenuses.
$\overline{AC}$ is a leg common to both triangles. By the Reflexive Property of Congruence, $\overline{AC} \cong \overline{AC}$.
Therefore, $\triangle ACB \cong \triangle ACD$ by HL.

In this example it is given that one pair of sides is congruent and that $\angle ACB$ is a right angle. Since $\angle ACB$ is a right angle, $\angle ACD$ is a right angle as well. Therefore, the pair of given congruent sides are the hypotenuses. If one pair of legs can be shown to be congruent, then the HL Theorem can be used to prove that these triangles are congruent.

Since the triangles contain a common side, and that side is not the hypotenuse of either triangle, then by the Reflexive Property of Congruence, the triangles contain a pair of congruent legs. So, the triangles are congruent by HL.

Triangle Congruence: CPCTC

Key Objectives

- Using CPCTC to prove parts of triangles are congruent.

Key Terms

- **CPCTC** is an abbreviation for the phrase "Corresponding Parts of Congruent Triangles are Congruent," which can be used as a justification in a proof after two triangles have been shown to be congruent.

Remember, congruent triangles are triangles with corresponding parts (angles and sides) that are congruent. The SSS, SAS, ASA, AAS, and HL theorems and postulates are used to prove that two triangles are congruent by using some combination of their corresponding parts. In other words, once some combination of corresponding parts has been shown to be congruent, then SSS, SAS, ASA, AAS, or HL can be used to justify that the two triangles are congruent.

CPCTC works in the opposite way. CPCTC is used to prove that corresponding parts of two triangles are congruent by using the congruence of two triangles. In other words, once two triangles are shown to be congruent, then CPCTC can be used to justify that any of their corresponding parts are congruent.

CPCTC is typically used in a proof after SSS, SAS, ASA, AAS, or HL has been used to show that the two triangles are congruent.

Example 1 Measuring Application

Manda wants to know the distance between her front door and her neighbor's. She located points *P*, *Q*, and *R*. How can she find *MN*?

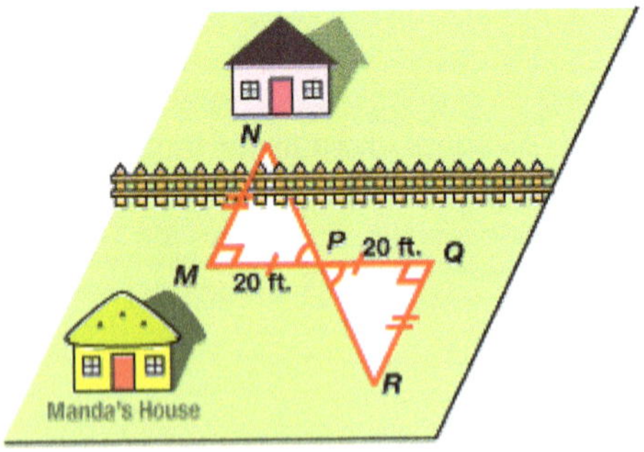

$\overline{MN} \cong \overline{QR}$ by CPCTC.
$MN = QR$ by the definition of congruent sides.
Manda can find the length of $\overline{MN}$ by measuring the length of $\overline{QR}$!

Indirect measurement is used in this example to show how the distance from M to N could be found using only measurements from Manda's side of the fence. Manda constructed the two triangles so that $\angle M$ and $\angle Q$ are right angles and $MP = PQ = 20$ ft. Use this given information to explain how she could find MN without actually measuring that distance.

First, prove that $\triangle NMP \cong \triangle RQP$ by using ASA since two angles ($\angle M$ and $\angle P$) and their included side in $\triangle NMP$ are congruent to two angles ($\angle Q$ and $\angle P$) and their included side in $\triangle RQP$.

Then, since the two triangles are congruent, CPCTC can be used to show that any pair of corresponding parts are congruent. So, use CPCTC to show that by finding QR, MN can also be found since QR and MN are the lengths of corresponding sides.

Example 2 Proving Corresponding Parts Congruent

Given: $\overline{AE} \cong \overline{EC}$, $\overline{BE} \cong \overline{ED}$

Prove: $\overline{AB} \cong \overline{DC}$

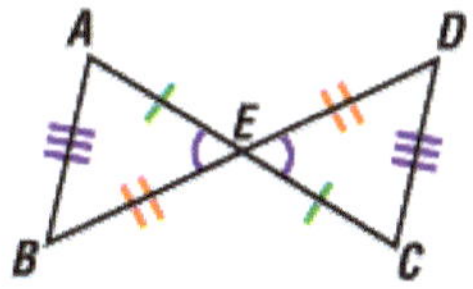

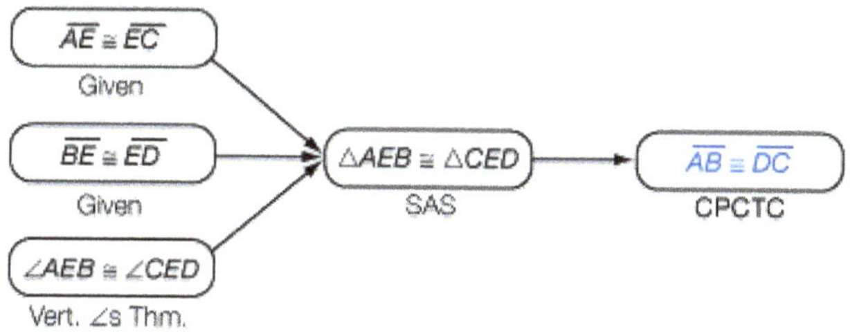

A flowchart proof is used in this example to prove that two sides are congruent. Remember, CPCTC can be used to prove the congruence of a pair of corresponding sides (or angles) in two triangles, but only after the two triangles are shown to be congruent. So, the first task in this proof is to show that $\triangle AEB \cong \triangle CED$.

Two pairs of congruent sides are given. Additionally, their included angles are congruent by the Vertical Angles Theorem (since $\angle AEB$ and $\angle CED$ are vertical angles). So, the two triangles must be congruent by SAS.

Now, since $\triangle AEB \cong \triangle CED$ by SAS, any of their corresponding parts can be shown to be congruent by CPCTC.

Example 3 Using CPCTC in a Proof

Given:
Y is the midpoint of $\overline{XZ}$,
$\angle XYW$ is a right angle.

Prove: $\angle XWY \cong \angle ZWY$

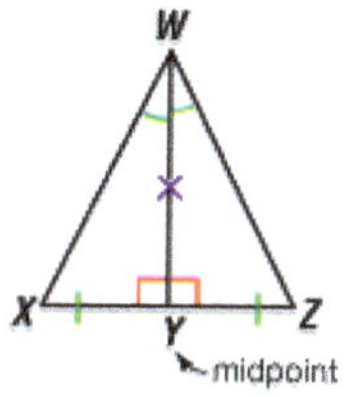

Statements:		Reasons:
1. $\angle XYW$ is a right angle.		1. Given
2. $m\angle XYW = 90°$		2. Def. of Rt. ∠
3. $m\angle ZYW = 90°$		3. Supp. ∠s Thm.
4. $\angle XYW \cong \angle ZYW$	*included angle*	4. Def. of ≅ ∠s
5. $\overline{XY} \cong \overline{ZY}$	*side*	5. Def. of mdpt.
6. $\overline{WY} \cong \overline{WY}$	*side*	6. Reflex. Prop. of ≅
7. $\triangle XWY \cong \triangle ZWY$		7. SAS
8. $\angle XWY \cong \angle ZWY$		8. CPCTC

In this proof there are two pieces of given information.
1. Y is the midpoint of XZ.
2. $\angle XYW$ is a right angle.

Notice that the angles to prove congruent, $\angle XWY$ and $\angle ZWY$, are corresponding angles in $\triangle XWY$ and $\triangle ZWY$. Therefore, first show that $\triangle XWY \cong \triangle ZWY$ and then use CPCTC to show that $\angle XWY \cong \angle ZWY$.

Example 4 Using CPCTC in a Coordinate Proof

Given: $A(-3,-6)$, $B(-3,6)$, $C(-7,4)$, $D(3,8)$, $E(3,-4)$, $F(7,-2)$

Prove: $\angle BCA \cong \angle DFE$

$\triangle ABC \cong \triangle DEF$ by SSS.

$\angle BCA \cong \angle DFE$ by CPCTC.

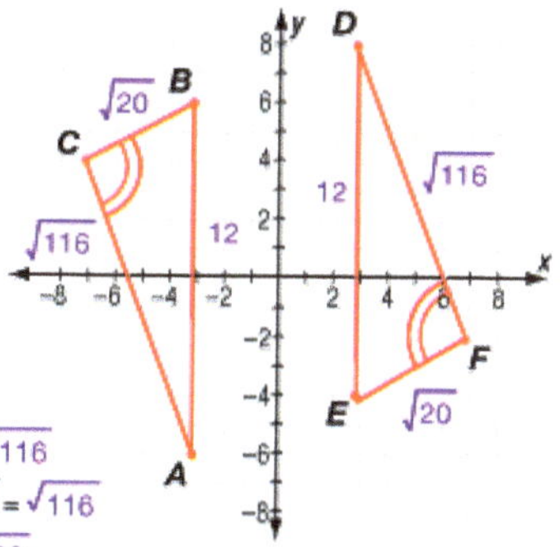

Use the Distance Formula to find the lengths of the sides of each triangle.

$DF = \sqrt{(7-3)^2 + (-2-8)^2} = \sqrt{16+100} = \sqrt{116}$

$AC = \sqrt{(-3-(-7))^2 + (-6-4)^2} = \sqrt{16+100} = \sqrt{116}$

$EF = \sqrt{(7-3)^2 + (-2-(-4))^2} = \sqrt{16+4} = \sqrt{20}$

$BC = \sqrt{(-7-(-3))^2 + (4-6)^2} = \sqrt{16+4} = \sqrt{20}$

$DE = \sqrt{(3-3)^2 + (8-(-4))^2} = \sqrt{0+144} = 12$

$AB = \sqrt{(-3-(-3))^2 + (6-(-6))^2} = \sqrt{0+144} = 12$

CPCTC can be used in a coordinate proof just as it was used in other proofs. First show that the two triangles are congruent, then use CPCTC to show that some pair of corresponding parts are congruent.

Remember, when given a figure on a coordinate plane, the lengths of the figure's sides can always be determined by using the distance formula.

For this proof, first find the length of each side of $\triangle ABC$ and of $\triangle DEF$ by using the distance formula. Then, since $DF = AC$, $EF = BC$, and $DE = AB$, the two triangles are congruent by SSS. So, since $\angle BCA$ and $\angle DFE$ are corresponding angles, $\angle BCA \cong \angle DFE$ by CPCTC.

Introduction to Coordinate Proof

Key Objectives

- Position figures in the coordinate plane for use in coordinate proofs.
- Prove geometric concepts by using coordinate proof.

Key Terms

- A **coordinate proof** is a style of proof that uses coordinate geometry and algebra.

Example 1 Positioning a Figure in the Coordinate Plane

The figure in a coordinate proof is either given or placed on a coordinate plane. If coordinates of the figure's vertices are not given in the conjecture, then the figure can be placed anywhere on the coordinate plane. However, there are some guidelines for positioning a figure that will result in easier steps in the proof.

Position a rectangle with a length of 6 units and a width of 4 units in the coordinate plane.

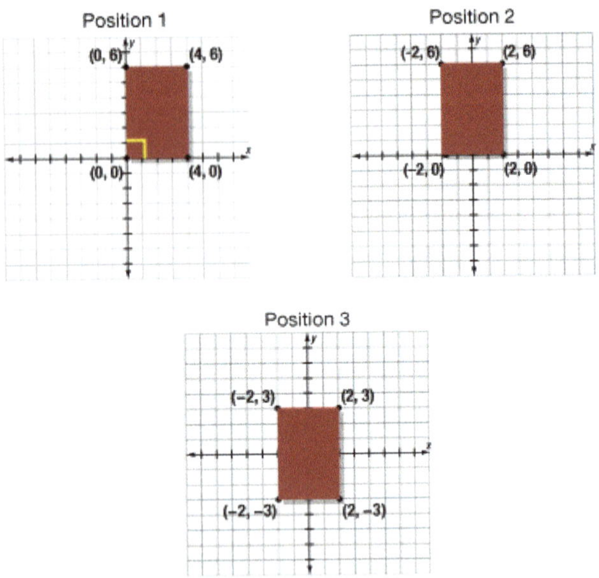

In this example, a figure is given. The figure is a rectangle with length 6 units and width 4 units. The rectangle can be positioned (placed) on the coordinate plane in any of the four quadrants. However, it is typically easiest to place a figure in the first quadrant so that the coordinates in each vertex's ordered pair will be a positive number.

The rectangle could be oriented in any way, but placing the rectangle so that its sides are vertical and horizontal, ideally with one side on the x-axis and one side on the y-axis, is best.

So, Position 1 is the best position for this rectangle because the coordinates of the vertices are all positive whole numbers and two of the rectangle's sides are placed along an axis.

Example 2 Writing a Proof Using Coordinate Geometry

Given:
Right $\triangle ABC$ has vertices $A(0,8)$, $B(0,0)$, and $C(5,0)$. D is the midpoint of $\overline{AC}$.

Prove:
The area of $\triangle ABC$ is twice the area of $\triangle DBC$.

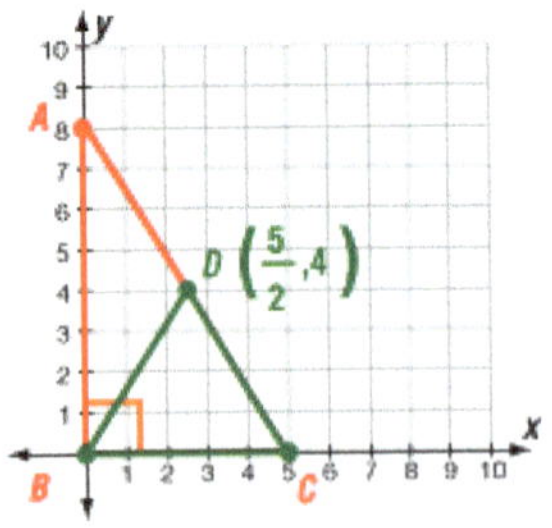

$\triangle ABC$ is a right triangle with height AB and base BC.

Area of $\triangle ABC = \frac{1}{2}bh$

$= \frac{1}{2}(5)(8) = 20$ square units

By the Midpoint Formula, the coordinates of D = $\left(\frac{0+5}{2}, \frac{0+8}{2}\right) = \left(\frac{5}{2}, 4\right)$.

$\triangle DBC$ has a base BC and height equal to the y-coordinate of D.

Area of $\triangle DBC = \frac{1}{2}bh$

$= \frac{1}{2}(5)(4) = 10$ square units

Since $20 = 2(10)$, the area of $\triangle ABC$ is twice the area of $\triangle DBC$.

Make a plan for proving that the area of $\triangle ABC$ is twice the area of $\triangle DBC$. Find the area of each triangle and then multiply the area of $\triangle DBC$ by 2 to determine whether the area of $\triangle ABC = 2$(the area of $\triangle DBC$).

Remember, the area of a triangle is 1/2 the product of the triangle's base and height (where the height is the perpendicular distance from the base to the opposite vertex). So, the area of $\triangle ABC$ is $(1/2)(5)(8) = 20$ units^2.

Before the area of $\triangle DBC$ can be found, the location of D must be determined. Since D is given to be the midpoint between A and C, the location of D can be found by using the midpoint formula. The y-coordinate of D is the height of $\triangle DBC$. Now the area of $\triangle DBC$ can be found.

Example 3 Assigning Coordinates to a Vertex

Position each figure in the coordinate plane and give the coordinates of each vertex.

A right triangle with leg lengths m and n

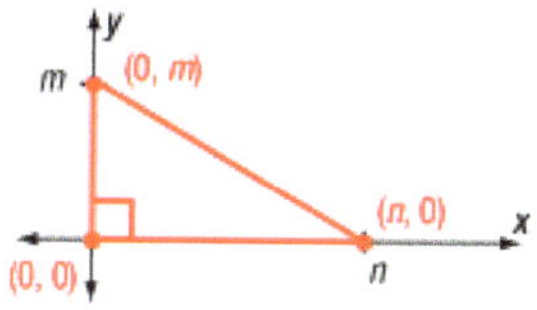

A rectangle with length q and width r

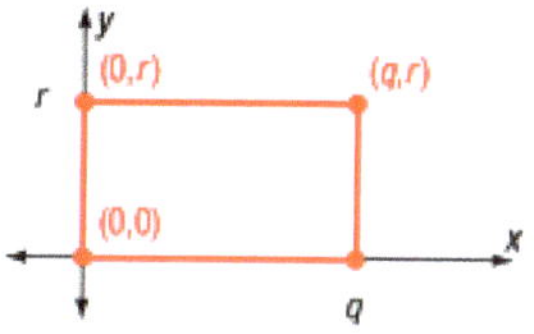

The first figure given here is a right triangle. Since a right triangle has one right angle, position that triangle so that the vertex at the right angle is at the origin. This triangle's legs have length m and n. So, assign one leg length m and the other length n, but it doesn't matter which leg has which length. Here, the vertical leg is assigned length m. Therefore, the vertical leg extends to m on the y-axis, or from (0, 0) to (0, m). And since the horizontal leg is assigned length n, this leg extends to n on the x-axis, or from (0, 0) to (n, 0).

The rectangle has four right angles, so place the rectangle so that one vertex is at the origin, one vertex is on the x-axis, and another vertex is on the y-axis.

Example 4 Writing a Coordinate Proof

In this example, two methods are demonstrated for proving that the area $\triangle DBE$ is one-fourth the area of $\triangle ABC$. The same conjecture is proven in each proof, but the figure's coordinates are identified differently in each proof.

In each proof, it is given that a triangle, $\triangle ABC$, contains a right angle at $\angle B$. Additionally, it is given that the midpoints of two of the triangle's sides are identified as D and E, where D is the midpoint between A and B and E is the midpoint between B and C.

The first step in each proof is to position $\triangle ABC$ on a coordinate plane. Since $\angle B$ is a right angle, position B at the origin. It follows that A must lie on the y-axis and C must lie on the x-axis. The lengths of the sides of $\triangle ABC$ are unknown, so the next step is to identify variables for the lengths of AB and BC. The difference in the two proofs below is based on the lengths chosen for AB and BC.

Given:
$\angle B$ is a right angle in $\triangle ABC$.
D is the midpoint of $\overline{AB}$.
E is the midpoint of $\overline{BC}$.

Prove: The area of $\triangle DBE$ is one-fourth the area of $\triangle ABC$.

Method 1

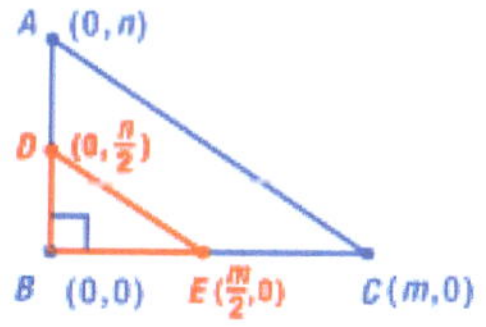

Proof:

By the Midpoint Formula, the coordinates of $D = \left(\frac{0+0}{2}, \frac{0+n}{2}\right) = \left(0, \frac{n}{2}\right)$.

By the Midpoint Formula, the coordinates of $E = \left(\frac{0+m}{2}, \frac{0+0}{2}\right) = \left(\frac{m}{2}, 0\right)$.

$\triangle ABC$ is a right triangle with base m and height n.

Area of $\triangle ABC = \frac{1}{2}bh$

$= \frac{1}{2}(m)(n) = \frac{mn}{2}$ square units

$\triangle DBE$ is a right triangle with base $\frac{m}{2}$ and height $\frac{n}{2}$.

Area of $\triangle DBE = \frac{1}{2}bh$

$= \frac{1}{2}\left(\frac{m}{2}\right)\left(\frac{n}{2}\right) = \frac{mn}{8}$ square units

Since $\frac{mn}{8} = \frac{1}{4}\left(\frac{mn}{2}\right)$, the area of $\triangle DBE$ is one fourth the area of $\triangle ABC$.

Let $AB = n$ and let $BC = m$. So, the coordinates of A must be (0, n) and the coordinates of C must be (0, m).

Now that the coordinates of A, B, and C are identified, the coordinates of vertices D and E in $\triangle DBE$ can be deduced (note that vertex B is the same point in both triangles). Since D is the midpoint of A and B, use the midpoint formula with (0, 0) and (0, n) to find the coordinates of D. The midpoint formula can also be used to find the coordinates of E, but use (0, 0) and (0, m) since E is the midpoint between B and C.

Now find the area of the two triangles. Since $\triangle ABC$ has base m and height n, the area of $\triangle ABC$ is $mn/2$ units^2. Since $\triangle DBE$ has base $m/2$ and height $n/2$, the area of $\triangle DBE$ is $mn/8$ units^2.

Therefore, since $mn/8 = (1/4)(mn/2)$, the area of $\triangle DBE$ is one-fourth the area of $\triangle ABC$.

Given:
$\angle B$ is a right angle in $\triangle ABC$.
D is the midpoint of $\overline{AB}$.
E is the midpoint of $\overline{BC}$.

Prove: The area of $\triangle DBE$ is one-fourth the area of $\triangle ABC$.

Method 2

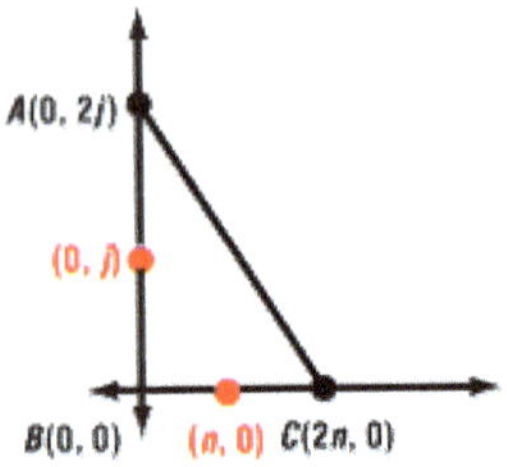

Proof:
By the Midpoint Formula, the coordinates of $D = \left(\frac{0+0}{2}, \frac{0+2j}{2}\right) = (0, j)$.
By the Midpoint Formula, the coordinates of $E = \left(\frac{2n+0}{2}, \frac{0+0}{2}\right) = (n, 0)$.
$\triangle ABC$ is a right triangle with base $2n$ and height $2j$.
Area of $\triangle ABC = \frac{1}{2}bh$
$= \frac{1}{2}2n2j = 2nj$
$\triangle DBE$ is a right triangle with base n and height j.
Area of $\triangle DBE = \frac{1}{2}bh$
$= \frac{1}{2}nj$
$\frac{\frac{1}{2}nj}{2nj} = \frac{1}{4}$

In Method 2, the triangle is again positioned on a coordinate plane such that B is at the origin, A is on the y-axis and C is on the x-axis. But in this proof, the lengths of the sides are not m and n. Instead, let $AB = 2j$ and let $BC = 2n$. The point of identifying these lengths as "twice the variable" is so that the coordinates of the midpoints will not contain fractions.

By using the midpoint formula, the base of $\triangle DBE$ is found to be n and the height of $\triangle DBE$ is found to be j.

Now find the area of each triangle and compare the areas to prove the conjecture.

Isosceles and Equilateral Triangles

Key Objectives

- Prove theorems about isosceles and equilateral triangles.
- Apply properties of isosceles and equilateral triangles.

Key Terms

- The congruent sides of an isosceles triangle are called the **legs**.
- The **vertex angle** is the angle formed by the legs (congruent sides) of an isosceles triangle.
- The side opposite of the vertex angle in an isosceles triangle is called the **base**.
- The **base angles** of an isosceles triangle are the two angles that have the base as a side.

Theorems, Postulates, Corollaries, and Properties

- **Isosceles Triangle Theorem** If two sides of a triangle are congruent, then the angles opposite the sides are congruent.
- **Converse of Isosceles Triangle Theorem** If two angles of a triangle are congruent, then the sides opposite those angles are congruent.
- **Equilateral Triangle Theorem** If a triangle is equilateral, then it is equiangular.
- **Equiangular Triangle Theorem** If a triangle is equiangular, then it is equilateral.

Example 1 Applying the Isosceles Triangle Theorem

Isosceles Triangle Theorem
If two sides of a triangle are congruent, then the angles opposite the sides are congruent.
(Isosc. △ Thm.)

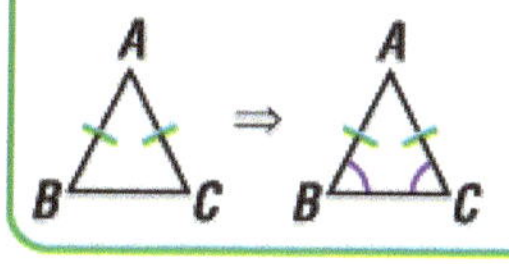

$\overline{AC} \cong \overline{AB} \Rightarrow \angle B \cong \angle C$

An isosceles triangle is a triangle with at least two congruent sides.

The Isosceles Triangle Theorem states that when a triangle has two congruent sides, it can be assumed that the angles opposite of those sides are also congruent.

Converse of Isosceles Triangle Theorem
If two angles of a triangle are congruent, then the sides opposite the angles are congruent.
(Conv. of the Isosc. △ Thm.)

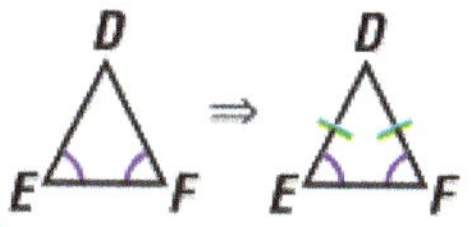

$\angle F \cong \angle E \Rightarrow \overline{DE} \cong \overline{DF}$

The Converse of Isosceles Triangle Theorem states the reverse of the Isosceles Triangle Theorem. By the converse, when a triangle has two congruent angles, it can be assumed that the sides opposite of those angles are also congruent.

Given:
C is the midpoint of $\overline{BD}$.
$m\angle ABE = 105°$
$m\angle CDE = 75°$

Explain why $BE = ED$.

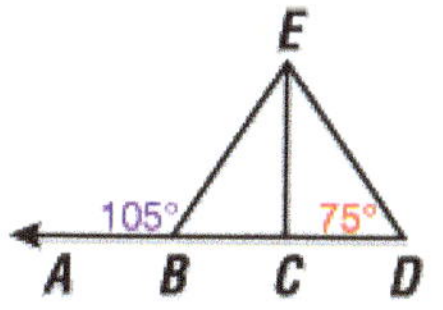

$m\angle ABE = 105°$	Given
$m\angle CDE = 75°$	Given
$\angle ABE$ and $\angle CBE$ are supplementary $\angle$s.	Given
$m\angle ABE + m\angle CBE = 180°$	Def. of Supp. $\angle$s
$m\angle CBE = 180° - m\angle ABE$	Subtract $m\angle ABE$ from both sides.
$m\angle CBE = 180° - 105°$	Substitute 105° for $m\angle ABE$.
$m\angle CBE = 75°$	Subtract 105° from 180°.
$m\angle CBE = m\angle CDE$	Trans. Prop. of =
$\angle CBE \cong \angle CDE$	Def. of $\cong$ $\angle$s
$\overline{BE} \cong \overline{ED}$	Conv. of the Isosc. △ Thm.
$BE = ED$	Def. of $\cong$ sides

The figure is given for this proof, along with $m\angle ABE$, $m\angle CDE$, and the fact that C is the midpoint between B and D. However, notice that this proof does not use the fact that C is a midpoint.

Make a plan for the proof. The conclusion to be proven is that $BE = ED$. Notice that BE and ED are the measures of two sides of $\triangle BED$. So, show that $\angle CBE$ is congruent to $\angle CDE$. From there, the Converse of Isosceles Triangle Theorem can be applied because $\triangle BED$ contains two congruent angles, $\angle CBE$ and $\angle CDE$. So, by the Converse of Isosceles Triangle Theorem, the sides of $\triangle BED$ that are opposite of $\angle CBE$ and $\angle CDE$ must also be congruent.

Example 2 Finding the Measure of an Angle

Find m∠Q.

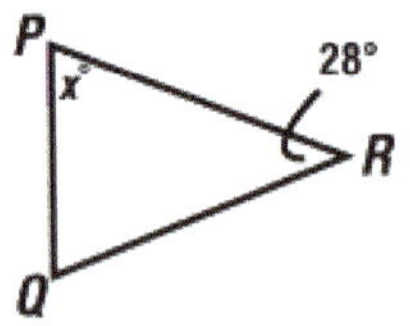

m∠R = 28°	Given
$\overline{PR} \cong \overline{QR}$	Given
∠P ≅ ∠Q	Isosc. △ Thm.
m∠Q = m∠P = x°	Def. of ≅ ∠s
180° = m∠P + m∠Q + m∠R	△ Sum Thm.
180 = 2x + 28	Substitute the given values.
152 = 2x	Subtract 28 from both sides.
76 = x	Divide both sides by 2.

Thus, m∠Q = 76°.

In this example the measure of only one angle is given.

$m\angle R = 28°$

Additionally, from the figure you know that two sides of $\triangle PQR$ are also congruent. The Isosceles Triangle Theorem can be applied to any triangle with two congruent sides. So, by the Isosceles Triangle Theorem, the angles opposite of the each congruent side must also be congruent. Therefore, if $m\angle P = x°$, then $m\angle Q = x°$ as well.

Now use the Triangle Sum Theorem to write an equation where the sum of the angles is equal to 180°. Then, solve this equation for x. Since $m\angle Q = x°$, the value of x is the answer.

Find m∠C.

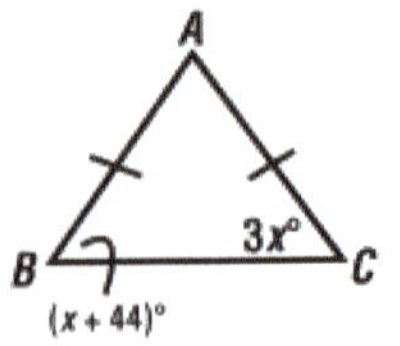

m∠B = m∠C	Isosc. △ Thm.
x + 44 = 3x	Substitute the given values.
44 = 2x	Subtract x from both sides.
22 = x	Divide both sides by 2.

m∠C = 3x = 3(22) = 66°

The given triangle, $\triangle ABC$, is an isosceles triangle since two of its sides, $\overline{AB}$ and $\overline{AC}$ are given to be congruent. So, the Isosceles Triangle Theorem can be applied. By the Isosceles Triangle Theorem, $\angle B$ is congruent to $\angle C$ since $\angle B$ and $\angle C$ are the angles opposite of the congruent sides. Then by the definition of congruence, the measure of the two angles must be equal. So, write an equation using the expression given for $m\angle B$, $(x + 44)°$, and the expression given for $m\angle C$, $3x°$. Solve this equation for x and then substitute the value of x back into the expression for $m\angle C$ to find that angle's measure.

Example 3 Using Properties of Equilateral Triangles

Equilateral Triangle Theorem

If a triangle is equilateral, then it is equiangular. (Equilateral △ Thm.)

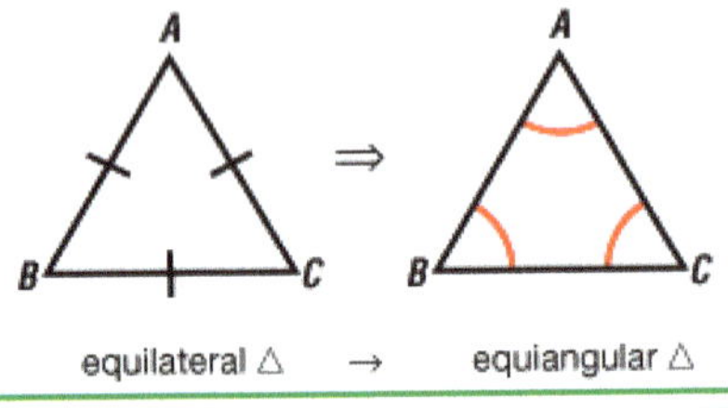

Equiangular Triangle Theorem

If a triangle is equiangular, then it is equilateral. (Equiangular △ Thm.)

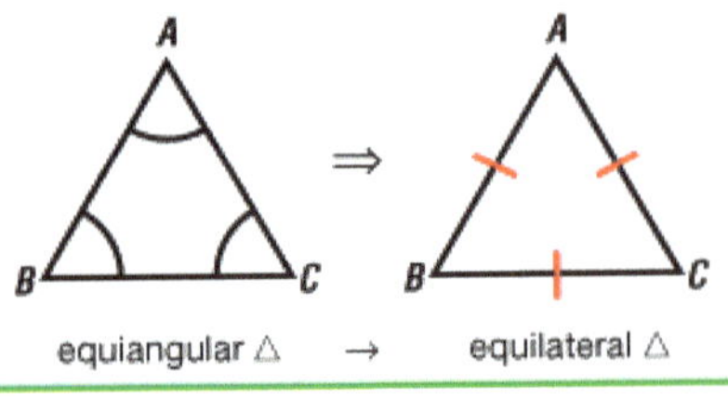

An equilateral triangle has three congruent sides and an equiangular triangle has three congruent angles.

The Equilateral Triangle Theorem and the Equiangular Triangle Theorem basically state that being one of those two types of triangles implies the other. Specifically, by the Equilateral Triangle Theorem, if it is known that a triangle is equilateral, then it can be assumed that the triangle is also equiangular. And by the Equiangular Triangle Theorem, if it is known that a triangle is equiangular, then it can be assumed that the triangle is also equilateral.

So, all equilateral triangles are equiangular and all equiangular triangles are equilateral.

Find the value of x. 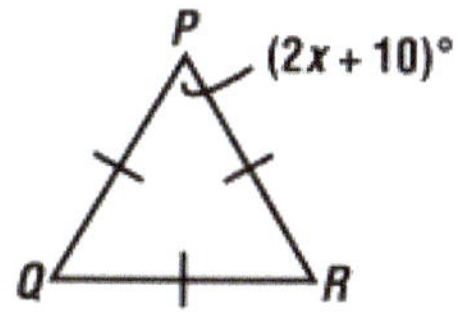 $\triangle PQR$ is equiangular. — Equilateral $\triangle \rightarrow$ equiangular $\triangle$ $m\angle P = 60°$ — The measure of each $\angle$ of an equiangular $\triangle$ is 60°. $2x + 10 = 60$ — Substitute the given values. $2x = 50$ — Subtract 10 from both sides. $x = 25$ — Divide both sides by 2.	In this example it is given that the three sides of $\triangle PQR$ are congruent. Therefore, $\triangle PQR$ is an equilateral triangle. So, by the Equilateral Triangle Theorem, it can be assumed that $\triangle PQR$ is also equiangular. It follows that the measure of each angle in $\triangle PQR$ must be equal to 60°, since $\triangle PQR$ is equiangular. So, set the expression given for $m\angle P$, $(2x + 10)°$, equal to 60° and solve the equation for x.
Find the value of t. 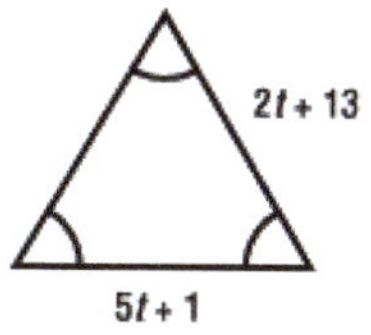 The $\triangle$ is equilateral. — Equiangular $\triangle \rightarrow$ equilateral $\triangle$ All of the sides of the $\triangle$ are congruent. — Def. of equilateral $\triangle$ $2t + 13 = 5t + 1$ — Substitute the given values. $12 = 3t$ — Subtract $2t$ and subtract 1 from both sides. $4 = t$ — Divide both sides by 3.	The triangle given here has three congruent angles. Therefore, the triangle is equiangular. So, by the Equiangular Triangle Theorem, it can be assumed that the triangle is also equilateral. Equilateral triangle must have three congruent sides, by definition. So, the expressions given for the lengths of two of the triangle's sides can be set equal to each other. $2t + 13 = 5t + 1$ Solve this equation for t.

Example 4 Using Coordinate Proof

Given: $\triangle ABC$ is right isosceles. D is the midpoint of $\overline{AC}$. $\overline{AB} \cong \overline{BC}$ **Prove:** $\triangle BDC$ is isosceles. 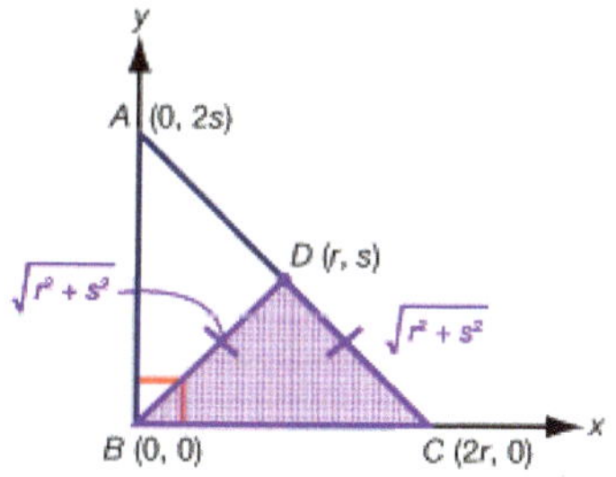 By the Midpoint Formula, the coordinates of D are $\left(\frac{2r+0}{2}, \frac{0+2s}{2}\right) = (r, s)$. By the Distance Formula, $BD = \sqrt{(r-0)^2 + (s-0)^2} = \sqrt{r^2+s^2}$, and $CD = \sqrt{(2r-r)^2 + (0-s)^2} = \sqrt{r^2+s^2}$. Since $BD = CD$, $\overline{BD} \cong \overline{CD}$ by definition. So $\triangle BDC$ is isosceles.	Remember, when writing a coordinate proof, the given figure can be placed anywhere on a coordinate plane. Since the figure given here is a right triangle, place the right angle at the origin where one side of the triangle is placed on the x-axis and another side is placed on the y-axis. And since it is given that $\overline{AB}$ and $\overline{BC}$ are congruent, the right angle must be $\angle B$. So, draw $\triangle ABC$ so that $\angle B$ is at the origin, A is on the y-axis, and C is on the x-axis. Here, the length of $\overline{AB}$ is chosen to be $2s$ and the length of $\overline{BC}$ is chosen to be $2r$. So, the coordinates of A and C are $(0, 2s)$ and $(2r, 0)$, respectively. D is given to be the midpoint of $\overline{AC}$. So, use the midpoint formula to find the coordinates of D. To prove that $\triangle BDC$ is isosceles, show that the lengths of two of the sides are equal. Use the distance formula to find the lengths of $\overline{BD}$ and $\overline{CD}$.

5

5 Properties and Attributes of Triangles

Perpendicular and Angle Bisector Theorems

Key Objectives

- Prove and apply theorems about perpendicular bisectors.
- Prove and apply theorems about angle bisectors.

Key Terms

- When a point is the same distance from two or more objects, the point is said to be **equidistant** from the objects.
- A **locus** is a set of points that satisfies a given condition.
- An **angle bisector** is a ray that divides an angle into two congruent angles.
- The **perpendicular bisector** of a segment is a line that passes through the midpoint of a segment that is perpendicular to the segment.

Theorems, Postulates, Corollaries, and Properties

- **Perpendicular Bisector Theorem** If a point is on the perpendicular bisector of a segment, then it is equidistant from the endpoints of the segment.
- **Converse of the Perpendicular Bisector Theorem** If a point is equidistant from the endpoints of a segment, then it is on the perpendicular bisector of the segment.
- **Angle Bisector Theorem** If a point is on the bisector of an angle, then it is equidistant from the sides of the angle.
- **Converse of the Angle Bisector Theorem** If a point in the interior of an angle is equidistant from the sides of the angle, then it is on the bisector of the angle.

Example 1 Using Perpendicular Bisector Theorems

Perpendicular Bisector Theorem

If a point is on the perpendicular bisector of a segment, then it is equidistant from the endpoints of the segment.
(⊥ Bisector Thm.)

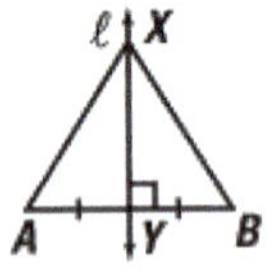

$\Rightarrow XA = XB$

$\overline{XY} \perp \overline{AB}$

$\overline{YA} \cong \overline{YB}$

By the Perpendicular Bisector Theorem, if a point is on the perpendicular bisector of a segment, then the point is equidistant from the endpoints of the segment.

Converse of the Perpendicular Bisector Theorem

If a point is equidistant from the endpoints of a segment, then it is on the perpendicular bisector of the segment.
(Conv. of ⊥ Bisector Thm.)

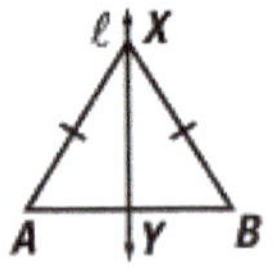

$\Rightarrow \overline{XY} \perp \overline{AB}$, $\overline{YA} \cong \overline{YB}$

$XA = XB$

By the Converse of the Perpendicular Bisector Theorem, if a point is equidistant from the endpoints of a segment, then the point is on the perpendicular bisector of the segment.

Find the measure.

ML

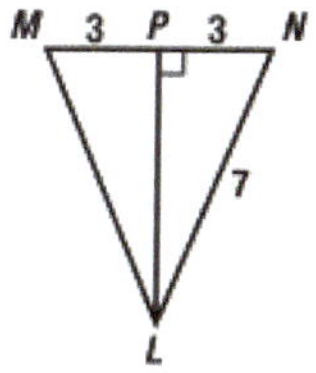

Since $MP = 3$ and $PN = 3$, $\overline{MP} \cong \overline{PN}$.

It is given that $\overline{MN} \perp \overline{PL}$.

Since $\overline{MP} \cong \overline{PN}$ and $\overline{MN} \perp \overline{PL}$, $\overline{PL}$ is a perpendicular bisector of $\overline{MN}$.

L is a point on $\overline{PL}$.

Since $\overline{PL}$ is a perpendicular bisector of $\overline{MN}$, L must be equidistant from the endpoints of $\overline{MN}$ by the Perpendicular Bisector Theorem.

Therefore, $ML = NL$.

Since $ML = NL$ and $NL = 7$, it follows that $ML = 7$.

$ML = 7$

The measure of a line segment is determined here using the Perpendicular Bisector Theorem.

In the figure, $\overline{PL}$ is given to be perpendicular to $\overline{MN}$. Additionally, $\overline{PL}$ bisects $\overline{MN}$ since $MP = PN$. Therefore, $\overline{PL}$ is a perpendicular bisector of $\overline{MN}$. So, the Perpendicular Bisector Theorem can be applied.

The Perpendicular Bisector Theorem states that any point on the perpendicular bisector of a segment must be equidistant from the endpoints of that segment.

Here, L is a point on perpendicular bisector $\overline{PL}$. So, L must be equidistant from the endpoints of $\overline{MN}$, M and N, by the Perpendicular Bisector Theorem. It follows that $ML = NL$. It is given that $NL = 7$. Thus, $ML = 7$ as well.

Find the measure.

PN

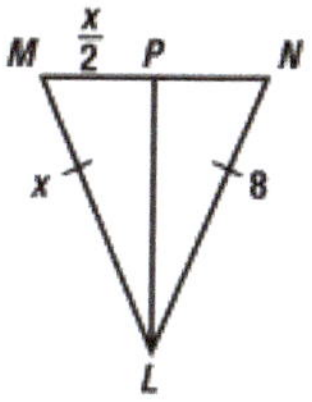

Since $\overline{ML} \cong \overline{NL}$, by the Converse of the Perpendicular Bisector Theorem, $\overline{PL}$ is a perpendicular bisector of $\overline{MN}$. Therefore, by the definition of perpendicular bisector, $MP = PN$.

Find x.

It is given that $\overline{ML} \cong \overline{NL}$. Therefore, since $ML = 8$, $x = 8$.

Use the value of x to find PN.

$MP = \frac{x}{2} = \frac{8}{2} = 4$

Since $MP = PN$, $PN = 4$.

$PN = 4$

Here, $\overline{PL}$ is *not* given to be perpendicular to $\overline{MN}$. However, it is given that $\overline{ML} \cong \overline{NL}$. It follows that $x = 8$ since $NL = 8$ and $ML = x$.

Since $x = 8$ and $MP = x/2$, it follows that $MP = 8/2 = 4$.

It would be easy to assume that since $MP = 4$, $PN = 4$ as well, but first the fact that $MP = PN$ must be confirmed since it is not given.

Use the Converse of the Perpendicular Bisector Theorem to confirm that $MP = PN$.

Example 2 Using Angle Bisector Theorems

Angle Bisector Theorem

If a point is on the bisector of an angle, then it is equidistant from the sides of the angle.
($\angle$ Bisector Thm.)

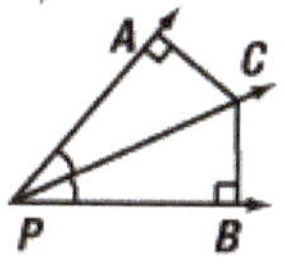

$\Rightarrow$ $AC = BC$

$\angle APC \cong \angle BPC$

By the Angle Bisector Theorem, if a point is on the bisector of an angle, then it is equidistant from the sides of the angle.

Notice that the distance between the point and the sides of the angle is the length of a line segment perpendicular to the side of the angle. This is the shortest distance.

Converse of the Angle Bisector Theorem

If a point in the interior of an angle is equidistant from the sides of the angle, then it is on the bisector of the angle.
(Conv. of $\angle$ Bisector Thm.)

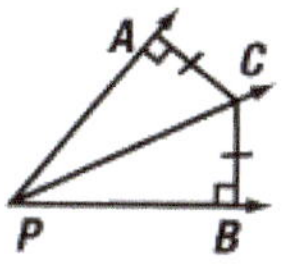

$\Rightarrow$ $\angle APC \cong \angle BPC$

$AC = BC$

By the Converse of the Angle Bisector Theorem, if a point in the interior of an angle is equidistant from the sides of the angle, then it is on the bisector of the angle.

Find the measure.

PB

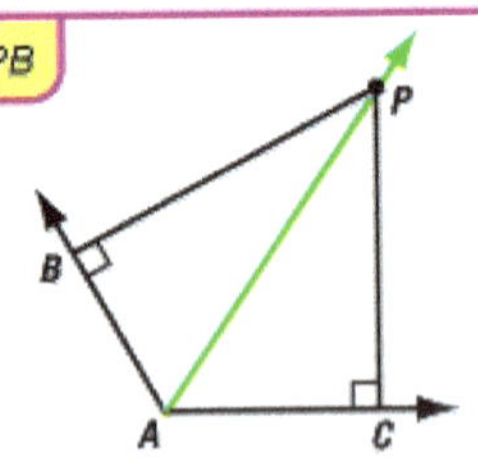

Given: $\angle BAP \cong \angle CAP$ and $PC = 10$

If $\angle BAP \cong \angle CAP$, then $\overline{AP}$ is an angle bisector of angle $\angle BAC$ by definition.

Since $\overline{AP}$ bisects $\angle BAC$, P is equidistant from the sides of angle $\angle BAC$ by the Angle Bisector Theorem.

Therefore, $PC = PB$.

It is given that $PC = 10$.

Since $PC = PB$, $PB = 10$.

$PB = 10$

In this example, PB is found. It is given that $\angle BAP \cong \angle CAP$. Therefore, $\overline{AP}$ bisects $\angle BAC$, and so $\overline{AP}$ is an angle bisector of $\angle BAC$. Since $\overline{AP}$ is an angle bisector, the Angle Bisector Theorem can be applied.

The Angle Bisector Theorem states that if a point is on the bisector of an angle, then that point is equidistant from the sides of the angle.

P is a point on angle bisector $\overline{AP}$. Therefore, P must be equidistant from the sides of $\angle BAC$. Since P must be equidistant from the sides of $\angle BAC$, it follows that $PC = PB$. It is given that $PC = 10$. Thus, $PB = 10$ as well.

Find the measure.

$m\angle CAP$

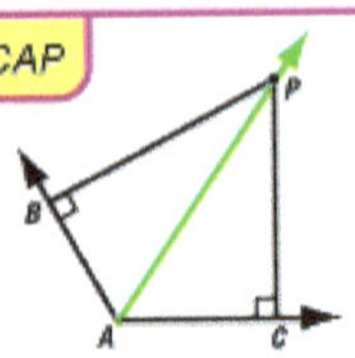

Given: $\overline{PB} \cong \overline{PC}$, $m\angle BAP = (7y + 14)°$ and $m\angle CAP = (9y - 2)°$

Since $\overline{PB} \cong \overline{PC}$, $\overline{PA}$ bisects $\angle BAC$ by the Converse of the Angle Bisector Theorem.

Since $\overline{PA}$ bisects $\angle BAC$, $\angle BAP \cong \angle CAP$ by definition of angle bisector. It follows that $m\angle BAP = m\angle CAP$.

Set the expression for $m\angle BAP$ equal to the expression for $m\angle CAP$ and solve for y.

$$9y - 2 = 7y + 14$$
$$2y = 16$$
$$y = 8$$

Substitute $y = 8$ into the expression for $m\angle CAP$ and simplify.

$m\angle CAP = (9(8) - 2)° = (72 - 2)° = 70°$ $m\angle CAP = 70°$

In this example, P is equidistant from the sides of $\angle CAP$ since $\overline{PB} \cong \overline{PC}$. Therefore, $\overline{PA}$ bisects $\angle BAC$ by the Converse of the Angle Bisector Theorem.

Since $\overline{PA}$ bisects $\angle BAC$, it follows that $m\angle BAP = m\angle CAP$.

Expressions for $m\angle BAP$ and $m\angle CAP$ are given. Set these expressions equal to each other and solve for y. Then, substitute the value of y into the expression for $m\angle CAP$ to find $m\angle CAP$.

Example 3 Writing Equations of Bisectors in the Coordinate Plane

In this example, the equation of a perpendicular bisector is written. Remember, the equation of a line can be found by using either two points that the line passes through or one point that the line passes through and the slope of the line.

Write an equation in point-slope form for the perpendicular bisector of the segment with endpoints $P(0, 3)$ and $Q(4, -1)$.

Step 1 Find midpoint of $\overline{PQ}$.

$$M = \left(\frac{0+4}{2}, \frac{3+-1}{2}\right) = (2, 1)$$ *Use the midpoint formula.*

Step 2 Find slope of $\overline{PQ}$.

$$m = \frac{-1-3}{4-0} = -1$$ *Use the slope formula.*

Step 3 Find the slope of the line perpendicular to $\overline{PQ}$.

The slope of $\overline{PQ}$ is -1. So, the slope of any line perpendicular to $\overline{PQ}$ is 1. *The slopes of perpendicular lines are opposite reciprocals.*

Step 4 Write the equation of the line.

Substitute the coordinates of the midpoint of $\overline{PQ}$, (2, 1), and the slope of the line perpendicular to $\overline{PQ}$, 1, into point-slope form.

$y - y_1 = m(x - x_1)$	*Point-Slope Form*
$y - 1 = 1(x - 2)$	*Substitute.*
$y - 1 = x - 2$	*Simplify.*
$y = x - 1$	*Write in slope-intercept form.*

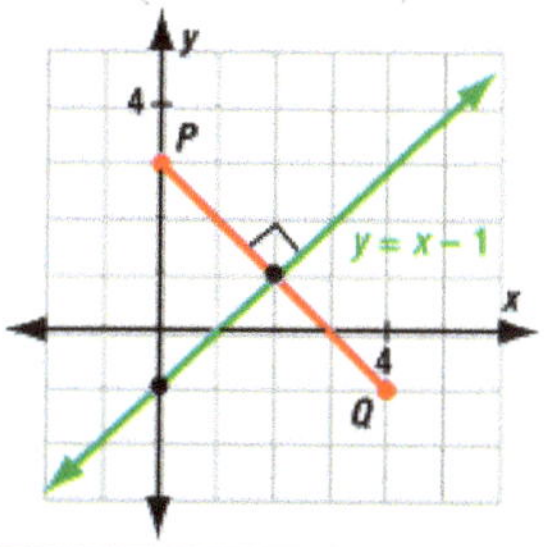

Remember

Point-Slope Form: $y - y_1 = m(x - x_1)$

Midpoint Formula: $M\left(\frac{x_1 + x_2}{2}, \frac{y_1 + y_2}{2}\right)$

The coordinates of the endpoints of $\overline{PQ}$ are given. Neither of these points can be used to write the equation of the perpendicular bisector of $\overline{PQ}$ because the perpendicular bisector does not pass through the segment's endpoints. Rather, since the perpendicular bisector of $\overline{PQ}$ *bisects* $\overline{PQ}$, it must pass through the midpoint of $\overline{PQ}$. So, find coordinates of the midpoint of $\overline{PQ}$ by using the Midpoint Formula: $M(2, 1)$.

The perpendicular bisector of $\overline{PQ}$ must be *perpendicular* to $\overline{PQ}$. Use this fact to find the slope of the perpendicular bisector. First, find the slope of $\overline{PQ}$. Then, find the slope of the line perpendicular to $\overline{PQ}$. The slope of $\overline{PQ}$ is -1. Since the slopes of perpendicular lines are opposite reciprocals, the slope of the perpendicular bisector is 1.

Now that a point on the perpendicular bisector is known, $M(2, 1)$, and the slope is known, $m = 1$, use point-slope form to write the equation of the perpendicular bisector.

Medians, Altitudes, and Midsegments in Triangles

Key Objectives

- Apply properties of medians, altitudes, and midsegments of a triangle.

Key Terms

- A **median of a triangle** is a segment whose endpoints are a vertex of the triangle and the midpoint of the other side.
- The point of concurrency of the medians of a triangle is the **centroid of the triangle**.
- An **altitude of a triangle** is a perpendicular segment from a vertex to the line containing the opposite side.
- The point of concurrency of the altitudes of a triangle is the **orthocenter of a triangle**.
- A **midsegment of a triangle** is a segment that joins the midpoints of the two sides of the triangle.
- Every triangle has three midsegments, which form the **midsegment triangle**.

Theorems, Postulates, Corollaries, and Properties

- **Centroid Theorem** A triangle's centroid is located 2/3 of the distance from each vertex to the midpoint of the opposite side.
- **Triangle Midsegment Theorem** A midsegment of a triangle is parallel to a side of the triangle, and its length is half the length of that side.

Medians, altitudes, and midsegments are types of line segments in triangles. Medians and altitudes are similar in that each has an endpoint at a vertex and the other endpoint is on the opposite side of the triangle (or on the line containing the opposite side). A midsegment does not have an endpoint at a vertex. Instead, both of a midsegment's endpoints are on sides of the triangle. A midsegment extends from the midpoint of one side of the triangle to the midpoint of another side of the triangle.

Each triangle has three medians, three altitudes, and three midsegments. A triangle's three medians intersect at one point, the centroid. Similarly, a triangle's three altitudes also intersect at one point, the orthocenter. However, a triangle's three midsegments do not intersect at one point. Each midsegment does share an endpoint with another midsegment and the three midsegments form another triangle, the midsegment triangle.

Example 1 Using Triangle Medians

Find *PB*.

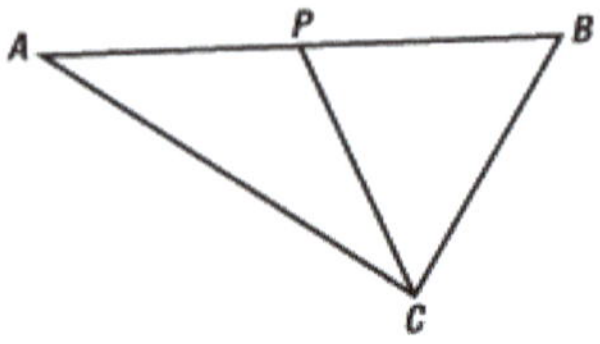

Given: $\overline{PC}$ is a median of $\triangle ABC$, $AP = 3x + 8$ and $PB = 5x$.

Since $\overline{PC}$ is a median of $\triangle ABC$ $\overline{AP} \cong \overline{PB}$. Therefore, $AP = PB$.

Set the expression for *AP* equal to the expression for *PB* and solve for *x*.

$$AP = PB$$
$$3x + 8 = 5x$$
$$8 = 2x$$
$$x = 4$$

Substitute $x = 4$ into the expression for *PB* and simplify.

$PB = 5x = 5(4) = 20$ $\quad$ $PB = 20$

$\overline{PC}$ is given to be a median. Therefore, by definition of median, P is the midpoint of $\overline{AB}$. It follows that $\overline{AP} \cong \overline{PB}$, and so $AP = PB$.

Expressions for AP and PB are given. Set these expressions equal to each other and solve for x. Then, substitute the value of x into the expression for PB to find PB.

Centroid Theorem

The centroid of a triangle is located 2/3 of the distance from each vertex to the midpoint of the opposite side. (Centroid Thm.)

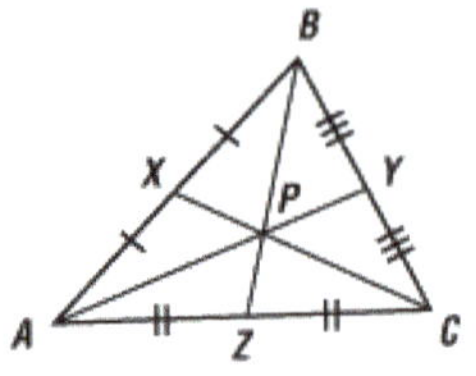

$\Rightarrow$

$AP = \frac{2}{3}AY \qquad BP = \frac{2}{3}BZ \qquad CP = \frac{2}{3}CX$

The centroid of the triangle is the point of concurrency of the medians of a triangle. In other words, the a triangle's centroid is the point at which the triangle's three medians intersect. By the Centroid Theorem, the centroid of a triangle is located 2/3 of the distance from each vertex to the midpoint of the opposite side.

This theorem can be used to find the length of a median. For example, suppose $BP = 12$. Then, since $BP = (2/3)BZ$, it follows that $12 = (2/3)BZ$, and so $BZ = 18$.

Example 2 Finding the Equation of a Median

ΔABC has vertices $A(-4, 4)$, $B(4, 0)$, and $C(-4, -1)$. Find the equation of the median through C in slope-intercept form.

Step 1 Plot the given vertices and draw ΔABC.

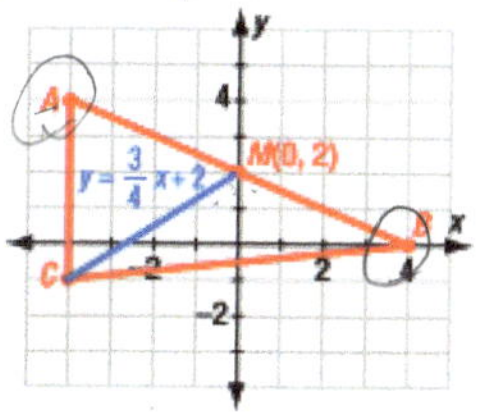

Step 2 Identify two points on the median through vertex C.
The median through vertex C must pass through the midpoint of the opposite side, $\overline{AB}$.
So, two points on the median are $C(-4, -1)$ and the midpoint of $\overline{AB}$.

Step 3 Find the coordinates of the midpoint of $\overline{AB}$.
Let $(x_1, y_1) = (-4, 4)$ and $(x_2, y_2) = (4, 0)$.
Substitute the coordinates into the midpoint formula.

$$M = \left(\frac{-4+4}{2}, \frac{4+0}{2}\right) = (0, 2)$$

Step 4 Write the equation of the median, the line that passes through $C(-4, -1)$ and $M(0, 2)$.

Find the slope of the line through $C(-4, -1)$ and $M(0, 2)$.

$$m = \frac{-1-2}{-4-0} = \frac{-3}{-4} = \frac{3}{4}$$

Identify the y-intercept.
Since the median passes through $M(0, 2)$, the y-intercept is 2.

Substitute $m = \frac{3}{4}$ and $b = 2$ into slope-intercept form to write the equation of the line. $y = mx + b \Rightarrow y = \frac{3}{4}x + 2$

Remember

Midpoint Formula	Slope Formula	Slope-Intercept Form
$M\left(\frac{x_1+x_2}{2}, \frac{y_1+y_2}{2}\right)$	$m = \frac{y_2-y_1}{x_2-x_1}$	$y = mx + b$

Remember, the equation of a line can be written using the coordinates of two points on that line.

A median is a segment whose endpoints are a vertex of the triangle and the midpoint of the other side. In this example, the median passes through vertex C. So, the median passes through $(-4, -1)$. The side opposite of C is $\overline{AB}$. So, the median must pass through the midpoint of $\overline{AB}$.

Use the midpoint formula to find the midpoint of $\overline{AB}$: (0, 2).

Now that two points on the median are known, $(-4, -1)$ and (0, 2), use the slope formula to find the slope of the line that passes through those two points: $m = 3/4$.

Next, identify the y-intercept. Notice that the median passes through (0, 2), which is a point on the y-axis. Therefore, the y-intercept of the median is 2.

Use the slope and the y-intercept to write the equation of the line.

Example 3 Finding the Equation of an Altitude

ΔABC has vertices $A(-4, 4)$, $B(4, 0)$, and $C(-4, -1)$. Find the equation of the altitude through C in slope intercept form.

Step 1 Plot the given vertices and draw ΔABC.

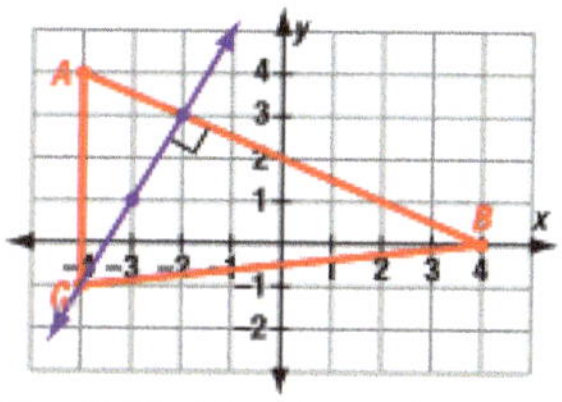

The altitude of ΔABC through vertex C must be perpendicular to the side opposite of C, which is $\overline{AB}$.

Step 2 Find the slope of the line perpendicular to $\overline{AB}$.
The slope of $\overline{AB}$ is $-\frac{1}{2}$.
So, since perpendicular lines have opposite reciprocal slopes, the slope of the line perpendicular to $\overline{AB}$ is 2.

Step 3 Write the equation of the altitude, the line that passes through $C(-4, -1)$ with slope 2.
Substitute $m = 2$ and the coordinates from the point $(-4, -1)$ into point-slope form and simplify to write the equation of the line.

$y - y_1 = m(x - x_1)$	*Point-Slope Form*
$y - (-1) = 2(x - (-4))$	*Substitute.*
$y + 1 = 2x + 8$	*Simplify.*
$y = 2x + 7$	*Write in slope-intercept form.*

An altitude of a triangle is a perpendicular segment from a vertex to the line containing the opposite side. So, an altitude is similar to a median since they both pass through a vertex and the opposite side. However, the median must pass through the midpoint of that opposite side, but the altitude does not necessarily pass through the midpoint of the opposite side. The altitude passes through the point which makes the segment perpendicular to the opposite side. The altitude and median are the same line when that line is the perpendicular bisector of the side opposite of the vertex.

As in the previous example, the coordinates of the vertex C, $(-4, -1)$, can be used to write the equation of the line since the altitude passes through C. Next, instead of finding the midpoint of $\overline{AB}$, find the slope of the line that is perpendicular to $\overline{AB}$. The slope of $\overline{AB}$ is $-1/2$. So, since perpendicular lines have opposite reciprocal slopes, the slope of the line perpendicular to $\overline{AB}$ is 2.

Now use point-slope form to write the equation of the altitude through C.

Example 4 Using the Triangle Midsegment Theorem

Triangle Midsegment Theorem

A midsegment of a triangle is parallel to a side of the triangle, and its length is half the length of that side.
(△ Midsegment Thm.)

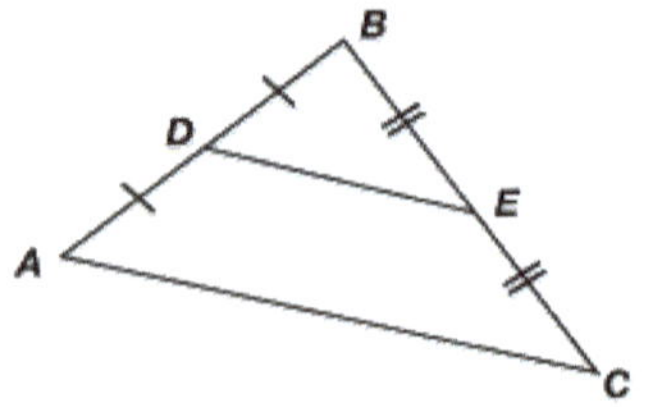

$\overline{DE} \parallel \overline{AC}$, $DE = \frac{1}{2}AC$

A midsegment of a triangle is a segment with endpoints at the midpoints of two sides of a triangle. Every triangle has three midsegments. These three midsegments form a triangle called the midsegment triangle.

By the Triangle Midsegment Theorem, if a segment is a midsegment, then the segment is parallel to a side of the triangle and its length is half the length of that side.

Find XA, XY, BY, and $m\angle YPC$.

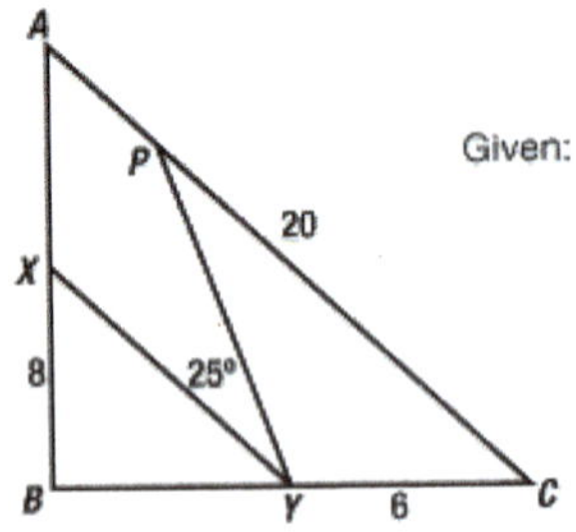

Given: $\overline{XY}$ is a midsegment of ΔABC.

In this example, it is given that $\overline{XY}$ is a midsegment of triangle ABC. Use the definition of midsegment and the Triangle Midsegment Theorem to find the missing lengths.

XA

Since $\overline{XY}$ is a midsegment of ΔABC, X is the midpoint of $\overline{AB}$ by the definition of midsegment.
Since X is the midpoint of $\overline{AB}$, $XA = XB$.
It is given that $XB = 8$. Therefore, $XA = 8$.

By the definition of midsegment, X is the midpoint of $\overline{AB}$. It follows that $XA = XB$, by the definition of midpoint. In the figure, it is given that $XB = 8$. So, $XA = 8$ as well.

XY

Since $\overline{XY}$ is a midsegment of ΔABC, $XY = \frac{1}{2}AC$ by the Triangle Midsegment Theorem.
It is given that $AC = 20$.
Therefore $XY = \frac{1}{2}(20) = 10$.

$\overline{XY}$ is a midsegment. So, by the Triangle Midsegment Theorem, XY is half the length of AC. In the figure, it is given that $AC = 20$. It follows that $XY = (1/2)(20) = 10$.

BY

Since $\overline{XY}$ is a midsegment of ΔABC, Y is the midpoint of $\overline{BC}$ by the definition of midsegment.
Since Y is the midpoint of $\overline{BC}$, $BY = YC$.
It is given that $YC = 6$. Therefore, $BY = 6$.

By the definition of midsegment, X is the midpoint of $\overline{BC}$. It follows that $BY = YC$, by the definition of midpoint. In the figure, it is given that $YC = 6$. So, $BY = 6$ as well.

$m\angle YPC$

Since $\overline{XY}$ is a midsegment of ΔABC, $\overline{XY} \parallel \overline{AC}$ by the definition of midsegment.
$\overline{PY}$ is a transversal. $\angle YPC$ and $\angle XYP$ are alternate interior angles.
Therefore, by the Alternate Interior Angles Theorem, $m\angle YPC = m\angle XYP$.
It is given than $m\angle XYP = 25°$. Therefore, $m\angle YPC = 25°$.

Here, the measure of an angle is to be found. By the Triangle Midsegment Theorem, midsegment $\overline{XY}$ is parallel to $\overline{AC}$. Notice that $\overline{PY}$ is a transversal that passes through parallel lines $\overline{XY}$ and $\overline{AC}$. Therefore, the theorems and postulates regarding corresponding, alternate interior, and alternate exterior angles can be applied. Consider angles XYP and YPC. These are alternate interior angles. Therefore, by the Alternate Interior Angles Theorem, $m\angle XYP = m\angle YPC$. It is given that $m\angle XYP = 25°$. So, $m\angle YPC = 25°$ as well.

Indirect Proof and Inequalities in One Triangle

Key Objectives

- Write indirect proofs.
- Apply inequalities in one triangle.

Key Terms

- In an **indirect proof**, or proof by contradiction, the conclusion is assumed to be false.

Theorems, Postulates, Corollaries, and Properties

- **Theorem** If two sides of a triangle are not congruent, then the larger angle is opposite of the longer side.
- **Theorem** If two angles of a triangle are not congruent, then the longer side is opposite of the larger angle.
- **Triangle Inequality Theorem** The sum of any two side lengths of a triangle is greater than the third side length.

Writing an Indirect Proof

1. Identify the conjecture to be proven.
2. Assume the opposite (the negative) of the conclusion is true.
3. Use direct reasoning to show that the assumption leads to a contradiction.
4. Conclude that, since the assumption is false, the original conjecture must be true.

Example 1 Writing an Indirect Proof

The first step in writing an indirect proof is to identify the conclusion of the conjecture to be proven and then to assume that the opposite of that conclusion is true. In other words, treat the negation of the conjecture's conclusion as given information, but do not negate the conjecture's hypothesis. Then, use direct reasoning to show a contradiction of either the given information (the conjecture's hypothesis), a theorem, a postulate, or a definition.

Write an indirect proof that an obtuse triangle cannot be a right triangle.

Given: $\triangle ABC$ is obtuse.

Prove: $\triangle ABC$ is not a right triangle.

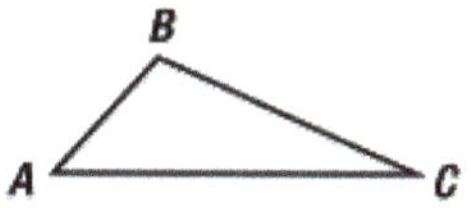

Assume $\triangle ABC$ is a right $\triangle$. Let $\angle B$ be obtuse and $\angle A$ be right.

Statements:	Reasons:
1. $m\angle B > 90°$ and $m\angle A = 90°$	1. Def. of obtuse and right $\angle$s
2. $m\angle A + m\angle B > 180°$	2. Add. Prop. of Inequalities
3. $m\angle C < 0°$	3. $\triangle$ Sum Theorem
4. A $\triangle$ cannot have an angle measure less than 0°.	4. Protractor Postulate
5. $\triangle ABC$ is not a right $\triangle$.	5. The assumption that $\triangle ABC$ is a right $\triangle$ is false.

In this example, the conclusion of the conjecture is that $\triangle ABC$ is *not* a right triangle. So, the negation of the conclusion is that $\triangle ABC$ is a right triangle. Assume that the negation is true. Use the negation of the conclusion along with the given information, that $\triangle ABC$ is obtuse, to find a contradiction.

Here, the Protractor Postulate is contradicted. Once the contradiction is found, the negation of the conclusion can be assumed to be false. If the statement "$\triangle ABC$ is a right triangle" is false, then the negation of that statement must be true. So, "$\triangle ABC$ is not a right triangle" must be true.

Example 2 Ordering Triangle Side Lengths and Angle Measures

Angle-Side Relationships in Triangles

Theorem
If two sides of a triangle are not congruent, then the larger angle is opposite the longer side.
(In △, larger ∠ is opp. longer side.)

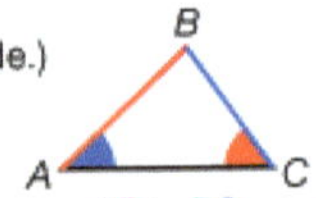

$AB > BC \Rightarrow m\angle C > m\angle A$

Theorem
If two angles of a triangle are not congruent, then the longer side is opposite the larger angle.
(In △, longer side is opp. larger ∠.)

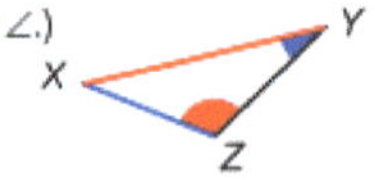

$m\angle Z > m\angle Y \Rightarrow XY > XZ$

There is a relationship between the lengths of a side of a triangle and the measures of the opposite angle.

The first theorem here states that for any two noncongruent sides of a triangle, the angle opposite of the longer of the two sides will be larger than the angle opposite of the shorter side. Therefore, if the lengths of the three sides of a triangle are known, then the measures of the three angles can be ordered from smallest to largest where the largest angle is opposite of the longest side.

The second theorem relates a triangle's angle measures to the lengths of its sides. For any two noncongruent angles in a triangle, the side opposite of the larger of the two angles will be longer than the side opposite of the smaller angle. Therefore, if a triangle's angle measures are known, then the lengths of the three sides can be ordered from shortest to longest where the longest side is opposite of the largest angle.

Write the angles in order from smallest to largest.

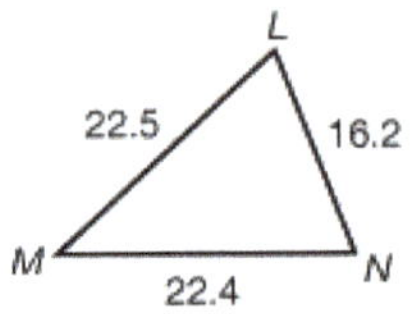

The shortest side is $\overline{LN}$, so the smallest angle is $\angle M$.

The second-to-longest side is $\overline{MN}$, so the second-to-largest angle is $\angle L$.

The longest side is $\overline{LM}$, so the largest angle is $\angle N$.

The angles from smallest to largest are $\angle M$, $\angle L$, and $\angle N$.

The triangle's three side lengths are given in this example. To relate the angle measures, first relate the side lengths. The side lengths from shortest to longest are $\overline{LN}$, $\overline{MN}$, and $\overline{LM}$, since $16.2 < 22.4 < 22.5$. So, the angles from smallest to largest must be the angle opposite of $\overline{LN}$, then the angle opposite of $\overline{MN}$, and then the angle opposite of $\overline{LM}$. Therefore, the angles from smallest to largest are $\angle M$, $\angle L$, and $\angle N$.

Write the sides in order from shortest to longest.

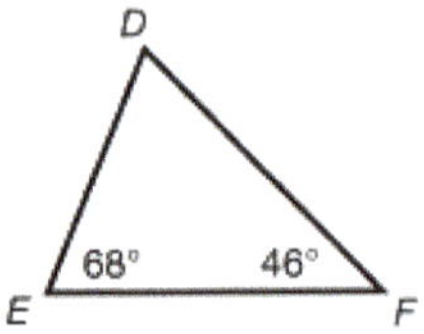

$m\angle D = 180° - (68° + 46°)$ △ Sum Thm.
$= 180° - 114° = 66°$

The smallest angle is $\angle F$, so the shortest side is $\overline{DE}$.

The second-to-largest angle is $\angle D$, so the second-to-longest side is $\overline{EF}$.

The largest angle is $\angle E$, so the largest side is $\overline{DF}$.

The sides from shortest to longest are $\overline{DE}$, $\overline{EF}$, and $\overline{DF}$.

In this example, only two of the triangle's angle measures are given. So, before the lengths of all three sides can be related, the measure of the third angle, $\angle D$, must be found.

By the Triangle Sum Theorem, the sum of the measures of the three angles must be 180°. Therefore, $68° + 46° + m\angle D = 180°$. Solve this equation to find $m\angle D$.

Once the measures of the three angles are known, use these measures to relate the lengths of the sides where the shortest side is opposite of the smallest angle and the longest side is opposite of the largest angle.

Example 3 Applying the Triangle Inequality Theorem

Triangle Inequality Theorem

The sum of any two side lengths of a triangle is greater than the third side length. (△ Inequal. Thm.)

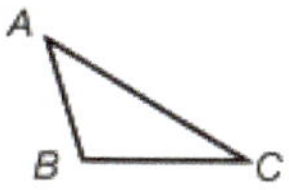

$AB + BC > AC$
$BC + AC > AB$
$AC + AB > BC$

The Triangle Inequality Theorem states that the sum of the lengths of any two sides of a triangle must be greater than the length of the third side. In other words, the length of any side of a triangle must be less than the sum of the other two sides.

When given the length of three segments, the Triangle Inequality Theorem can be used to determine whether the three segments can form a triangle. If the three lengths do not satisfy the Triangle Inequality Theorem, then those three segments cannot form a triangle.

Tell whether a triangle can have sides with the given lengths. Explain.

6, 10, 14

$6 + 10 = 16 > 14$ ✓
$6 + 14 = 20 > 10$ ✓
$10 + 14 = 24 > 6$ ✓

Yes—the sum of each pair of lengths is greater than the third length.

5, 9, 14

$5 + 9 = 14 \not> 14$ ✗

No—by the Triangle Inequality Theorem, a triangle cannot have these side lengths.

$a - 1$, $a^2 - 2$, and $3a$, when $a = 4$

$a - 1$	$a^2 - 2$	$3a$
$4 - 1$	$4^2 - 2$	$3(4)$
3	14	12

$3 + 14 = 17 > 12$ ✓
$3 + 12 = 15 > 14$ ✓
$14 + 12 = 26 > 3$ ✓

Yes—the sum of each pair of lengths is greater than the third length.

Use the Triangle Inequality Theorem to determine whether each given trio of side lengths can form a triangle. The Triangle Inequality Theorem must hold for all three sides. In other words, it is not sufficient to show that the length of one side is less than the sum of the other two sides. This fact must be demonstrated for all three sides, so it takes three inequalities to show that three given side lengths form a triangle.

However, having one side length that is greater than the sum of the other two sides is sufficient information for failing the Triangle Inequality Theorem.

The third trio of side lengths is given as expressions and a value of the variable is given, $a = 4$. So, first substitute 4 into each expression to find the side length and then compare each side length to the sum of the other two sides to determine whether a triangle is formed.

Example 4 Finding Side Lengths

When the lengths of two sides of a triangle are set, the third side cannot be any length. The third side can be a range of values that is limited by its relationship to the lengths of the other two sides.

The lengths of two sides of a triangle are 5 inches and 9 inches. Find the range of possible lengths for the third side.

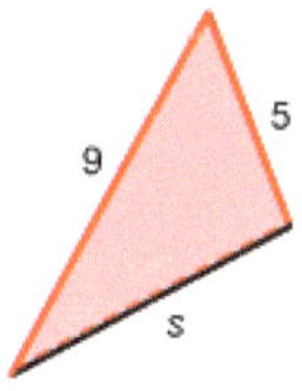

$s + 5 > 9$ $\quad$ $5 + 9 > s$
$s > 4$ $\quad$ $14 > s$

$4 < s < 14$

The length of the third side is greater than 4 inches and less than 14 inches.

To find the range of lengths for the third side of a triangle when two side lengths are known, first assign a variable for the length of the third side. Here, the length of the unknown side is represented by s.

Then write the three inequalities where each relates the sum of two sides to the other side. Here, those three inequalities are $s + 5 > 9$, $s + 9 > 5$, and $5 + 9 > s$.

Solve each inequality.
$s > 4$
$s > -4$
$14 > s$, or $s < 14$

Now find the range of values that satisfies all three inequalities. The range between 4 and 14 satisfies all 3 inequalities. Therefore, the range of side lengths for this triangle is $4 < s < 14$.

Example 5 Travel Application

The distance from Austin to Mason is 108 miles, and the distance from Mason to San Antonio is 111 miles. What is the range of distances from Austin to San Antonio?

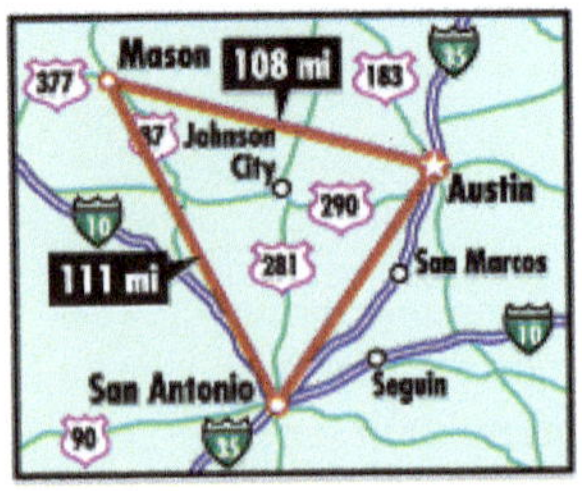

Let d be the distance from Austin to San Antonio.

$d + 108 > 111$ $d > 3$	$111 + 108 > d$ $219 > d$	*Triangle Inequality Theorem*
$3 < d < 219$		*Combine the inequalities.*

The distance from Austin to San Antonio is greater than 3 miles and less than 219 miles.

A triangle can be formed on this map where the three cities, Austin, Mason, and San Antonio, are the triangle's vertices. Since the distance from Austin to Mason and the distance from Mason to San Antonio are known, the lengths of two of the triangle's sides are known. Let d represent the distance from Austin to San Antonio. Use the two known lengths and d to write the inequalities. However, note that the inequality $d + 111 > 108$ can be ignored because that solution set, $d > -3$, includes negative values and distance can never be negative.

Inequalities in Two Triangles

Key Objectives

- Apply inequalities in two triangles.

Theorems, Postulates, Corollaries, and Properties

- **Hinge Theorem** If two sides of one triangle are congruent to two sides of another triangle and the included angles are not congruent, then the longer third side is across from the larger included angle.
- **Converse of the Hinge Theorem** If two sides of one triangle are congruent to two sides of another triangle and the third sides are not congruent, then the larger included angle is across from the longer third side.

The Hinge Theorem and its converse give information about the relative size of either two angles in a pair of triangles or two sides in a pair of triangles. With these theorems, inequalities involving the lengths of sides or the measures of angles in two triangles can be written to give information about the two triangles.

Example 1 Using the Hinge Theorem and Its Converse

In these examples, the Hinge Theorem and its converse are used to write inequalities involving the lengths of sides of two triangles or the measures of angles in two triangles. Begin by determining that the Hinge Theorem or the Converse of the Hinge Theorem apply to the two triangles, then write an appropriate inequality.

Hinge Theorem

If two sides of one triangle are congruent to two sides of another triangle and the included angles are not congruent, then the longer third side is across from the larger included angle.
(Hinge Thm.)

$m\angle A > m\angle D \Rightarrow BC > EF$

Converse of the Hinge Theorem

If two sides of one triangle are congruent to two sides of another triangle and the third sides are not congruent, then the larger included angle is across from the longer third side.
(Conv. Hinge Thm.)

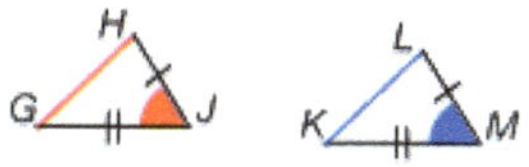

$GH > KL \Rightarrow m\angle J > m\angle M$

According to the Hinge Theorem, when two sides of one triangle are congruent to two sides of another triangle, the triangle with the larger included angle will have the longer third side.

According to the Converse of the Hinge Theorem, when two sides of one triangle are congruent to two sides of another triangle, the triangle with the longer third side has the larger included angle.

Compare $m\angle ACB$ and $m\angle BCD$.

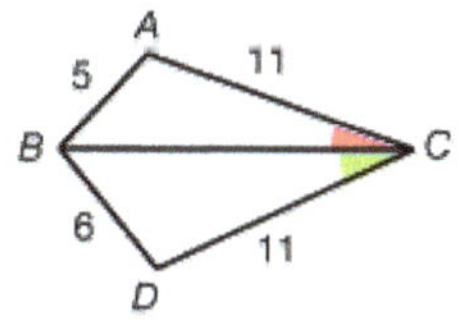

Compare the side lengths in $\triangle ABC$ and $\triangle BDC$.

$DC = AC \quad BC = BC \quad DB > AB$

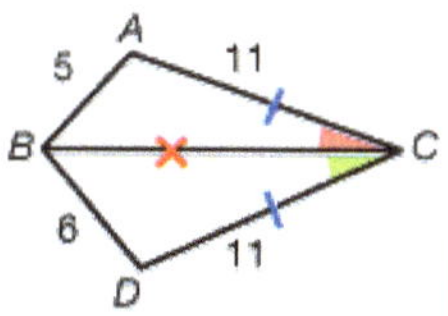

By the Converse of the Hinge Theorem, $m\angle BCD > m\angle ACB$.

The measures of two angles in two triangles are compared in this example.

Begin by comparing the side lengths of the two triangles. That comparison shows that $DC = AC$ (both equal 11 units), $BC = BC$ (because they are the same side) and $DB > AB$ (since $6 > 5$). So, $\triangle ABC$ has two sides congruent to $\triangle BDC$.

The third side of $\triangle BDC$ is larger than the third side of $\triangle ABC$, so by the Converse of the Hinge Theorem, the measure of the included angle BCD is larger than the measure of the included angle ACB.

Compare *PQ* and *RS*.

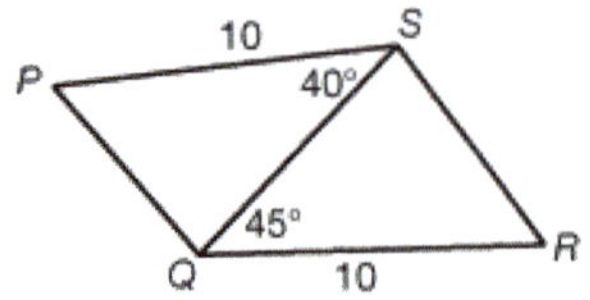

Compare the sides and angles in $\triangle PQS$ and $\triangle RSQ$.

$RQ = PS$ $\quad QS = SQ$ $\quad m\angle RQS > m\angle PSQ$

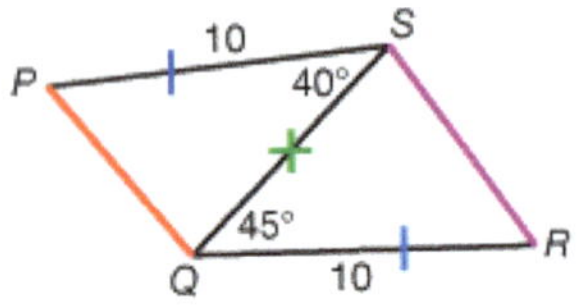

By the Hinge Theorem, $RS > PQ$.

The lengths of two sides in two triangles are compared in this example.

Begin by comparing the side lengths of the two triangles. That comparison shows that $RQ = PS$ (both equal to 10 units), and $QS = SQ$ (they are the same side). Therefore, $\triangle PQS$ has two sides congruent to two sides of $\triangle RSQ$.

The included angle RQS in the one triangle is given to be greater than the included angle RSQ in the other triangle, so by the Hinge Theorem, the opposite side RS is greater than the opposite side PQ.

Find the range of values for *z*.

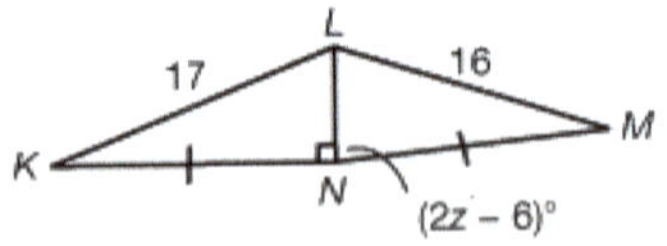

Compare the side lengths in $\triangle KLN$ and $\triangle MLN$.

$MN = KN$ $\quad LN = LN$ $\quad ML < KL$

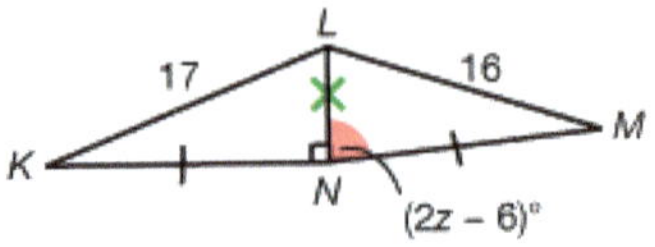

Since $\angle LNM$ is in a triangle, $m\angle LNM > 0°$.

$2z - 6 > 0$	*Substitute 2z – 6 for m∠LNM.*
$2z > 6$	*Add 6 to both sides.*
$z > 3$	*Divide both sides by 2.*

By the Converse of the Hinge Theorem, $m\angle LNM < m\angle LNK$.

$2z - 6 < 90$	*Substitute 2z – 6 for m∠LNM and 90 for m∠LNK.*
$2z < 96$	*Add 6 to both sides.*
$z < 48$	*Divide both sides by 2.*

The range of values for z is $3 < z < 48$.

The Hinge Theorem and its converse produce inequalities that can be used to determine the range of values for the measure of an angle.

To find the range of values for z, first compare the side lengths of the two triangles. $MN = KN$, because the sides are congruent. Also, $LN = LN$ because they are the same side. Finally, $ML < KL$ (since $16 < 17$).

Since $\angle LNM$ is in a triangle, $\angle LNM$ must be greater than zero. Therefore, $m\angle LNM > 0$. Substitute the expression given in the figure for $m\angle LNM$ into this inequality.

$m\angle LNM > 0$
$2z - 6 > 0.$
Solve the inequality for z.
The result is $z > 3$.

By the converse of the Hinge Theorem, $\angle LNM$ must be less than $\angle KNL$ and, as given, $m\angle KNL = 90°$. Therefore, $m\angle LNM < 90°$. Substitute the expression given in the figure for $m\angle LNM$ into this inequality.

$m\angle LNM < 90$
$2z - 6 > 90.$
Solve the inequality for z.
The result is $z < 48$.

Combining the inequalities yields $3 < z < 48$ as the answer.

Example 2 Construction Site Application

The boom of a crane makes an angle with the main body of the crane. The angle changes based on the position of the boom. The diagram shows the position of the boom at two different settings. Which worker is farther from the base of the crane? Explain.

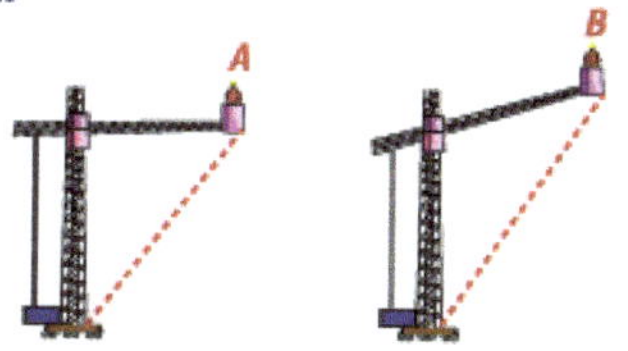

The heights of the towers are the same and the lengths of the booms are the same.

The included angle of crane *A* is smaller than the included angle of crane *B*.

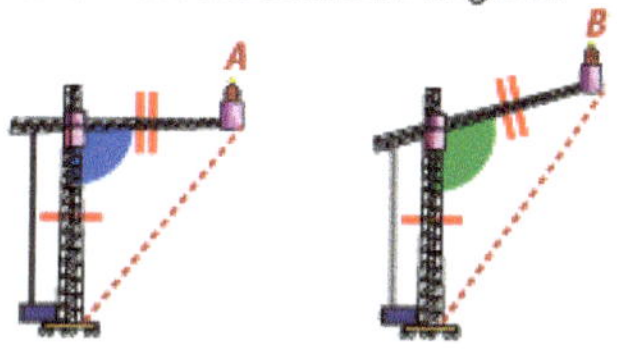

By the Hinge Theorem, the worker in crane *B* is farther from the base.

In this example, the Hinge Theorem is applied to a real-world case.

In the example the boom of the same crane rests at two different positions, so the length of the boom and the height of the tower in each case are the same. These are the lengths of two of the sides of the two triangles formed by the crane. The third sides of the triangles are the distance between the worker at the end of the boom and the base of the tower.

For crane B the measure of the included angle made by the boom and the tower is greater than the measure of the included angle made by crane A. Therefore, by the Hinge Theorem, the third side, or the distance between the worker and the crane base, is longer in crane B than crane A. Worker B is farther from the base than worker A.

Example 3 Proving Triangle Relationships

Write a two-column proof.

Given: $\overline{AB} \cong \overline{AC}$

Prove: $BD > CD$

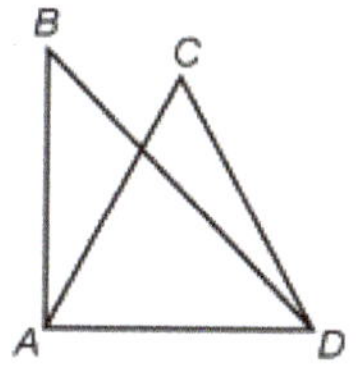

Statements:	Reasons:
1. $\overline{AB} \cong \overline{AC}$	1. Given
2. $\overline{AD} \cong \overline{AD}$	2. Reflex. Prop. of ≅
3. $m\angle DAB = m\angle DAC + m\angle CAB$	3. Segment Addition Postulate
4. $m\angle DAB > m\angle DAC$	4. Comparison Prop. of Inequal.
5. $BD > CD$	5. Hinge Theorem

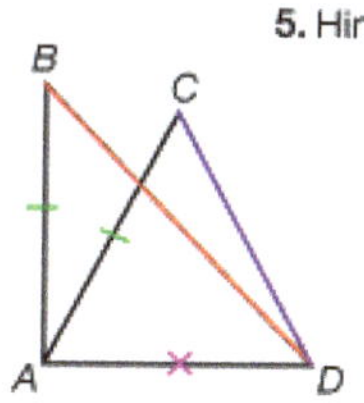

The conjecture to prove in this example is that the length of one side of one triangle is greater than the length of one side of another triangle. The given information is that $AB = AC$.

Begin by making a plan for the proof. Then use the plan to set out the specific steps of the proof and their justifications.

Statement 1 is the given information. Statement 2 is true because the two lines are the same line. Notice that two sides of $\triangle ABD$ are congruent to two sides of $\triangle ACD$. Now consider the included angle.

According to the Segment Addition Postulate $m\angle DAB = m\angle DAC + m\angle CAB$. This means, by the Comparison Property of Inequality, that $m\angle DAB > m\angle DAC$.

Since $m\angle DAB$ is greater than $m\angle DAC$, the length of the side opposite $\angle DAC$, BD, must be greater than the length of the side opposite $\angle DAC$, CD, thus proving the conjecture.

The Pythagorean Theorem

Key Objectives

- Use the Pythagorean Theorem and its converse to solve problems.
- Use Pythagorean inequalities to classify triangles.

Key Terms

- A set of three nonzero whole numbers a, b, and c such that $a^2 + b^2 = c^2$ is called a **Pythagorean triple**.

Theorems, Postulates, Corollaries, and Properties

- **Pythagorean Theorem** In a right triangle, the sum of the squares of the lengths of the legs is equal to the square of the length of the hypotenuse.
- **Converse of the Pythagorean Theorem** If the sum of the squares of the lengths of two sides of a triangle is equal to the square of the length of the third side, then the triangle is a right triangle.
- **Pythagorean Inequalities Theorem** In $\triangle ABC$, c is the length of the longest side. If $c^2 > a^2 + b^2$, then $\triangle ABC$ is an obtuse triangle. If $c^2 < a^2 + b^2$, then $\triangle ABC$ is an acute triangle.

The Pythagorean Theorem

The sum of the squares of the lengths of the legs of a right triangle equals the square of the length of the hypotenuse.
(Pyth. Thm.)

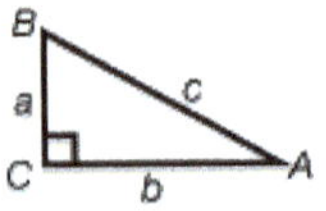

$\triangle ABC$ is a right triangle. $\Rightarrow$ $a^2 + b^2 = c^2$

Converse of the Pythagorean Theorem

If the sum of the squares of the lengths of the two shorter sides of a triangle is equal to the square of the length of the third side, then the triangle is a right triangle.
(Conv. Pyth. Thm.)

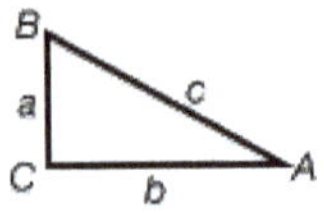

$a^2 + b^2 = c^2$ $\Rightarrow$ $\triangle ABC$ is a right triangle.

The Pythagorean Theorem is one of the most useful relationships in geometry. It states that in a right triangle, the sum of the squared side lengths of the legs of the triangle (the sides that include the right angle) is equal to the square of the length of the other side (the side opposite the right angle). The side opposite the right angle is called the hypotenuse. The Pythagorean Theorem can be used to find the length of one side of a right triangle when the lengths of the other two sides are known.

The Converse of the Pythagorean Theorem states that if the sum of the squares of the two shorter sides of a triangle are equal to the square of the third angle, then that triangle is a right triangle. The Converse of the Pythagorean Theorem can be used to determine whether a triangle is a right triangle when the lengths of a triangle's three sides are known.

Example 1 Using the Pythagorean Theorem

Find the value of x. Give your answer in simplest radical form.

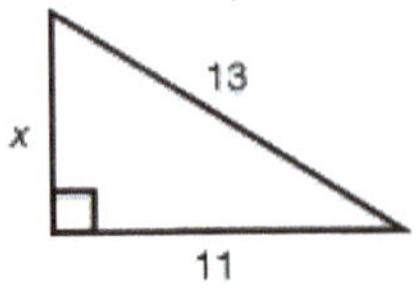

$x^2 + 11^2 = 13^2$	*Pythagorean Theorem*
$x^2 + 121 = 169$	*Simplify.*
$x^2 = 48$	*Subtract 121 from both sides.*
$x = \sqrt{48}$	*Take the positive square root of both sides.*
$= \sqrt{16 \cdot 3}$	*Simplify.*
$= 4\sqrt{3}$	*Simplify.*

The Pythagorean Theorem is used to find the length of a leg of a right triangle in this example.

The length of one leg and the hypotenuse are given. Substitute the values of the length of the leg and the hypotenuse of the triangle into the Pythagorean formula. Remember that the side opposite the right angle is the hypotenuse, or c in $a^2 + b^2 = c^2$.

Solve the equation for x, the length of the unknown leg.

Find the value of *x*. Give your answer in simplest radical form.

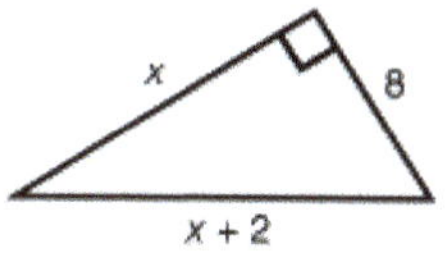

$x^2 + 8^2 = (x + 2)^2$	*Pythagorean Theorem.*
$x^2 + 64 = (x + 2)^2$	*Simplify.*
$x^2 + 64 = x^2 + 4x + 4$	*Simplify (FOIL).*
$64 = 4x + 4$	*Subtract x^2 from both sides.*
$60 = 4x$	*Subtract 4 from both sides.*
$15 = x$	*Divide both sides by 4.*

The Pythagorean Theorem is used to find an unknown value related to the lengths of the sides of a right triangle in this example.

The length of one leg and the hypotenuse are given as algebraic expressions involving one unknown, x.

Substitute the values of the lengths of the sides of the triangle into $a^2 + b^2 = c^2$.

Solve the equation for x. The solution yields $x = 15$. This value for x could also be used to calculate the lengths of the two unknown sides.

Example 2 Safety Application

To prevent a ladder from shifting, safety experts recommend that the ratio of *a*:*b* be 4:1. How far from the base of the wall should you place the foot of a 12-foot ladder? Round to the nearest inch.

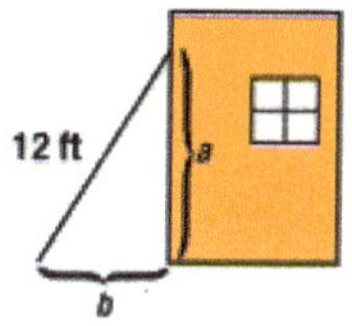

$a^2 + b^2 = 144$	*Pythagorean Theorem*
$(4b)^2 + b^2 = 144$	$\frac{a}{b} = \frac{4}{1} \rightarrow a = 4b$. *Substitute 4b for a.*
$16b^2 + b^2 = 144$	*Simplify.*
$17b^2 = 144$	*Combine like terms.*
$b^2 = \frac{144}{17}$	*Divide both sides by 17.*
$b = \sqrt{\frac{144}{17}}$	*Take the positive square root of both sides.*
$= \frac{12}{\sqrt{17}}$	*Simplify.*
$\approx$ 2 ft, 11 in.	*Simplify and round.*

The Pythagorean Theorem is used to find the length of two legs of a right triangle in this application example. There are two unknown legs, but the ratio of the lengths of the legs is given, which allows the length of one leg to be expressed in terms of the length of the other leg.

To express the value of one leg length in terms of the other, set up the ratio a is to b as 4 is to 1 and cross multiply. The lengths a, b, and c can now be expressed as 12, b, and $4b$. Substitute the values of the lengths of the sides of the triangle into $a^2 + b^2 = c^2$.

Solve the equation for b. The solution yields $b \approx$ 2 feet 11 inches. This is the value for the recommended distance of the foot of the ladder from the wall.

Example 3 Identifying Pythagorean Triples

Pythagorean triples are sets of three nonzero whole numbers that satisfy the Pythagorean formula.

Find the missing side length. Tell if the side lengths form a Pythagorean triple. Explain.

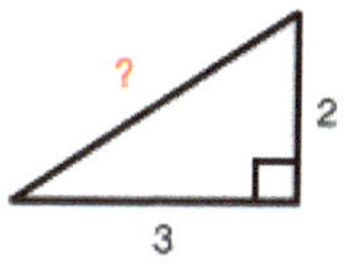

$?^2 = 3^2 + 2^2$	*Pythagorean Theorem*
$?^2 = 13$	*Simplify.*
$? = \sqrt{13}$	*Take the positive square root of both sides.*

The side lengths do not form a Pythagorean triple because $\sqrt{13}$ is not a whole number.

The Pythagorean Theorem is used to find the length of the hypotenuse of a right triangle in this example. Then, it is determined whether the side lengths form a Pythagorean triple.

Leg lengths of 2 and 3 are given. Substitute the values of the lengths of the sides of the triangle into $a^2 + b^2 = c^2$.

Solve the equation for the hypotenuse.

The unknown side has a length of square root of 13, which is not a whole number. Therefore, these sides do not form a Pythagorean triple.

Find the missing side length. Tell if the side lengths form a Pythagorean triple. Explain.

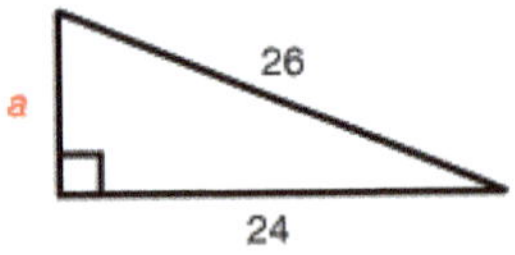

$a^2 + 24^2 = 26^2$	*Pythagorean Theorem*
$a^2 = 26^2 - 24^2$	*Subtract 24^2 from both sides.*
$a^2 = 100$	*Simplify.*
$a = 10$	*Take the positive square root of both sides.*

The side lengths are nonzero whole numbers that satisfy the equation $a^2 + b^2 = c^2$, so they form a Pythagorean triple.

The Pythagorean Theorem is used to find the length of a leg of a right triangle in this example. Then it is determined whether the side lengths form a Pythagorean triple.

The length of one leg and the hypotenuse are given. Substitute the values of the lengths of the sides of the triangle into $a^2 + b^2 = c^2$. Be sure to substitute the correct length (26) for c, the hypotenuse.

Solve the equation for the unknown leg.

The unknown leg has a length of 10, which is a whole number. Therefore, these sides do form a Pythagorean triple.

Example 4 Classifying Triangles

Pythagorean Inequalities Theorem

In $\triangle ABC$, c is the length of the longest side.

If $c^2 > a^2 + b^2$, then $\triangle ABC$ is an **obtuse** triangle.

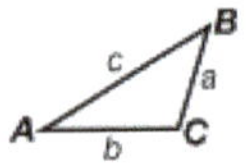

$c^2 > a^2 + b^2$

If $c^2 < a^2 + b^2$, then $\triangle ABC$ is an **acute** triangle.

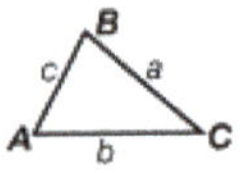

$c^2 < a^2 + b^2$

(Pyth. Inequal. Thm.)

The Pythagorean Inequalities Theorem is used to determine whether a triangle is obtuse, acute, or right.

When a triangle is a right triangle, the sum of the squares of its two legs, the shorter sides, are equal to the square of the hypotenuse.

When a triangle is obtuse, the sum of the squares of the two shorter sides of the triangle is greater than the square of the length of the longer side.

When a triangle is acute, the sum of the squares of the shorter two sides in the triangle is less than the square of the third side.

Tell if the measures can be the side lengths of a triangle. If so, classify the triangle as acute, obtuse, or right.

5, 8, 10 $5 + 8 = 13 > 10$ $5 + 10 = 15 > 8$ $8 + 10 = 18 > 5$

By the Triangle Inequality Theorem, 5, 8, and 10 can be the side lengths of a triangle.

$a^2 + b^2 \stackrel{?}{=} c^2$	*Compare c^2 to $a^2 + b^2$.*
$5^2 + 8^2 \stackrel{?}{=} 10^2$	*Substitute the longest side length for c.*
$25 + 64 \stackrel{?}{=} 100$	*Multiply.*
$89 < 100$	*Add and compare.*

By the Pythagorean Inequalities Theorem, since c^2 is greater than $a^2 + b^2$, the triangle is obtuse.

2.3, 7.2, 9.6

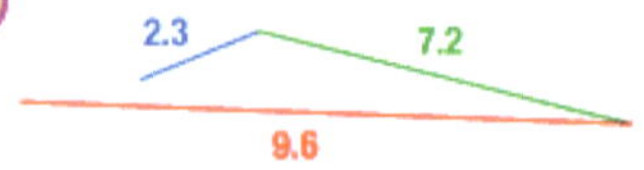

Since $2.3 + 7.2 = 9.5$ and $9.5 \not> 9.6$, these cannot be the side lengths of a triangle.

The Triangle Inequality Theorem is used here to determine whether three measures can be the side lengths of a triangle. If it is a triangle, it is classified as acute, obtuse, or right.

To determine if the measures can belong to the sides of a triangle, apply the Triangle Inequality Theorem. Sum the values in pairs and test whether the sum is greater than the third number. The values 5, 8, and 10 can be the sides of a triangle, since the sum of each pair is greater than the third number.

Use the Pythagorean Inequalities Theorem to classify the triangle. The sum of the squares of the two shorter sides is $5^2 + 8^2 = 25 + 64 = 89$. Since 89 is less than the square of the third side ($10^2 = 100$), the triangle is obtuse.

In the second example, the sum of the two shorter measures is less than the third measure, so these values cannot be the lengths of a triangle.

Applying Special Right Triangles

Key Objectives

- Justify and apply properties of 45°-45°-90° triangles.
- Justify and apply properties of 30°-60°-90° triangles.

Theorems, Postulates, Corollaries, and Properties

- **45°-45°-90° Triangle Theorem** In a 45°-45°-90° triangle, both legs are congruent, and the length of the hypotenuse is the length of a leg times $\sqrt{2}$.
- **30°-60°-90° Triangle Theorem** In a 30°-60°-90° triangle, the length of the hypotenuse is 2 times the length of the shorter leg, and the length of the longer leg is the length of the shorter leg times $\sqrt{3}$.

Example 1

45°-45°-90° Theorem

In a 45°-45°-90° triangle, both legs are congruent, and the length of the hypotenuse is the length of a leg times $\sqrt{2}$.

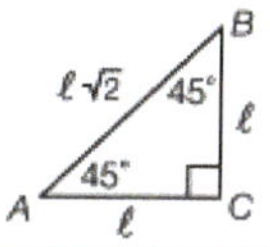

$AC = BC = \ell$
$AB = \ell\sqrt{2}$

The 45°-45°-90° Triangle Theorem describes the relationship between the lengths of the legs and the length of the hypotenuse in right triangles with angles 45°, 45°, and 90°.

In these triangles, the two legs are the same length (congruent), and the length of the hypotenuse is the length of a leg times $\sqrt{2}$.

Find the value of x. Give your answer in simplest radical form.

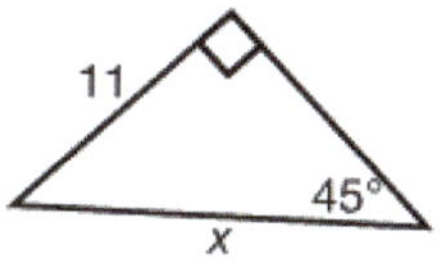

By the Triangle Sum Theorem, the measure of the third angle of the triangle is 45°.

$x = 11\sqrt{2}$ — *Hypotenuse = leg* $\sqrt{2}$

The 45°-45°-90° Triangle Theorem is used to determine the unknown length of the hypotenuse of a triangle in this example.

The length of one leg is given as 11 units and the measure of its opposite angle is given as 45°. First, use the Triangle Sum Theorem to determine that the unknown angle in the triangle is 45° (180° − 90° − 45° = 45°).

It follows from the 45°-45°-90° Triangle Theorem that the length of the hypotenuse, x, is the length of a leg times $\sqrt{2}$, or $11\sqrt{2}$.

Find the value of x. Give your answer in simplest radical form.

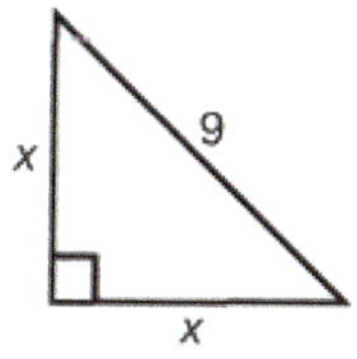

The triangle is an isosceles right triangle, which is a 45°-45°-90° triangle.

$x\sqrt{2} = 9$ — *Leg* $\sqrt{2}$ *= hypotenuse*

$x = \frac{9}{\sqrt{2}}$ — *Divide both sides by* $\sqrt{2}$.

$= \frac{9}{\sqrt{2}} \cdot \frac{\sqrt{2}}{\sqrt{2}} = \frac{9\sqrt{2}}{2}$ — *Multiply the numerator and the denominator by* $\sqrt{2}$.

The 45°-45°-90° Triangle Theorem is used here to determine the unknown lengths of the legs of a triangle.

It is given that the triangle is an isosceles right triangle. This means it is a 45°-45°-90° triangle. The hypotenuse is given as 9 units. By the 45°-45°-90° Theorem, the length of the hypotenuse is the length of a leg times $\sqrt{2}$. Thus, $x\sqrt{2} = 9$. Solve for x, the length of a leg, to yield $x = 9\sqrt{2}/2$.

Example 2

Tessa wants to make a bandana for her dog by folding a square of cloth into a 45°-45°-90° triangle. Her dog's neck has a circumference of about 10 in. The folded bandana needs to be an extra 6 in. long so Tessa can tie it around her dog's neck. What should the side length of the square be? Round to the nearest inch.

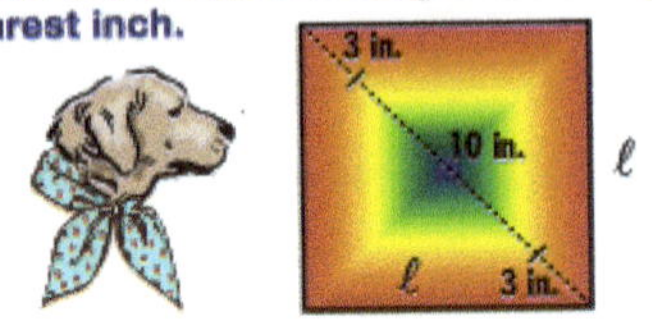

Tessa needs a 45°-45°-90° triangle with a hypotenuse of 16 in.

$\ell\sqrt{2} = 16$	*Leg $\sqrt{2}$ = hypotenuse (45°-45°-90° Thm.)*
$\ell = \frac{16}{\sqrt{2}}$	*Divide both sides by $\sqrt{2}$.*
$= \frac{16}{\sqrt{2}} \cdot \frac{\sqrt{2}}{\sqrt{2}}$	*Multiply the numerator and denominator by $\sqrt{2}$.*
$= \frac{16\sqrt{2}}{2}$	*Simplify.*
$= 8\sqrt{2} \approx 11$ in.	*Simplify and round.*

Each side of the bandana should be about 11 in. long.

The 45°-45°-90° Triangle Theorem is used to determine the unknown lengths of the legs of a triangle in this application example.

The hypotenuse is given to be 16 inches. To find the lengths of the sides of the square, which are the legs of a 45°-45°-90° triangle, set the hypotenuse equal to the unknown leg length times $\sqrt{2}$, or $\ell\sqrt{2} = 16$. Solve for ℓ. Use a calculator to determine the approximate length of the side of the square to be 11 inches.

Example 3

30°-60°-90° Theorem

In a 30°-60°-90° triangle, the length of the hypotenuse is 2 times the length of the shorter leg, and the length of the longer leg is the length of the shorter leg times $\sqrt{3}$.

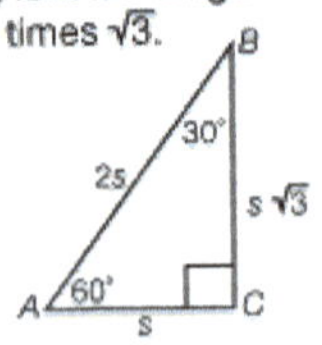

$AC = s$
$AB = 2s$
$BC = s\sqrt{3}$

The 30°-60°-90° Triangle Theorem describes the relationship between the lengths of the legs and hypotenuse in right triangles with angles 30°, 60°, and 90°.

In these triangles, if the short leg is length s, the longer leg is length $s\sqrt{3}$, and the hypotenuse is length $2s$.

Find the values of *x* and *y*. Give your answer in simplest radical form.

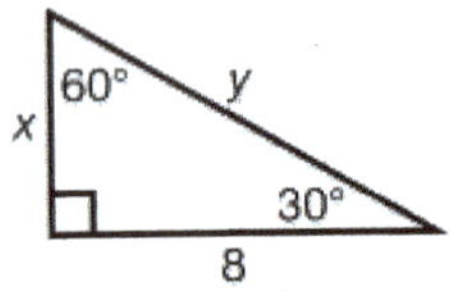

$x\sqrt{3} = 8$	*(Shorter leg) $\sqrt{3}$ = longer leg*
$x = \frac{8}{\sqrt{3}}$	*Divide both sides by $\sqrt{3}$.*
$= \frac{8}{\sqrt{3}} \cdot \frac{\sqrt{3}}{\sqrt{3}} = \frac{8\sqrt{3}}{3}$	*Multiply the numerator and denominator by $\sqrt{3}$ and simplify.*
$y = 2x$	*Hypotenuse = 2 (shorter leg)*
$= 2\left(\frac{8\sqrt{3}}{3}\right) = \frac{16\sqrt{3}}{3}$	*Substitute $\frac{8\sqrt{3}}{3}$ for x and simplify.*

The 30°-60°-90° Triangle Theorem is used here to determine the unknown length of the hypotenuse and one leg of a triangle.

The given triangle is a 30°-60°-90° triangle. The length of one leg is given as 8 units. Begin by expressing the unknown length of the short leg in terms of the known length of the longer leg. This yields a length for the shorter leg of $8\sqrt{3}/3$. The length of the hypotenuse is 2 times the length of the shorter leg, or $16\sqrt{3}/3$.

Find the values of x and y. Give your answer in simplest radical form.

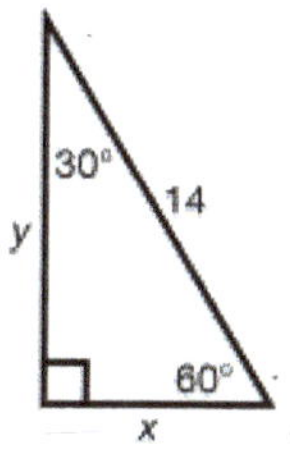

$2x = 14$	*2(shorter leg) = hypotenuse*
$x = 7$	*Divide both sides by 2.*
$y = x\sqrt{3}$	*Longer leg = (shorter leg)* $\sqrt{3}$
$= 7\sqrt{3}$	*Substitute 7 for x.*

The 30°-60°-90° Triangle Theorem is used here to determine the unknown lengths of the legs of a triangle.

The given triangle is a 30°-60°-90° triangle. The length of the hypotenuse is given as 14 units. Begin by expressing the unknown length of the short leg in terms of the known length of the hypotenuse. This yields a length for the shorter leg of 7. The length of the longer leg is the length of the shorter leg times $\sqrt{3}$, or $7\sqrt{3}$.

Example 4

A painting is framed in the shape of an equilateral triangle. The length of one side of the frame is 54 in. Will the painting fit on a wall above the mantle that is 50 in. below the ceiling?

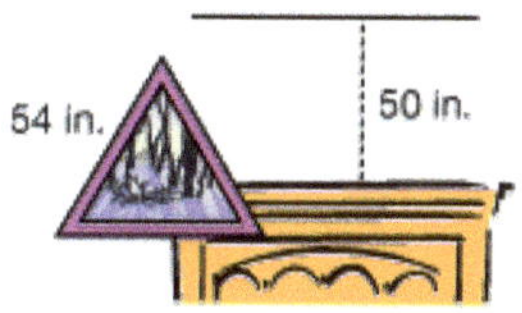

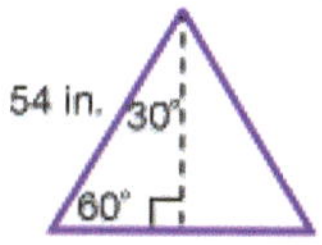

2(shorter leg) = hypotenuse	
2(shorter leg) = 54 in.	*Substitute 54 for the hypotenuse.*
shorter leg = 27 in.	*Divide both sides by 2.*
longer leg = (shorter leg) $\sqrt{3}$	
longer leg = $27\sqrt{3}$ in. ≈ 46.8 in.	*Substitute 27 for the shorter leg and round.*

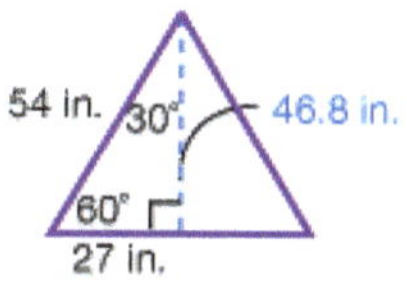

The frame is approximately 46.8 inches tall, which is less than 50 inches, so the frame will fit on the wall.

The 30°-60°-90° Triangle Theorem is used to determine the unknown length of the long leg of a triangle in this application example.

Is the height of the equilateral triangle less than the allowed space of 50 inches? The line segment that defines the height divides the triangle into two congruent 30°-60°-90° triangles. The length of the hypotenuse for these triangles, the length of the side of the equilateral triangle, is given as 54 inches.

According to the 30°-60°-90° Triangle Theorem, the length of the short leg is half the length of the hypotenuse, or 54/2 = 27 inches.

By the 30°-60°-90° Triangle Theorem, the length of the longer leg is $27\sqrt{3}$. Use a calculator to find the approximate length of the longer leg, also the height of the original equilateral triangle, to be approximately 46.8 inches. This is less than the available wall height, so the painting will fit within the space.

6

6 Polygons and Quadrilaterals

Properties and Attributes of Polygons

Key Objectives

- Classify polygons based on their sides and angles.
- Find and use the measures of interior and exterior angles of polygons.

Key Terms

- Each segment that forms a polygon is a **side of a polygon**.
- The common endpoint of two sides is a **vertex of the polygon**.
- A segment that connects any two nonconsecutive vertices is a **diagonal**.
- A **regular polygon** is one that is both equilateral and equiangular.
- A polygon is **concave** if any part of a diagonal contains points in the exterior of the polygon.
- If no diagonal contains points in the exterior, then the polygon is **convex**.

Theorems, Postulates, Corollaries, and Properties

- **Polygon Angle Sum Theorem** The sum of the interior angle measures of a convex polygon with n sides is $(n-2)180°$.
- **Polygon Exterior Angle Sum Theorem** The sum of the exterior angle measures, one angle at each vertex, of a convex polygon is 360°.

Example 1 Identifying Polygons

A polygon is a closed figure on a plane formed by three or more line segments that intersect only at their endpoints.

Tell whether each figure is a polygon. If it is a polygon, name it by the number of its sides.

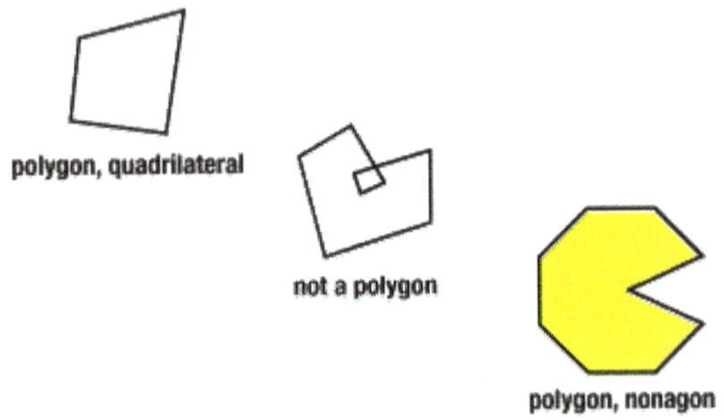

Three figures are determined to be polygons or not polygons. The polygons are named according to the number of sides they have.

The first and third figures are closed figures with only one interior and one exterior. The line segments defining each figure intersect only at their endpoints. These two figures are polygons. The second figure contains line segments that intersect away from their endpoints. It is not a polygon.

The first figure has four sides; it is called a quadrilateral. The third figure has nine sides; it is called a nonagon.

Example 2 Classifying Polygons

In a regular polygon, every side has the same length and every angle has the same measure. In a convex polygon, the diagonals do not include points exterior to the polygon. In a concave polygon, diagonals do include exterior points. Regular polygons are always convex.

Tell whether each polygon is regular or irregular. Tell whether it is concave or convex.

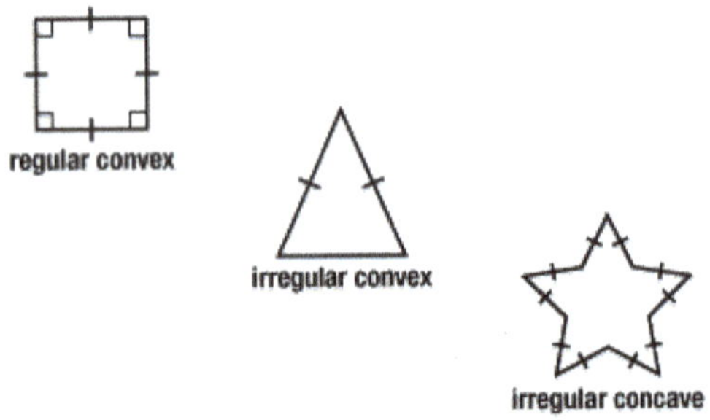

Three figures are identified as regular polygons or irregular polygons, and as concave or convex.

The hatch marks indicate congruent side lengths. In the first and third polygons all sides are the same length. In the first figure all the angles are congruent, so this is a regular polygon. In the third figure, the angles are not all congruent, so this is an irregular polygon. The first figure is convex, because its diagonals do not include points on the exterior. Figure three is concave because some diagonals include exterior points.

The second figure is an irregular polygon, because one of the sides is a different length from the other two. It is a convex polygon because the diagonals do not include exterior points.

Example 3 Finding Interior Angle Measures and Sums in Polygons

Demonstrate the polygon angle sum theorem.

Polygon Angle Sum Theorem
The sum of the interior angle measures of a convex polygon with n sides is $(n - 2)180°$.

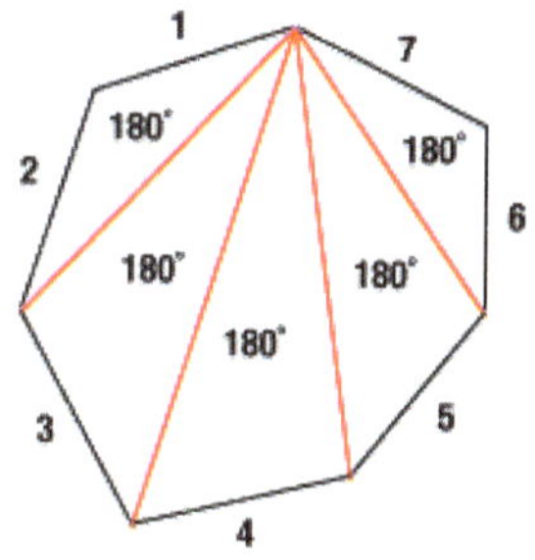

A seven-sided polygon is called a septagon.

$7 - 2 = 5$

The sum of interior angle measures of a convex septagon = 5(180°)

According to the Polygon Angle Sum Theorem, the sum of the interior angle measures of a convex polygon with n sides is $(n - 2)180°$.

To see why the theorem is true, draw diagonals from one vertex in a polygon. This forms $n - 2$ triangles in a polygon with n sides. For example, in the septagon, there are 5 triangles formed. Each triangle contains 180° of angles. The total measure of the angles within the polygon is the sum of the angles contained in all the triangles, or 5 times 180°.

Find the sum of the interior angle measures of a convex hexagon.

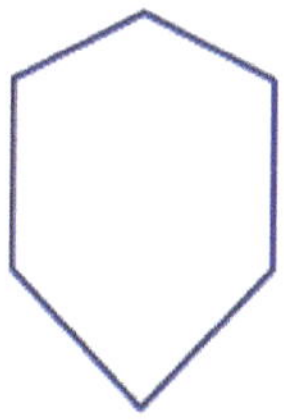

$(n - 2)180°$	*Polygon ∠ Sum Thm.*
$(6 - 2)180°$	*A hexagon has 6 sides, so substitute 6 for n.*
$720°$	*Simplify.*

The sum of the interior angles of a hexagon is determined in this example.

The hexagon has 6 sides. According to the Polygon Angle Sum Theorem, the sum of the interior angles is $(n - 2)180°$. Substitute 6 for n and simplify. The sum of the interior angles of a hexagon is 720°.

Find the measure of each interior angle of a regular octagon.

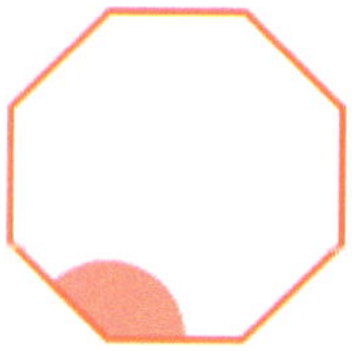

$(n - 2)180°$	*Polygon ∠ Sum Thm.*
$(8 - 2)180° = 1080°$	*Substitute 8 for n and simplify.*
$\frac{1080°}{8} = 135°$	*The int. ∠s are ≅, so divide by 8.*

The measure of the interior angles of a regular octagon is determined in this example.

The octagon has 8 sides. According to the Polygon Angle Sum Theorem, the sum of the interior angles is $(n - 2)180°$. Substitute 8 for n and simplify. The sum of the interior angles of the octagon is 1080°.

Since the octagon is regular, all the interior angles are congruent. The size of each angle is $1080°/8 = 135°$.

Find the measure of each interior angle of quadrilateral *PQRS*.

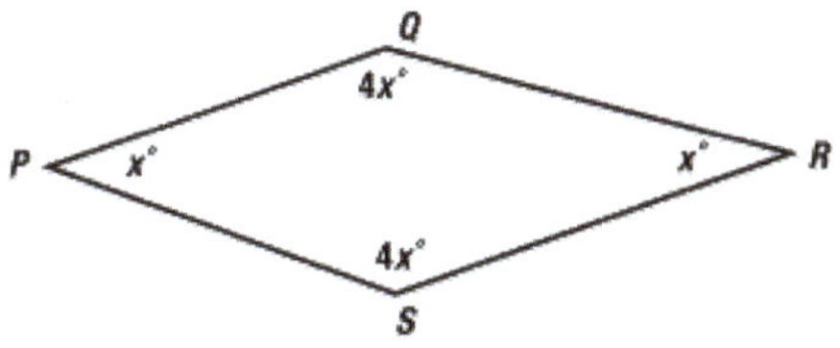

$m\angle P + m\angle Q + m\angle R + m\angle S = (4 - 2)180°$	*Polygon ∠ Sum Thm.*
$m\angle P + m\angle Q + m\angle R + m\angle S = 360°$	*Simplify.*
$x + 4x + x + 4x = 360°$	*Substitute.*
$10x = 360°$	*Simplify.*
$x = 36°$	*Divide both sides by 10.*
$m\angle P = m\angle R = 36°$	*Substitute 36° for x.*
$m\angle Q = m\angle S = 4(36°) = 144°$	*Substitute 36° for x.*

The measure of the interior angles of a quadrilateral is determined in this example. The measures of the angles of the quadrilateral are given as multiples of an unknown, x.

According to the Polygon Angle Sum Theorem, the sum of the interior angles is $(n - 2)180°$. Equate this value with the sum of the expressions we are given for the angle measures. Substitute 4 for n and solve for x. The value of x is 36°. This is the measure of angles P and R. The other two angles measure $4x$, or 144°.

Example 4 Finding Exterior Angle Measures in Polygons

Polygon Exterior Angle Sum Theorem

The sum of the exterior angle measures, one angle at each vertex, of a convex polygon is 360°.
(Poly. Ext. ∠ Sum Thm.)

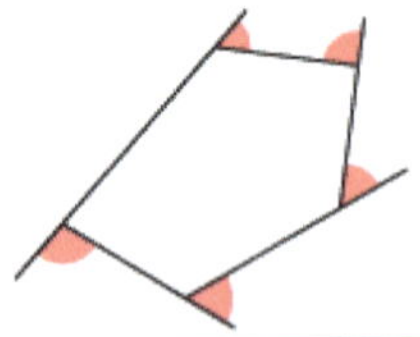

According to the Polygon Exterior Angle Sum Theorem, the sum of the exterior angle measures of a convex polygon is 360°. Only one exterior angle per vertex is counted. The angle is made by extending the line segment that forms a side of the polygon.

Find the measure of each exterior angle of a regular hexagon.

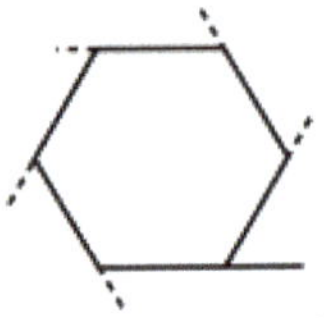

A hexagon has 6 sides and 6 vertices.

sum of ext. ∠s : 360°

measure of one ext. ∠ : $\frac{360^\circ}{6} = 60^\circ$

The measure of each exterior angle of a regular hexagon is 60°.

The measure of the exterior angles of a regular hexagon is determined in this example.

According to the Polygon Exterior Angle Sum Theorem, the sum of the interior angles is 360°. The hexagon has 6 sides. Since the hexagon is regular, all the interior angles are congruent. The size of each angle is 360°/6 = 60°.

Find the value of x in polygon *RSTU*.

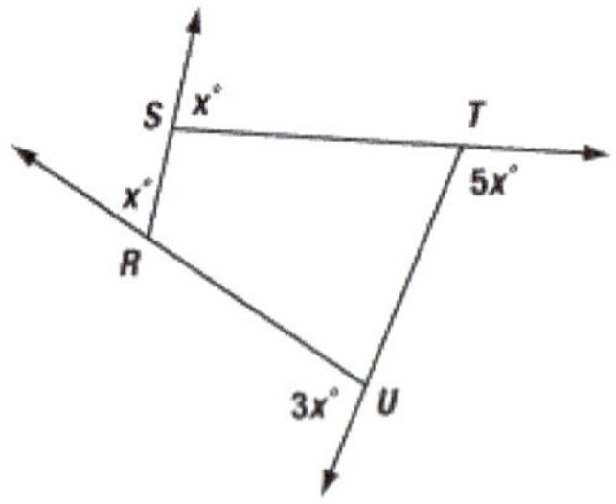

$x^\circ + x^\circ + 3x^\circ + 5x^\circ = 360^\circ$	*Polygon Ext. ∠ Sum Thm.*
$10x = 360$	*Combine like terms.*
$x = 36$	*Divide both sides by 10.*

The measures of unknown exterior angles of a polygon are determined in this example. The measures of the angles of the quadrilateral are given as multiples of an unknown x.

According to the Polygon Exterior Angle Sum Theorem, the sum of the exterior angles is 360°. Equate the sum of the expressions given for the angle measures to 360°. Solve for x. The value of x is 36°. This is the measure of the exterior angles at S and R. The measures of the other two angles can be determined by substitution.

Example 5 Problem-Solving Application

The Pentagon near Washington DC is a regular pentagon. What is the measure of an exterior angle?

By the Polygon Exterior Angle Sum Theorem, the sum of the exterior angle measures is 360°

measure of an exterior ∠ : $\frac{360^\circ}{5} = 72^\circ$

The measure of an exterior angle of a regular pentagon is determined in this application example.

The building shape is given as a regular pentagon. According to the Polygon Exterior Angle Sum Theorem, the sum of the exterior angles is 360°. Since the pentagon is regular, all the exterior angles are congruent. The size of each angle is 360°/5 = 72°.

Properties of Parallelograms

Key Objectives

- Prove and apply properties of parallelograms.
- Use properties of parallelograms to solve problems.

Key Terms

- A quadrilateral with two pairs of parallel sides is a **parallelogram**.

Theorems, Postulates, Corollaries, and Properties

- **Theorem** If a quadrilateral is a parallelogram, then its opposite sides are congruent.
- **Theorem** If a quadrilateral is a parallelogram, then its opposite angles are congruent.
- **Theorem** If a quadrilateral is a parallelogram, then its consecutive angles are supplementary.
- **Theorem** If a quadrilateral is a parallelogram, then its diagonals bisect each other.

Example 1 Problem-Solving Application

Theorem
If a quadrilateral is a parallelogram, then its opposite sides are congruent.
(▱ → opp. sides ≅)

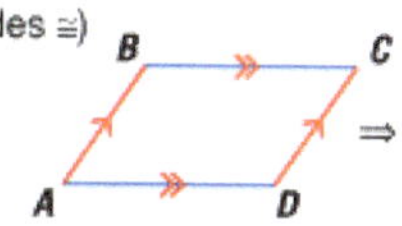

⇒ $\overline{AB} \cong \overline{CD}$
$\overline{BC} \cong \overline{DA}$

The sides opposite each other in a parallelogram are congruent.

Theorem
If a quadrilateral is a parallelogram, then its consecutive angles are supplementary.
(▱ → cons. ∠s supp.)

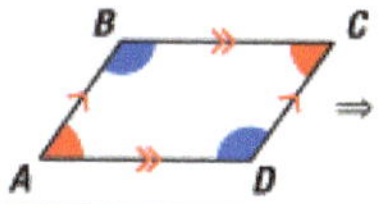

⇒ $m\angle A + m\angle B = 180°$
$m\angle B + m\angle C = 180°$
$m\angle C + m\angle D = 180°$
$m\angle D + m\angle A = 180°$

Consecutive angles, those on either end of a single side, in a parallelogram are supplementary, meaning their measures sum to 180°.

An inexpensive picnic table begins to lean over and forms a parallelogram. The diagram shows the picnic table. In ▱*ABCD*, *AD* = 60 inches, *AE* = 35 inches, and m∠*CDA* = 81°. Find each measure.

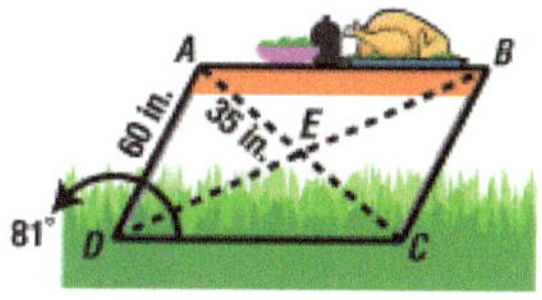

BC

$\overline{BC} \cong \overline{AD}$	▱ → opp. sides ≅
$BC = AD$	Def. of ≅ segs.
$BC = 60$ in.	Substitute 60 for AD.

m∠DAB

$m\angle DAB + m\angle CDA = 180°$	▱ → cons. ∠s supp.
$m\angle DAB + 81 = 180$	Substitute 81 for m∠CDA.
$m\angle DAB = 99°$	Subtract 81 from both sides.

The length of a side and measure of an angle in a parallelogram are determined in this example. AD, $m\angle CDA$, and AE are given.

To find BC, recognize that $\overline{BC} \cong \overline{AD}$ by the Properties of Parallelograms (opposite sides are congruent). Then, by the definition of congruent line segments, $BC = AD$. Substitute the known length for AD to find $BC = 60$ in.

To find $m\angle DAB$, recognize that $\angle CDA$ is consecutive with $\angle DAB$. Therefore, the sum of their measures equals 180°, according to the Properties of Parallelograms. Substitute the known value of 81° for $m\angle CDA$. The measure of $\angle DAB$ is found to be 99°.

Example 2 Using Properties of Parallelograms to Find Measures

Figure ABCD is a parallelogram. Find each measure.

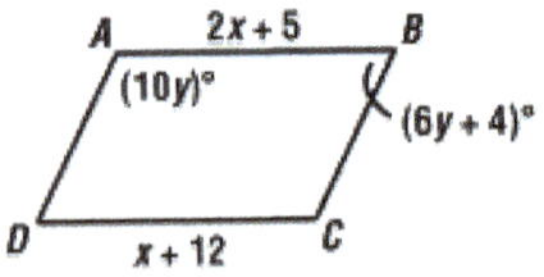

AB

$\overline{AB} \cong \overline{DC}$	▱ → opp. sides ≅
$AB = DC$	Def. of ≅ segs.
$2x + 5 = x + 12$	Substitute the given values.
$x + 5 = 12$	Subtract x from both sides.
$x = 7$	Subtract 5 from both sides.

$AB = 2x + 5 = 2(7) + 5 = 19$

m∠B

$m\angle A + m\angle B = 180°$	▱ → cons. ∠s supp.
$(10y) + (6y + 4) = 180$	Substitute the given values.
$16y + 4 = 180$	Combine like terms.
$16y = 176$	Subtract 4 from both sides.
$y = 11$	Divide both sides by 16.

$m\angle B = (6y + 4)° = [6(11) + 4]° = 70°$

The measures of a side and an angle in a parallelogram are determined in this example. AB and DC, $m\angle B$, and $m\angle A$ are given as algebraic expressions.

AB and DC are equal according to the Properties of Parallelograms. Set the expressions for the lengths of the two sides equal to each other and solve for x. The solution yields $x = 7$, and substituting this value for x in the expression for AB yields a length of 19 units.

$\angle B$ and $\angle A$ are consecutive angles in a parallelogram, so according to the Properties of Parallelograms their measures sum to 180°. Substitute the expressions for $m\angle B$ and $m\angle A$ into the equation for their sum and solve for the unknown, y. The solution yields $y = 11$, and substituting this value for y in the expression for the measure of $\angle B$ yields 70°.

Example 3 Parallelograms in the Coordinate Plane

Three vertices of a parallelogram are A(−3, 3), B(−4, 0), and C(2, −4). Find the coordinates of vertex D.

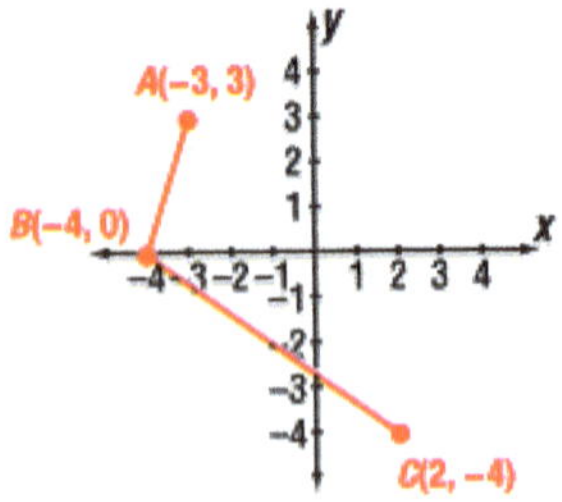

Find the slope of $\overline{BA}$ by counting the units from B to A.
The rise from 0 to 3 is 3 units.
The run from −4 to −3 is 1 unit.

Start at C and count the same number of units.
A rise of 3 units from −4 is −1.
A run of 1 unit from 2 is 3.

The coordinates of vertex D are (3, −1).

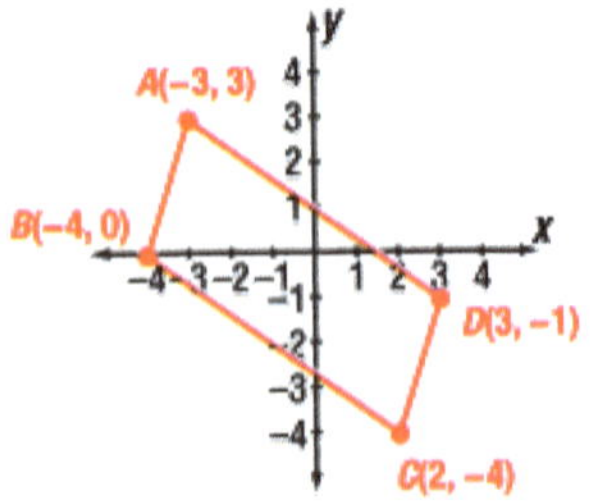

The coordinates are determined here for the unknown vertex of a parallelogram on the coordinate plane. The coordinates of three of the vertices for the parallelogram are given.

The fourth vertex, D, will form the endpoint of a line segment, $\overline{CD}$, that is parallel to line segment $\overline{AB}$. The slope of the two sides will be equal, since this is a parallelogram. The length of the two sides will be equal according to the Properties of Parallelograms.

Use the slope of $\overline{AB}$, a rise of 3 units and a run of 1 unit, to find point D relative to point C. For the x coordinate, $2 + 1 = 3$, and for the y coordinate, $-4 + 3 = -1$. The coordinates for D are $(3, -1)$.

Example 4 Using Properties of a Parallelogram in a Proof

Theorem

If a quadrilateral is a parallelogram, then its opposite angles are congruent.
(▱ → opp. ∠s ≅)

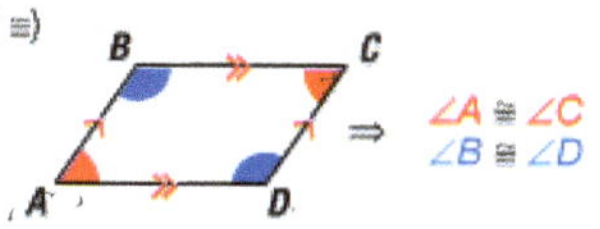

The angles opposite each other in a parallelogram are congruent.

Theorem

If a quadrilateral is a parallelogram, then its diagonals bisect each other.
(▱ → diags. bisect each other)

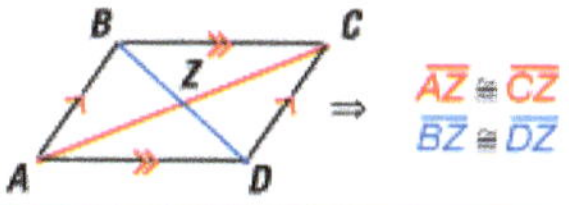

The two diagonals of a parallelogram bisect each other.

Given:
ABCD, *DEFG*, and *FIHJ* are parallelograms,
A, *D* and *E* are collinear,
C, *D* and *G* are collinear,
J, *F* and *G* are collinear,
and *E*, *F* and *I* are collinear.

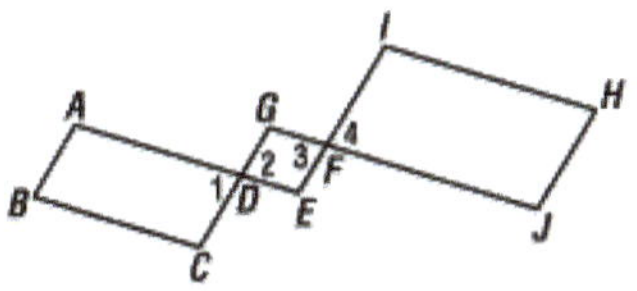

Prove:
∠*B* ≅ ∠4

Statements:	Reasons:
1. *ABCD*, *DEFG*, *FIHJ* are parallelograms.	1. Given
2. ∠*B* ≅ ∠1, ∠2 ≅ ∠3	2. ▱ → opp. ∠s ≅
3. ∠1 ≅ ∠2, ∠3 ≅ ∠4	3. Vert. ∠s Thm.
4. ∠*B* ≅ ∠4	4. Trans. Prop. of ≅

A proof that ∠*B* is congruent to ∠4 is developed in this example. It is given that there are three parallelograms *ABCD*, *DEFG*, and *FIHJ*, and that $\overline{AE}$, $\overline{CG}$, $\overline{JG}$, and $\overline{EI}$ are line segments.

According to the Properties of Parallelograms, ∠*B* and ∠1 are congruent, and ∠2 and ∠3 are congruent, because they are opposite angles. According to the Vertical Angles Theorem ∠1 and ∠2 are congruent, and ∠3 and ∠4 are congruent. Therefore, because of the Transitive Property of Congruence, ∠*B* ≅ ∠4.

Conditions for Parallelograms

Key Objectives

- Prove that a given quadrilateral is a parallelogram.

Theorems, Postulates, Corollaries, and Properties

- **Theorem** If one pair of opposite sides of a quadrilateral is parallel and congruent, then the quadrilateral is a parallelogram.
- **Theorem** If both pairs of opposite sides of a quadrilateral are congruent, then the quadrilateral is a parallelogram.
- **Theorem** If both pairs of opposite angles of a quadrilateral are congruent, then the quadrilateral is a parallelogram.
- **Theorem** If an angle of a quadrilateral is supplementary to both of its consecutive angles, then the quadrilateral is a parallelogram.
- **Theorem** If the diagonals of a quadrilateral bisect each other, then the quadrilateral is a parallelogram.

Example 1 Verifying Figures are Parallelograms

Theorem

If both pairs of opposite sides of a quadrilateral are congruent, then the quadrilateral is a parallelogram.
(Quad. with opp. sides ≅ → ▱)

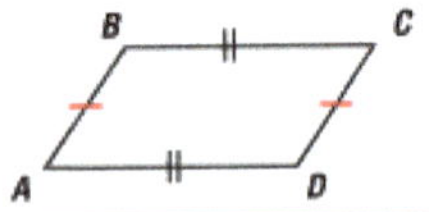

This is a condition of a quadrilateral being a parallelogram. If both pairs of opposite sides of the quadrilateral are congruent, then it is a parallelogram. Note the symbols for congruent sides in the figure.

Theorem

If both pairs of opposite angles of a quadrilateral are congruent, then the quadrilateral is a parallelogram.
(Quad. with opp. ∠s ≅ → ▱)

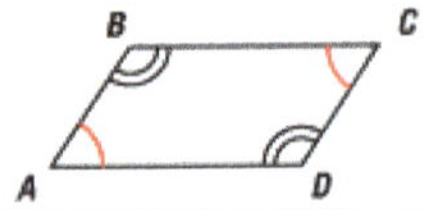

This is a condition of a quadrilateral being a parallelogram. If both pairs of the opposite angles of a quadrilateral are congruent, then that quadrilateral is a parallelogram. Note the symbols for congruent angles in the figure.

Theorem

If an angle of a quadrilateral is supplementary to both of its consecutive angles, then the quadrilateral is a parallllelogram.
(Quad. with ∠s supp. to cons. ∠s → ▱)

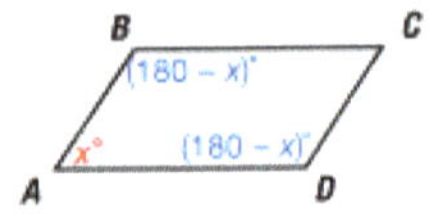

This is a condition of a quadrilateral being a parallelogram. If an angle of a quadrilateral is supplementary to both of its consecutive angles, then that quadrilateral is a parallelogram.

Show that *ABCD* is a parallelogram for $x = 5$ and $y = 2$.

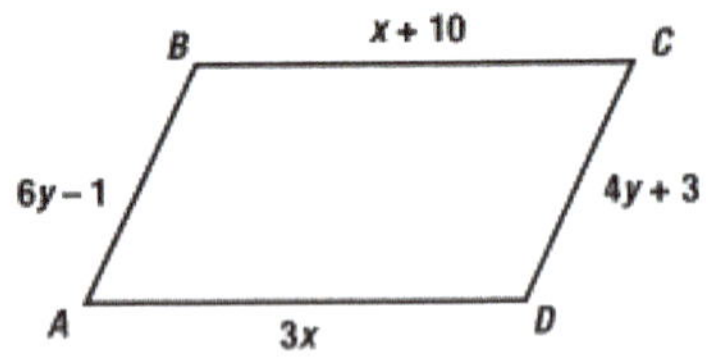

Find BC and DA.

$BC = x + 10$	$DA = 3x$	Given
$BC = 5 + 10 = 15$	$DA = 3(5) = 15$	Substitute and simplify.

Find AB and CD.

$AB = 6y - 1$	$CD = 4y + 3$	Given
$AB = 6(2) - 1 = 11$	$CD = 4(2) + 3 = 11$	Substitute and simplify.

$BC = DA$ and $AB = CD$.
The quadrilateral is a parallelogram. *quad. with opp. sides ≅ → ▱*

In this example, the quadrilateral *ABCD* is shown to be a parallelogram when the unknowns x and y have certain values.

The lengths of the sides of the quadrilateral are given as algebraic expressions in x and y. Substitute the given values for x and y into the expressions for the lengths of the sides, solve for the lengths, and compare opposite sides.

Since $BC = DA = 15$, and $AB = CD = 11$, then by the quadrilateral with congruent opposite sides condition for parallelograms, the figure is a parallelogram.

Show that *EFGH* is a parallelogram for $z = 10$ and $w = 15$.

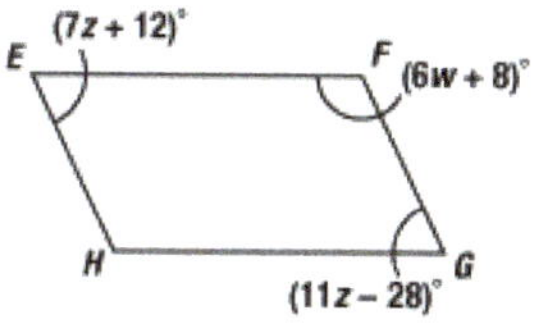

$m\angle E = (7z + 12)^\circ$	*Given*
$m\angle E = [7(10) + 12]^\circ = 82^\circ$	*Substitute 10 for z and simplify.*
$m\angle G = (11z - 28)^\circ$	*Given*
$m\angle G = [11(10) - 28]^\circ = 82^\circ$	*Substitute 10 for z and simplify.*
$m\angle F = (6w + 8)^\circ$	*Given*
$m\angle F = [6(15) + 8]^\circ = 98^\circ$	*Substitute 15 for w and simplify.*
$\angle F$ and $\angle G$ are supplementary	*$82^\circ + 98^\circ = 180^\circ$*
$\overline{EF} \parallel \overline{GH}$	*Conv. of Same-Side Int. ∠s Thm.*
$\angle E$ and $\angle H$ are supplementary	*Same-Side Int. ∠s Thm.*
$m\angle H = 98^\circ$	*$82^\circ + 98^\circ = 180^\circ$*
$m\angle F = m\angle H$ and $m\angle E = m\angle G$	*Trans Prop of =*
$\angle F \cong \angle H$ and $\angle E \cong \angle G$	*Def. of ≅ ∠s*
The quadrilateral is a parallelogram.	*quad. with ∠ supp. to cons. ∠s → ▱*

The quadrilateral *EFGH* is shown here to be a parallelogram when the unknowns z and w have certain values.

The measures of three of the angles of the quadrilateral are given as algebraic expressions in z and w. Substitute the given values for z and w into the expressions for the angle measures, and solve for the measures.

This proof is more direct than the one shown. $\angle F$ and $\angle G$ are consecutive angles. Since $m\angle F + m\angle G = 98^\circ + 82^\circ = 180^\circ$, $\angle F$ and $\angle G$ are supplementary. Therefore, *EFGH* is a parallelogram by the Properties of Parallelograms (Quad. with ∠s supp. to cons. ∠s → ▱). The same proof can be obtained using $\angle F$ and $\angle E$.

Example 2 Applying Conditions for Parallelograms

Theorem

If one pair of opposite sides of a quadrilateral are parallel and congruent, then the quadrilateral is a parallelogram.
(Quad. with pair of opp. sides ∥ and ≅ → ▱)

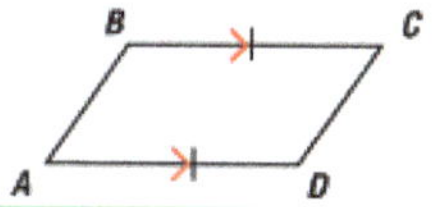

This is a condition of a quadrilateral being a parallelogram. If a quadrilateral has one pair of opposite congruent and parallel sides, then it is a parallelogram. Note the symbols for congruent sides and those for parallel sides in the figure.

Determine if each quadrilateral must be a parallelogram. Justify your answer.

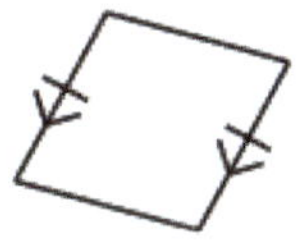

✓

One pair of opposite sides are congruent and parallel.

Yes, the quadrilateral is a parallelogram.

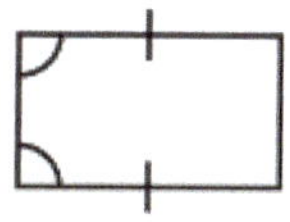

✗

One pair of opposite sides are congruent, and one pair of consecutive angles are congruent.

No, the quadrilateral is not necessariliy a parallelogram.

Quadrilateral figures are determined to be parallelograms or not.

It is given in the first figure that one of the pairs of opposite sides of the quadrilateral is both congruent and parallel. Therefore, this is a parallelogram by a condition for parallelograms.

It is given in the second figure that one pair of opposite sides is congruent and one pair of consecutive angles is congruent. This is not enough information to determine that the quadrilateral is a parallelogram. Remember not to make any assumptions based on what the figure looks like. For example, if the congruent angles are not right angles, then the congruent sides are not parallel.

Example 3 Proving Parallelograms in the Coordinate Plane

Show that the given quadrilateral is a parallelogram by using the given definition or theorem.

$A(-3, 3)$, $B(1, 1)$, $C(2, -2)$, $D(-2, 0)$

slope of $\overline{AD} = \frac{\text{rise}}{\text{run}} = \frac{-3}{1} = -3$

slope of $\overline{BC} = \frac{\text{rise}}{\text{run}} = \frac{-3}{1} = -3$

slope of $\overline{DC} = \frac{\text{rise}}{\text{run}} = \frac{-2}{4} = -\frac{1}{2}$

slope of $\overline{AB} = \frac{\text{rise}}{\text{run}} = \frac{-2}{4} = -\frac{1}{2}$

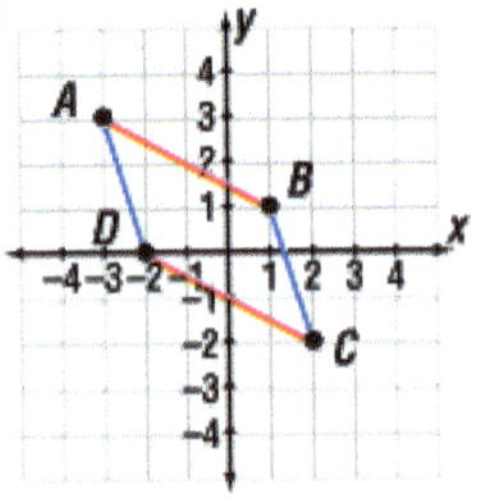

Since both pairs of opposite sides are parallel, *ABCD* is a parallelogram by definition.

A quadrilateral in the coordinate plane is proven here to be a parallelogram. The coordinates of the vertices of the parallelogram are given.

Calculate the slopes of the sides of the quadrilateral. For example the slope of $\overline{AB} = (y_B - y_A)/(x_B - x_A) = (1 - 3)/(1 - (-3)) = -2/4 = -1/2$. Calculating all slopes shows that $\overline{AB} \parallel \overline{DC}$ and $\overline{AD} \parallel \overline{BC}$. Since both pairs of opposite sides are parallel, by the conditions for parallelograms, the quadrilateral *ABCD* is a parallelogram.

Show that the given quadrilateral is a parallelogram by using the given definition or theorem.

$F(-3, 5)$, $G(2, 5)$, $H(4, 0)$, $J(0, 0)$

$FG = 4$
$JH = 4$

slope of $\overline{FG} = 0$
slope of $\overline{JH} = 0$

$\overline{FG}$ and $\overline{JH}$ have the same slope, so $\overline{FG} \parallel \overline{JH}$.

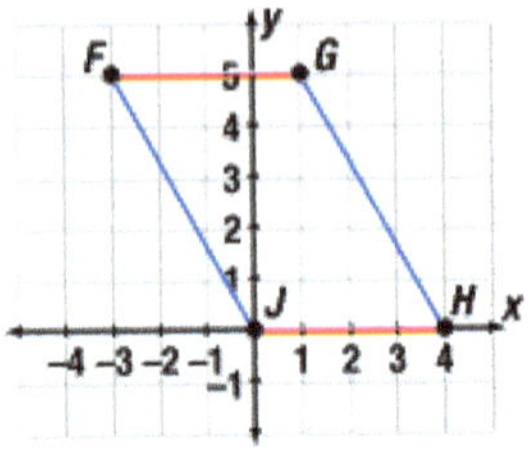

Since $FG = JH$, $\overline{FG} \cong \overline{JH}$.

So *FGHJ* is a parallelogram. *quad. with pair of opp. sides ∥ and ≅ → ▱*

A quadrilateral in the coordinate plane is proven here to be a parallelogram. The coordinates of the vertices of the parallelogram are given.

Calculate the slopes of two of the sides of the quadrilateral. For example, the slope of $\overline{FG} = (y_G - y_F)/(x_G - x_F) = (5 - 5)/(1 - (-3)) = 0/4 = 0$. The slope of $\overline{JH}$ is also 0. Therefore, $\overline{FG} \parallel \overline{JH}$. Calculate the lengths *FG* and *JH*. For example, the length *FG* is $(1 - (-3)) = 4$. This is also the length *JH*. Since the opposite sides are congruent and parallel, by a condition for parallelograms, the quadrilateral *FGHJ* is a parallelogram.

Example 4 Problem-Solving Application

Theorem
If the diagonals of a quadrilateral bisect each other, then the quadrilateral is a parallelogram.
(Quad. with diags. bisecting each other → ▱)

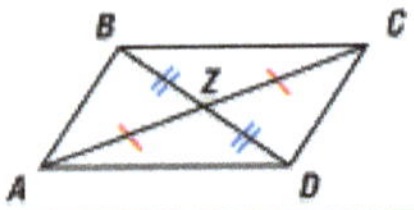

This is a condition of a quadrilateral being a parallelogram. If a quadrilateral's two diagonals bisect each other, then the quadrilateral is a parallelogram. Notice the symbols for congruent line segments in the figure.

In the stained glass, *AE* = *EC* and *BE* = *ED*. Why is *ABCD* a parallelogram?

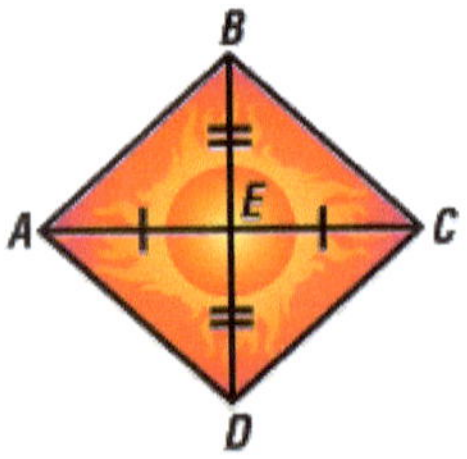

Since the diagonals bisect each other, *ABCD* is a parallelogram.

A quadrilateral is determined to be a parallelogram in this application example. It is given that $\overline{AE}$ and $\overline{EC}$ are equal in length and $\overline{BE}$ and $\overline{ED}$ are equal in length.

The givens indicate that the diagonals of the quadrilateral bisect each other. Therefore, this is a parallelogram according to the conditions for parallelograms.

Properties of Special Parallelograms

Key Objectives

- Prove and apply properties of rectangles, rhombuses, and squares.
- Use properties of rectangles, rhombuses, and squares to solve problems.

Key Terms

- A **rectangle** is a quadrilateral with four right angles.
- A **rhombus** is a quadrilateral with four congruent sides.
- A **square** is a quadrilateral with four right angles and four congruent sides.

Theorems, Postulates, Corollaries, and Properties

- **Theorem** If a quadrilateral is a rectangle, then it is a parallelogram.
- **Theorem** If a parallelogram is a rectangle, then its diagonals are congruent.
- **Theorem** If a quadrilateral is a rhombus, then it is a parallelogram.
- **Theorem** If a parallelogram is a rhombus, then its diagonals are perpendicular.
- **Theorem** If a parallelogram is a rhombus, then each diagonal bisects a pair of opposite angles.

Example 1 Craft Application

Theorem
If a quadrilateral is a rectangle, then it is a parallelogram.
(Rect. → ▱)

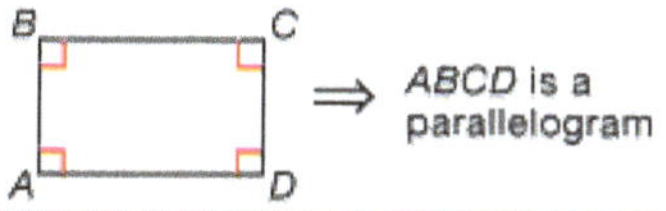

This is a property of a rectangle. If a quadrilateral is a rectangle, then it is a parallelogram.

Theorem
If a parallelogram is a rectangle, then its diagonals are congruent.
(Rect. → diags. ≅)

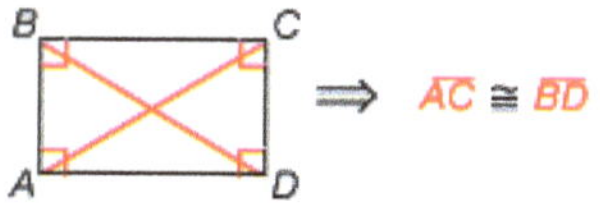

This is a property of a rectangle. If a parallelogram is a rectangle, then its diagonals are congruent.

In this rectangular patch, the quilter cut the strips so that *EH* = 12 in. and *EG* = 20 in. Find *FG* and *FJ*.

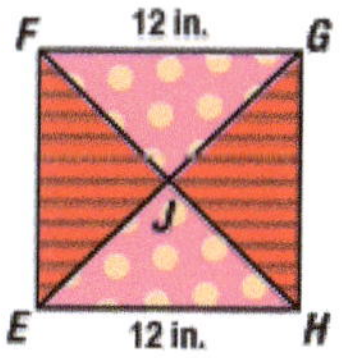

FG

$\overline{FG} \cong \overline{EH}$ — ▱ → opp. sides ≅

$FG = EH = 12$ in. — Def. of ≅ segs.

A property of rectangles is used to find the length of the sides of a quilt patch in this application example. It is given that the patch is a rectangle. The length of one side, $\overline{EH}$, and the length of a diagonal, $\overline{EG}$, are given.

By the properties of rectangles, opposite sides are congruent, so $\overline{FG} \cong \overline{EH}$. *EH* is given as 12 in., so by the definition of congruent line segments, *FG* is also 12 in.

In this rectangular patch, the quilter cut the strips so that $EH = 12$ in. and $EG = 20$ in. Find FG and FJ.

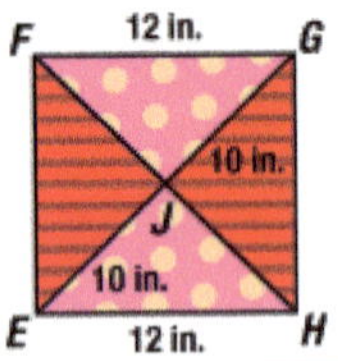

FJ

$EJ = GJ = \frac{EG}{2}$	*▱ → diags. bisect each other*
$EJ = GJ = 10$ in.	*Substitute 20 for EG.*
$\overline{FH} \cong \overline{EG}$	*rect. → diags. ≅*
$FH = EG = 20$ in.	*Def. of ≅ segs.*
$FJ = \frac{FH}{2}$	*▱ → diags. bisect each other*
$FJ = 10$ in.	*Substitute 20 for FH.*

A property of rectangles is used to find the length of half the diagonal of a quilt patch in this application example. It is given that the patch is a rectangle. The length of one side, $\overline{EH}$, and the length of a diagonal, $\overline{EG}$, are given.

Here is a more direct proof than the one shown; try comparing them. By the properties of rectangles, the diagonals $\overline{FH}$ and $\overline{EG}$ are congruent. EG is given to be 20 in., so FH is 20 in. by the definition of congruent line segments.

The diagonals of a rectangle bisect each other, so $FJ = FH/2$. Substitute 20 in. for FH to yield 10 in. for FJ.

Example 2 Using Properties of Rhombuses to Find Measures

Theorem

If a parallelogram is a rhombus, then its diagonals are perpendicular.
(Rhombus → diags. ⊥)

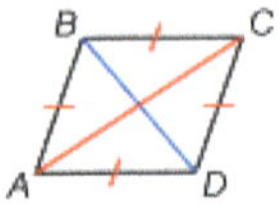

$\Rightarrow \overline{AC} \perp \overline{BD}$

This is a property of a rhombus. If a parallelogram is a rhombus, then its diagonals are perpendicular.

Theorem

If a parallelogram is a rhombus, then each diagonal bisects a pair of opposite angles.
(Rhombus → each diag. bisects opp. ∠s)

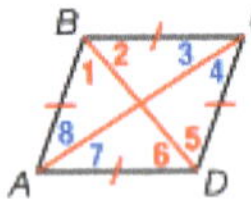

$\Rightarrow$ $\angle 1 \cong \angle 2$, $\angle 3 \cong \angle 4$, $\angle 5 \cong \angle 6$, $\angle 7 \cong \angle 8$

This is a property of a rhombus. If a parallelogram is a rhombus, then each of its diagonals bisects a pair of opposite angles.

Theorem

If a quadrilateral is a rhombus, then it is a parallelogram.
(Rhombus → ▱)

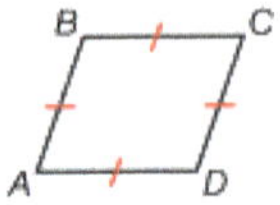

$\Rightarrow$ $ABCD$ is a parallelogram.

This is a property of a rhombus. If a quadrilateral is a rhombus, then it is a parallelogram.

$ABCD$ is a rhombus. Find each measure.

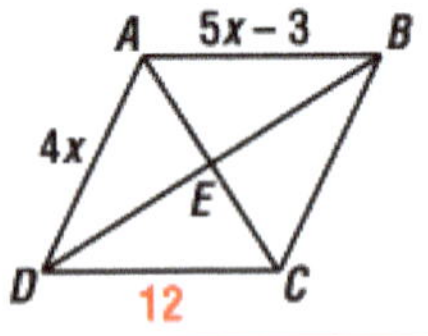

DC

$AD = AB$	*Def. of rhombus*
$4x = 5x - 3$	*Substitute the given values.*
$0 = x - 3$	*Subtract 4x from both sides.*
$3 = x$	*Add 3 to both sides.*
$DC = AD$	*Def. of rhombus*
$DC = 4x$	*Substitute 4x for AD.*
$= 4(3) = 12$	*Substitute 3 for x and simplify.*

The properties of a rhombus are used here to determine the length of a side of a rhombus. The figure is given to be a rhombus with the lengths of two sides given as algebraic expressions.

To find DC, begin by solving for x. By the definition of a rhombus, all sides are congruent, so $AD = AB$. Substitute the given algebraic expressions for AD and AB and solve for x. The solution yields $x = 3$.

Again, by the definition of a rhombus all sides are congruent, so to find DC substitute the value for x into either of the expressions for the length of the two given sides. $DC = AD = 4x = 4(3) = 12$.

ABCD is a rhombus. Find each measure.

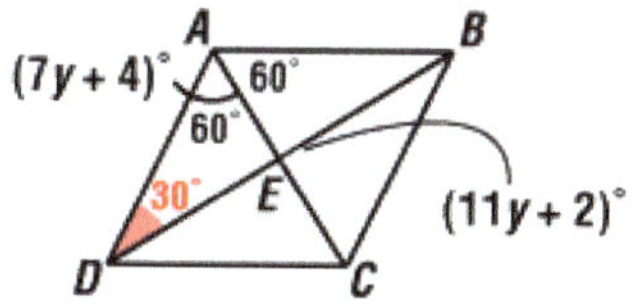

m∠ADB

$11y + 2 = 90$	*rhombus → diags. ⊥*
$11y = 88$	*Subtract 2 from both sides.*
$y = 8$	*Divide both sides by 11.*
$m\angle DAC = 7(8) + 4 = 60^\circ$	*Substitute 8 for y and simplify.*
$m\angle CAB = 60^\circ$	*rhombus → each diag. bisects opp. ∠s*
$m\angle DAB = m\angle DAC + m\angle CAB$	
$= 60 + 60 = 120^\circ$	*Substitute 60° for m∠DAC and m∠CAB.*
$m\angle ADC + m\angle DAB = 180^\circ$	*Same-side Int. ∠s Thm.*
$m\angle ADC + 120^\circ = 180^\circ$	*Substitute 120° for m∠DAC.*
$m\angle ADC = 60^\circ$	*Subtract 120° from both sides.*
$m\angle ADB = \frac{1}{2}m\angle ADC = \frac{1}{2}(60^\circ) = 30^\circ$	*rhombus → each diag. bisects opp. ∠s*

The properties of a rhombus are used here to determine the measure of an angle of a rhombus. The figure is given to be a rhombus with the measures of two angles given as algebraic expressions with the unknown y.

To find $m\angle ADB$ begin by solving for y. By the properties of a rhombus the diagonals are perpendicular, so

$m\angle BEC = (11y + 2)^\circ = 90^\circ$. Solve for y to yield $y = 8$.

Substitute the value of y to find $m\angle DAC$.

$m\angle DAC = (7y + 4)^\circ = (7(8) + 4)^\circ = 60^\circ$

By the properties of a rhombus each diagonal bisects opposite angles, so $m\angle CAB = m\angle DAC = 60^\circ$. Substitute to find $m\angle DAB$.

$m\angle DAB = m\angle DAC + m\angle CAB = 60^\circ + 60^\circ = 120^\circ$.

By the Same Side Interior Angles Theorem,

$m\angle ADC + m\angle DAB = m\angle ADC + 120^\circ = 180^\circ$. Therefore,

$m\angle ADC = 60^\circ$. According to the properties of a rhombus each diagonal bisects opposite angles, so

$m\angle ADB = (1/2)m\angle ADC = (1/2)60^\circ = 30^\circ$.

Example 3 Verifying Properties of Squares

Show that the diagonals of square ABCD are congruent perpendicular bisectors of each other.

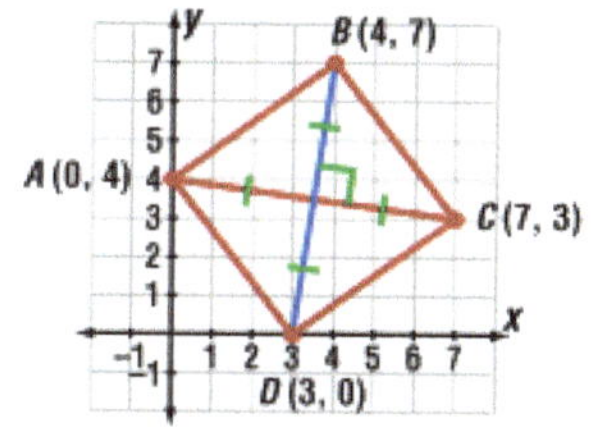

$AC = \sqrt{(7-0)^2 + (4-3)^2} = \sqrt{49 + 1} = \sqrt{50} = 5\sqrt{2}$

$BD = \sqrt{(4-3)^2 + (7-0)^2} = \sqrt{1 + 49} = \sqrt{50} = 5\sqrt{2}$

Since $AC = BD$, $\overline{AC} \cong \overline{BD}$.

mdpt. of $\overline{AC} = \left(\frac{0+7}{2}, \frac{4+3}{2}\right) = \left(\frac{7}{2}, \frac{7}{2}\right)$

mdpt. of $\overline{BD} = \left(\frac{4+3}{2}, \frac{7+0}{2}\right) = \left(\frac{7}{2}, \frac{7}{2}\right)$

Since $\overline{AC}$ and $\overline{BD}$ have the same mdpt., they bisect each other.

slope of $\overline{AC} = \frac{\text{rise}}{\text{run}} = \frac{-1}{7} = -\frac{1}{7}$

slope of $\overline{BD} = \frac{\text{rise}}{\text{run}} = \frac{7}{1} = 7$

Since $\left(-\frac{1}{7}\right)7 = -1$, $\overline{AC} \perp \overline{BD}$.

The diagonals of a square in the coordinate plane are shown to be congruent and to be perpendicular bisectors of each other.

To show the diagonals are congruent, calculate their lengths using the Distance Formula. The lengths are both found to be $5\sqrt{2}$. By the definition of congruent line segments, the diagonals are congruent.

To show the diagonals bisect each other, calculate their midpoints using the Midpoint Formula. Both midpoints are the same point on the coordinate plane. This means the two diagonals intersect at their midpoints, or bisect each other.

To show the diagonals are perpendicular, calculate their slopes. The slope of $\overline{AC}$ is $-1/7$ and the slope of $\overline{BD}$ is 7. The product of the slopes is equal to -1, indicating that the lines are perpendicular.

Example 4 Using Properties of Special Parallelograms in Proofs

Given:
ABCD is a rectangle.
E is the midpoint of $\overline{AD}$.

Prove:
$\triangle BCE$ is isosceles.

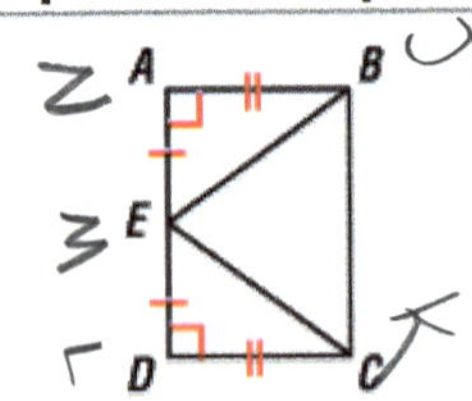

Statements:	Reasons:
1. *ABCD* is a rectangle. *E* is the midpoint of $\overline{AD}$.	1. Given
2. $\angle A$ and $\angle D$ are rt. $\angle$s.	2. Def. of rect.
3. $\angle A \cong \angle D$	3. rt. $\angle \cong$ Thm.
4. *ABCD* is a parallelogram.	4. rect. $\rightarrow \square$
5. $\overline{AB} \cong \overline{DC}$	5. $\square \rightarrow$ opp. sides $\cong$
6. $\overline{AE} \cong \overline{ED}$	6. Def. of mdpt.
7. $\triangle ABE \cong \triangle DCE$	7. SAS $\cong$ Thm.
8. $\overline{BE} \cong \overline{CE}$	8. CPCTC
9. $\triangle BCE$ is isosceles.	9. Def. of isosceles $\triangle$

A triangle is proven to be isosceles using properties of rectangles and parallelograms. *ABCD* is given to be a rectangle with *E* the midpoint of $\overline{AD}$.

By the definition of a rectangle, $\angle A$ and $\angle D$ are right angles and are congruent by the Right Angle Congruence Theorem. *ABCD* is a parallelogram since a rectangle is a parallelogram. $\overline{AB} \cong \overline{DC}$ according to the Properties of Parallelograms.

$\overline{AE} \cong \overline{ED}$ by the definition of a midpoint. Therefore, $\triangle ABE \cong \triangle DCE$ by the SAS Congruence Postulate. $\overline{BE} \cong \overline{CE}$ by the Corresponding Parts of Congruent Triangles are Congruent Theorem. Thus by the definition of an isosceles triangle, $\triangle BCE$ is isosceles.

Conditions for Special Parallelograms

Key Objectives

- Prove that a given quadrilateral is a rectangle, rhombus, or square.

Theorems, Postulates, Corollaries, and Properties

- **Theorem** If one angle of a parallelogram is a right angle, then the parallelogram is a rectangle.
- **Theorem** If the diagonals of a parallelogram are congruent, then the parallelogram is a rectangle.
- **Theorem** If one pair of consecutive sides of a parallelogram are congruent, then the parallelogram is a rhombus.
- **Theorem** If the diagonals of a parallelogram are perpendicular, then the parallelogram is a rhombus.
- **Theorem** If one diagonal of a parallelogram bisects a pair of opposite angles, then the parallelogram is a rhombus.

Example 1 Carpentry Application

Theorem
If the diagonals of a parallelogram are congruent, then the parallelogram is a rectangle.
(▱ with diags. ≅ → rect.)

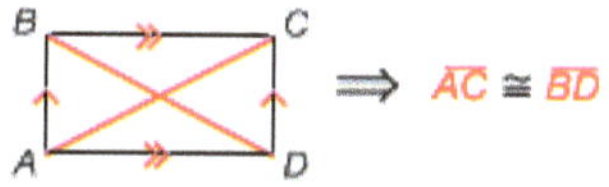

This is a condition for a rectangle. If the diagonals of a parallelogram are congruent, then the parallelogram is a rectangle.

Theorem
If one pair of consecutive sides of a parallelogram are congruent, then the parallelogram is a rhombus.
(▱ with one pair cons. sides ≅ → rhombus)

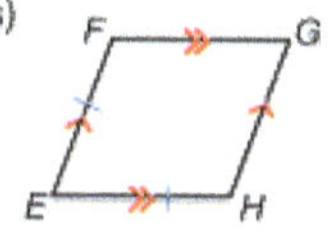

This is a condition for a rhombus. If one pair of consecutive sides of a parallelogram are congruent, then the parallelogram is a rhombus.

Theorem
If one diagonal of a parallelogram bisects a pair of opposite angles, then the parallelogram is a rhombus.
(▱ with diag. bisecting opp. ∠s → rhombus)

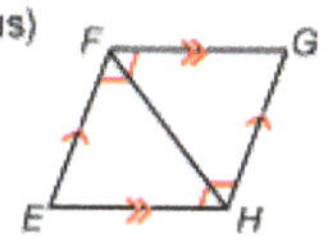

This is a condition for a rhombus. If one diagonal of a parallelogram bisects a pair of opposite angles, then the parallelogram is a rhombus.

A carpenter built a wood frame for a window so that $\overline{AB} \cong \overline{CD}$, $\overline{AC} \cong \overline{BD}$, and $\overline{AD} \cong \overline{BC}$. Why must the frame be a rectangle?

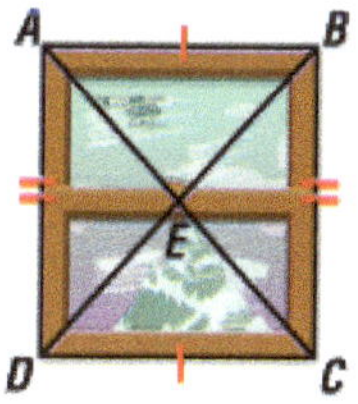

$\overline{AC} \cong \overline{BD}$:
The diagonals are congruent.

Both pairs of opposite sides of *ABCD* are congruent, so *ABCD* is a parallelogram.

Since $\overline{AC} \cong \overline{BD}$, the diagonals of ▱*ABCD* are congruent.

Therefore, the window is a rectangle (▱ with diags ≅ → rect.).

A quadrilateral *ABCD* is proved to be a rectangle in this application example. It is given that opposite sides of the quadrilateral are congruent, as are the diagonals.

By the conditions for parallelograms, *ABCD* is a parallelogram because its opposite sides are congruent.

By the conditions for rectangles, *ABCD* is a rectangle because the diagonals are congruent.

In carpentry, the way to ensure a frame is "square," meaning it has four right angles, is to make sure the lengths of the two diagonals are the same.

Example 2 Applying Conditions for Special Parallelograms

Theorem
If one angle of a parallelogram is a right angle, then the parallelogram is a rectangle.
(▱ with one rt. ∠ → rect.)

This is a condition for a rectangle. If one angle of a parallelogram is a right angle, then the parallelogram is a rectangle.

Theorem

If the diagonals of a parallelogram are perpendicular, then the parallelogram is a rhombus.
(▱ with diags. ⊥ → rhombus)

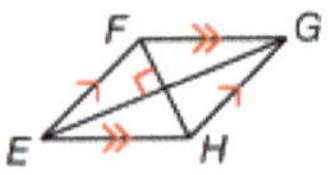

This is a condition for a rhombus. If the diagonals of a parallelogram are perpendicular, then the parallelogram is a rhombus.

Determine if the conclusion is valid. If not, tell what additional information is needed to make it valid.

Given:
$\overline{PQ} \cong \overline{SR}$, $\overline{PS} \cong \overline{QR}$, $\overline{PQ} \perp \overline{QR}$, and $\overline{PR} \perp \overline{QS}$.

Conclusion:
PQRS is a square.

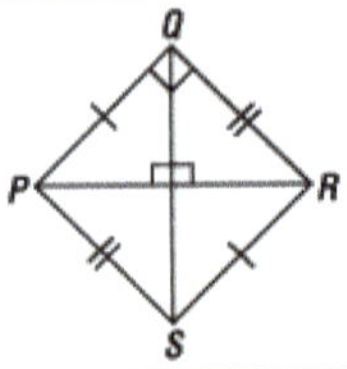

$\overline{PQ} \cong \overline{SR}$ and $\overline{PS} \cong \overline{QR}$	*Given*
PQRS is a parallelogram.	*Quad. with opp. sides ≅ → ▱.*
$\overline{PQ} \perp \overline{QR}$, so ∠*PQR* is a right angle.	*Def. of ⊥ segments*
PQRS is a rectangle.	*▱ with one rt. ∠ → rect.*
$\overline{PR} \perp \overline{QS}$	*Given*
PQRS is a rhombus.	*▱ with diags. ⊥ → rhombus*

Since *PQRS* is a rectangle and a rhombus, it has four right angles and four congruent sides. So *PQRS* is a square by definition.

It is determined that a quadrilateral satisfies the conditions to be a square in this example. It is given that two pairs of opposite sides are congruent, that two sides are perpendicular, and that the diagonals are perpendicular.

PQRS is a parallelogram by the conditions for parallelograms, because the opposite sides are congruent.

PQRS is a rectangle by the conditions for rectangles, because two of the sides form a right angle.

PQRS is a rhombus by the conditions for a rhombus, because the diagonals are perpendicular.

PQRS is a rectangle, so it has four right angles, and a rhombus, so it has four congruent sides. Therefore, by definition, *PQRS* is also a square.

Determine if the conclusion is valid. If not, tell what additional information is needed to make it valid.

Given:
$\overline{PR} \perp \overline{QS}$

Conclusion:
PQRS is a rhombus.

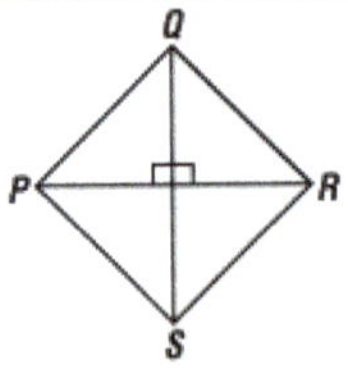

While *PQRS* could be a rhombus, it is not necessarily a rhombus.

A counterexample to the conclusion is a kite whose diagonals are perpendicular.

PQRS satisfies $\overline{PR} \perp \overline{QS}$, but it is not a rhombus.

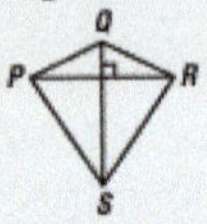

It is determined whether a quadrilateral satisfies the conditions to be a rhombus in this example.

It is given that the diagonals are perpendicular.

There is insufficient information to determine if the quadrilateral is a rhombus. For example, a kite has perpendicular diagonals like the given figure. But a kite is not a rhombus because it does not have four congruent sides.

Example 3 Identifying Special Parallelograms in the Coordinate Plane

Use the diagonals to determine whether a parallelogram with the given vertices is a rectangle, rhombus, or square. Give all the names that apply.

$A(0, 2)$, $B(2, 7)$, $C(7, 5)$, $D(5, 0)$

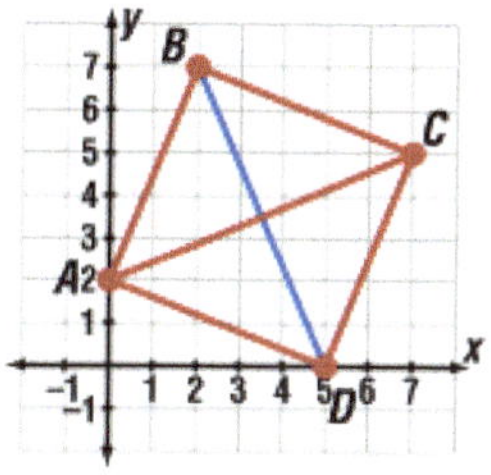

$AC = \sqrt{7^2 + 3^2} = \sqrt{58}$
$BD = \sqrt{3^2 + 7^2} = \sqrt{58}$

Since $\sqrt{58} = \sqrt{58}$, the diagonals are congruent.

ABCD is a rectangle (▱ with diags. ≅ → rect.).

slope of $\overline{AC} = \frac{\text{rise}}{\text{run}} = \frac{3}{7}$

slope of $\overline{BD} = \frac{\text{rise}}{\text{run}} = \frac{-7}{3} = -\frac{7}{3}$ Since $\left(\frac{3}{7}\right)\left(-\frac{7}{3}\right) = -1$, $\overline{AC} \perp \overline{BD}$.

Since $\overline{AC} \perp \overline{BD}$, ▱*ABCD* is a rhombus (▱ with diags. ⊥ → rhombus).

Since *ABCD* is a rectangle and a rhombus, it has four right angles and four congruent sides. So *ABCD* is a square by definition.

The properties of the diagonals of a parallelogram in the coordinate plane are used here to determine whether the parallelogram is a rectangle, a rhombus, or a square. The coordinates of the parallelogram are given.

First, test whether the parallelogram is a rectangle by determining if the diagonals are the same length. Calculate the length of the diagonals using the Pythagorean Theorem. The diagonals are congruent, so the figure is a rectangle by the conditions for rectangles.

Second, test whether the parallelogram is a rhombus by determining if the slopes of the diagonals are perpendicular. The slopes of the diagonals are 3/7 and −7/3, and the product of the slopes is −1, so the diagonals are perpendicular and the figure is a rhombus by the conditions for a rhombus.

The parallelogram is also a square because it is both a rectangle (four right angles) and a rhombus (four congruent sides).

Properties of Kites and Trapezoids

Key Objectives

- Use properties of kites to solve problems.
- Use properties of trapezoids to solve problems.

Key Terms

- A **kite** is a quadrilateral with exactly two pairs of congruent consecutive sides.
- A **trapezoid** is a quadrilateral with exactly one pair of parallel sides.
- Each of the parallel sides of a trapezoid is called a **base of a trapezoid**.
- The nonparallel sides of a trapezoid are called **legs**.
- **Base angles** of a trapezoid are two consecutive angles whose common side is a base.
- If the legs of a trapezoid are congruent, the trapezoid is an **isosceles trapezoid**.
- The **midsegment of a trapezoid** is the segment whose endpoints are the midpoints of the legs.

Theorems, Postulates, Corollaries, and Properties

- **Theorem** If a quadrilateral is a kite, then its diagonals are perpendicular.
- **Theorem** If a quadrilateral is a kite, then exactly one pair of opposite angles is congruent.
- **Theorem** If a quadrilateral is an isosceles trapezoid, then each pair of base angles is congruent.
- **Theorem** If a trapezoid has one pair of congruent base angles, then the trapezoid is isosceles.
- **Theorem** A trapezoid is isosceles if and only if its diagonals are congruent.
- **Theorem** The midsegment of a trapezoid is parallel to each base, and its length is one-half the sum of the lengths of the bases.

Example 1 Problem-Solving Application

Theorem

If a quadrilateral is a kite, then its diagonals are perpendicular. (Kite → diags. ⊥)

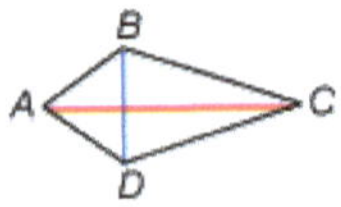

$\Rightarrow \overline{AC} \perp \overline{BD}$

This is a property of kites. If a quadrilateral is a kite, then its diagonals are perpendicular.

Alicia is using a pattern to make a kite and she must cover the outer edges with a cloth binding. There are 2 feet of binding in one package. What is the total amount of binding needed to cover the edges of the kite? How many packages of binding must Alicia buy?

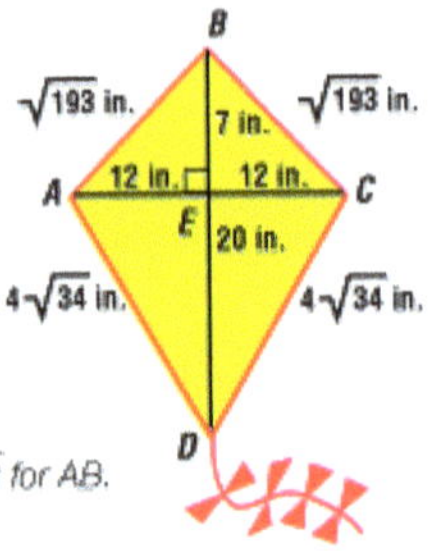

$AB = \sqrt{12^2 + 7^2} = \sqrt{193}$ in. — *Pyth. Thm.*

$BC = AB$ — *$\overline{BC} \cong \overline{AB}$*

$BC = \sqrt{193}$ in. — *Substitute $\sqrt{193}$ for AB.*

$CD = \sqrt{12^2 + 20^2}$ — *Pyth. Thm.*

$= \sqrt{544} = 4\sqrt{34}$ in. — *Simplify.*

$AD = CD$ — *$\overline{AD} \cong \overline{CD}$*

$AD = 4\sqrt{34}$ in. — *Substitute $4\sqrt{34}$ for CD.*

perimeter of ABCD $= 2(\sqrt{193}) + 2(4\sqrt{34}) \approx 74.4$ in.

$\frac{74.4}{24} \approx 3.1$ packages — *There are 24 in. of binding per package.*

Alicia will need to buy 4 packages of binding in order to cover the edges of the kite.

The perimeter of a kite is determined in this application example. The following lengths are given: $AE = CE = 12$ inches, $BE = 7$ inches, and $DE = 20$ inches.

Since the figure is a kite, the diagonals are perpendicular and form right triangles with their hypotenuses forming the sides of the kite. The length of $\overline{AB}$ is calculated using the Pythagorean Theorem to be $\sqrt{193}$ inches. By the definition of a kite, $\overline{AB} \cong \overline{BC}$. So BC is also equal to $\sqrt{193}$ inches. The length of $\overline{CD}$ is calculated using the Pythagorean Theorem to be $4\sqrt{34}$ inches. By the definition of a kite, this is also the length of $\overline{AD}$.

The perimeter of the kite is the sum of the four sides, which can be approximated using a calculator to be 74.4 inches. Since each package of binding contains 24 inches, 4 packages are needed to completely cover the perimeter.

Example 2 Using Properties of Kites

Theorem

If a quadrilateral is a kite, then exactly one pair of opposite angles is congruent.
(Kite → one pair opp. ∠s ≅)

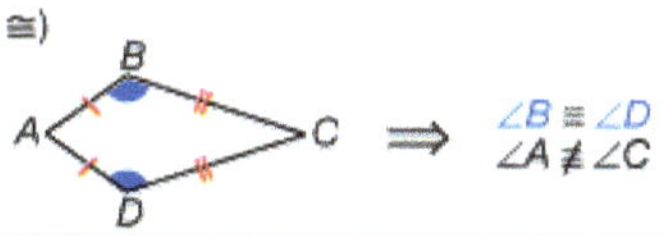

This is a property of a kite. If a quadrilateral is a kite, then exactly one pair of opposite angles is congruent.

In kite *PQRS*, m∠*RSW* = 20°, and m∠*RQW* = 49°. Find the measure.

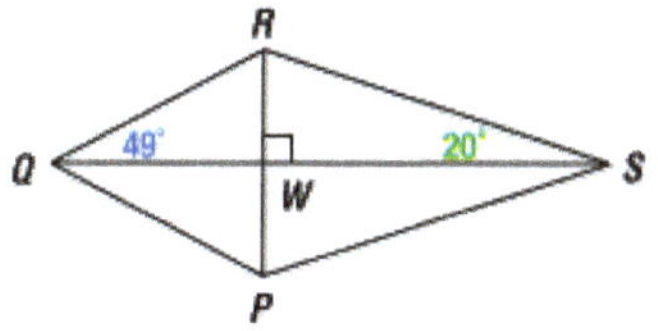

m∠WRS

$m\angle RWS = 90°$	*Kite → diags ⊥*
$m\angle WRS + m\angle RSW = 90°$	*Acute ∠s of rt. △ are comp.*
$m\angle WRS + 20° = 90°$	*Substitute 20° for m∠RSW.*
$m\angle WRS = 70°$	*Subtract 20° from both sides.*

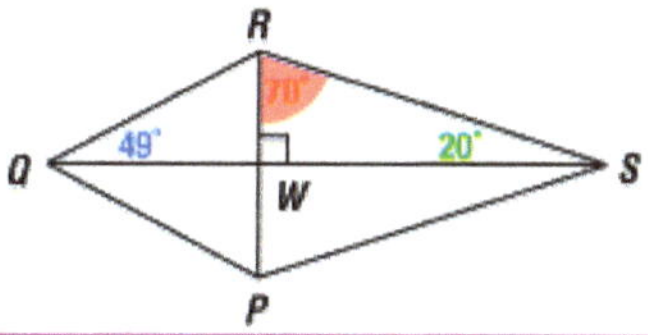

The measure of an angle in a kite, *PQRS*, is determined in this example. The measures of $\angle RQW$ and $\angle RSW$ are given.

By the properties of kites, the diagonals are perpendicular, so $m\angle RWS = 90°$. Since the acute angles of a right triangle are complementary, $m\angle WRS + m\angle RSW = 90°$. Substituting the given 20° for $m\angle RSW$ yields $m\angle WRS = 70°$.

In kite *PQRS*, m∠*RSW* = 20°, and m∠*RQW* = 49°. Find the measure.

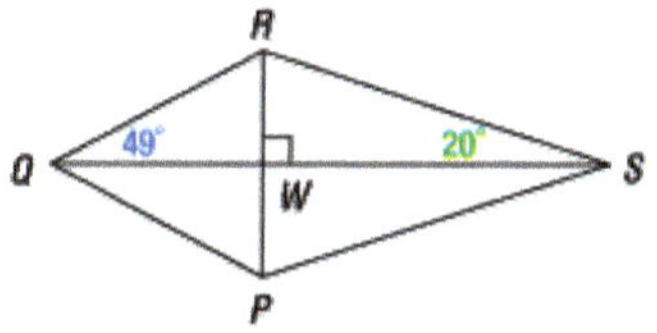

m∠QPS

$\overline{QR} \cong \overline{QP}$ and $\overline{RS} \cong \overline{SP}$	*Def. of kite*
$\angle QRS \cong \angle QPS$	*kite → one pair opp. ∠s ≅*
$\triangle QRS \cong \triangle QPS$	*SAS ≅*
$\angle PQW \cong \angle RQW$ and $\angle PSW \cong \angle RSW$	*CPCTC*
This proves that diagonal $\overline{QS}$ bisects ∠PQR and ∠PSR.	
$m\angle PQW = m\angle RQW$ and $m\angle PSW = m\angle RSW$	*Def. of ≅ ∠s*
$m\angle PQW = 49°$ and $m\angle PSW = 20°$	*Substitute.*
$m\angle QPS + m\angle PQW + m\angle PSW = 180°$	*△ Sum Thm.*
$m\angle QPS + 49° + 20° = 180°$	*Substitute.*
$m\angle QPS + 69° = 180°$	*Simplify.*
$m\angle QPS = 111°$	*Subtract 69° from both sides.*

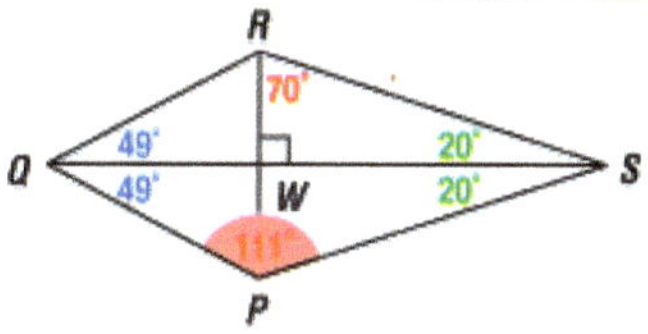

The measure of an angle in a kite, *PQRS*, is determined in this example. The measures of $\angle RQW$ and $\angle RSW$ are given.

By the definition of a kite, the short sides are congruent, as are the long sides. By the properties of kites, $\angle QRS$ and $\angle QPS$ are congruent. Thus, $\triangle QRS \cong \triangle QPS$ by Side-Angle-Side Congruence. It follows from CPCTC that $\angle PQW \cong \angle RQW$ and $\angle PSW \cong \angle RSW$ (and therefore that $\overline{QS}$ bisects both $\angle PQR$ and $\angle PSR$).

Therefore, $m\angle PQW = 49°$ and $m\angle PSW = 20°$. By the Triangle Sum Theorem, $m\angle QPS + m\angle PQW + m\angle PSW = 180°$. Substituting the known angles yields $m\angle QPS = 180° - (49° + 20°) = 111°$.

In kite $PQRS$, $m\angle RSW = 20°$, and $m\angle RQW = 49°$. Find the measure.

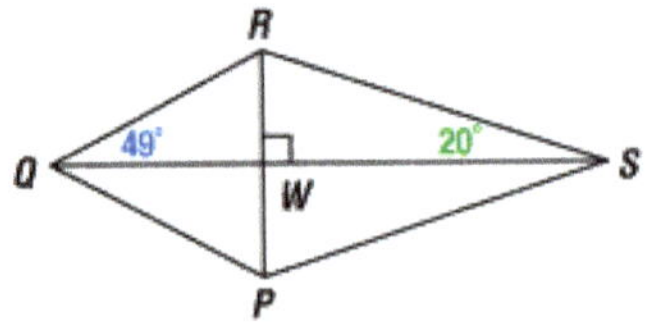

$m\angle QRW$

$\triangle RWQ$ is a right triangle, so $m\angle QRW + m\angle RQW = 90°$.

Substitute 49° for $m\angle RQW$ to find $m\angle QRW$.

$m\angle QRW = 90° - 49° = 41°$

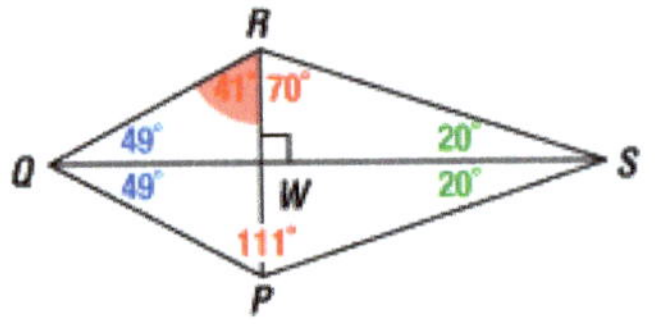

The measure of an angle in a kite, $PQRS$, is determined in this example. The measures of $\angle RQW$ and $\angle RSW$ are given.

Since $\overline{QS}$ and $\overline{PR}$ are perpendicular by the properties of kites, $\triangle RWQ$ is a right triangle. Therefore, $m\angle QRW + m\angle RQW = 90°$. Substitute 49° for $m\angle RQW$ to find $m\angle QRW$.

$m\angle QRW = 90° - 49° = 41°$.

Example 3 Using Properties of Isosceles Trapezoids

Theorem

If a quadrilateral is an isosceles trapezoid, then each pair of base angles is congruent.
(Isosc. trap. → base ∠s ≅)

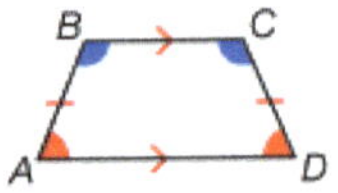

$\Rightarrow$ $\angle A \cong \angle D$, $\angle B \cong \angle C$

This is a property of an isosceles trapezoid. If a quadrilateral is an isosceles trapezoid, then each pair of base angles is congruent.

Theorem

A trapezoid is isosceles if and only if its diagonals are congruent.
(Isosc. trap. ↔ diags. ≅)

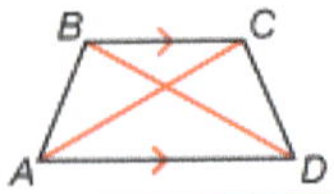

$\overline{AC} \cong \overline{DB}$
⇔
$ABCD$ is isosceles.

This is a property of an isosceles trapezoid. A trapezoid is isosceles if and only if its diagonals are congruent. Since this is a biconditional statement, it is true in both directions.

Find $m\angle C$.

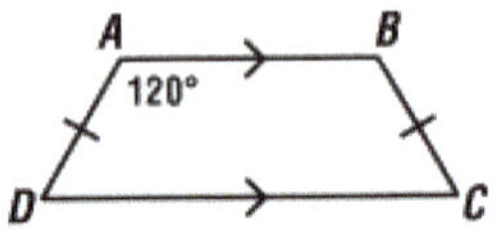

$m\angle A + m\angle D = 180°$	*Same-Side Int. ∠s Thm.*
$120 + m\angle D = 180$	*Substitute 120 for m∠A.*
$m\angle D = 60°$	*Subtract 120 from both sides.*
$\angle C \cong \angle D$	*Isosc. trap. → base ∠s ≅*
$m\angle C = m\angle D$	*Def. of ≅ ∠s*
$m\angle C = 60°$	*Substitute 60 for m∠D.*

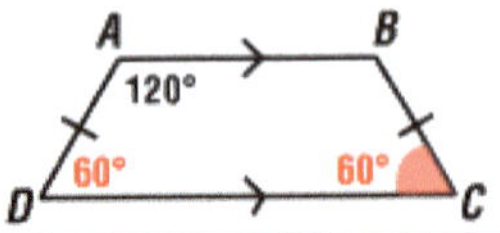

The measure of an angle in an isosceles trapezoid, $ABCD$, is determined in this example. The figure is given to be an isosceles trapezoid and the measure of $\angle DAB$ is given. It is also given that $\overline{AB} \parallel \overline{DC}$ and $\overline{AD} \cong \overline{BC}$.

Since $\overline{AB} \parallel \overline{DC}$, $m\angle A + m\angle D = 180°$ by the Same-Side Interior Angles Theorem. Substitute the given value for $m\angle A$ and solve to find $m\angle D = 60°$. By the Properties of Isosceles Trapezoids, $\angle C \cong \angle D$ and therefore $m\angle C = m\angle D = 60°$.

AC = 12.7 and BE = 7.1. Find ED.

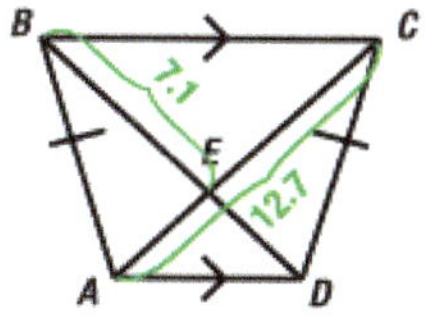

$\overline{BD} \cong \overline{AC}$	*Isosc. trap. → diags. ≅*
$BD = AC$	*Def. of ≅ segs.*
$BD = 12.7$	*Substitute 12.7 for AC.*
$BE + ED = BD$	*Seg. Add. Post.*
$7.1 + ED = 12.7$	*Substitute 7.1 for BE and 12.7 for BD.*
$ED = 5.6$	*Subtract 7.1 from both sides.*

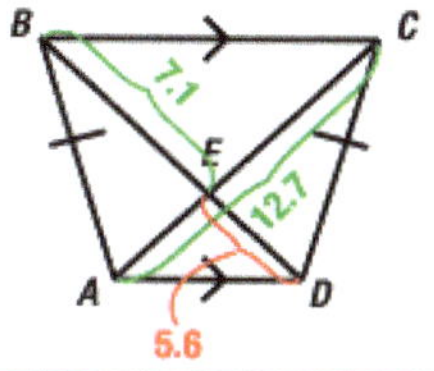

The length of half of a diagonal in an isosceles trapezoid, $ABCD$, is determined in this example. The figure is given to be an isosceles trapezoid. It is given that the diagonal $\overline{AC}$ has length 12.7 units, and $\overline{BE}$ has length 7.1 units. It is also given that $\overline{AD} \parallel \overline{BC}$ and $\overline{AB} \cong \overline{DC}$.

Because this is an isosceles trapezoid, the diagonals are congruent, and therefore $BD = 12.7$ by the definition of congruent line segments. By the Segment Addition Postulate, $BE + ED = BD$. Substituting gives $7.1 + ED = 12.7$. Solving gives $ED = 5.6$.

Example 4 Applying Conditions for Isosceles Trapezoids

Theorem

If a trapezoid has one pair of congruent base angles, then the trapezoid is isosceles.
(Trap. with pair base ∠s ≅ → isosc. trap.)

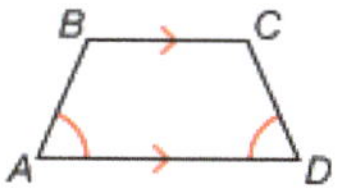

⇒ ABCD is isosceles.

This is a condition of isosceles trapezoids. If a trapezoid has one pair of congruent base angles, then it is an isosceles trapezoid.

Find the values of *a* so that *WXYZ* is isosceles.

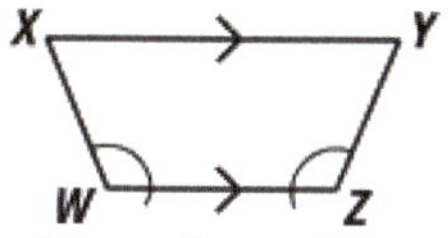

$\angle W \cong \angle Z$	*Trap. with pair base ∠s ≅ → isosc. trap.*
$m\angle W = m\angle Z$	*Def. of ≅ ∠s*
$3a^2 + 20 = 2a^2 + 45$	*Substitute $3a^2 + 20$ for m∠W, and $2a^2 + 45$ for m∠Z.*
$a^2 + 20 = 45$	*Subtract $2a^2$ from both sides.*
$a^2 = 25$	*Subtract 20 from both sides.*
$a = 5$ or $a = -5$	*Take the square root of both sides.*

The values of an unknown, a, in expressions for the base angles of a trapezoid that will make the trapezoid isosceles are determined in this example. It is given that two sides are parallel and the measures of two base angles are given as algebraic expressions with one unknown, a.

For the trapezoid to be isosceles, one pair of base angles must be congruent. Set $\angle W \cong \angle Z$ so that $m\angle W = m\angle Z$ and substitute the algebraic expressions for the angle measures. This gives $3a^2 + 20 = 2a^2 + 45$. Solving for a gives $a = 5$ or $a = -5$.

JL = 6x + 3, and KM = 2x + 31. Find the value of *x* so that *JKLM* is isosceles.

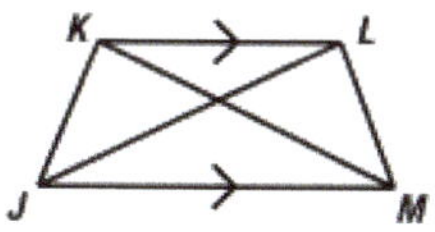

$\overline{JL} \cong \overline{KM}$	*Diags ≅ → isosc. trap.*
$JL = KM$	*Def. of ≅ segs.*
$6x + 3 = 2x + 31$	*Substitute 6x + 3 for JL and 2x + 31 for KM.*
$4x = 28$	*Subtract 2x from both sides and subtract 3 from both sides.*
$x = 7$	*Divide both sides by 4.*

The values of an unknown, x, in expressions for the lengths of the diagonals of a trapezoid that will make the trapezoid isosceles are determined in this example. It is given that two sides are parallel and the measures of the diagonals are given as algebraic expressions with one unknown, x.

For the trapezoid to be isosceles, the diagonals must be congruent. Set $\overline{JL} \cong \overline{KM}$ so that $JL = KM$ and substitute the algebraic expressions for the lengths. This gives $6x + 3 = 2x + 31$. Solving for the expression gives $x = 7$.

Example 5 Finding Lengths Using Midsegments

Trapezoid Midsegment Theorem

The midsegment of a trapezoid is parallel to each base, and its length is one half the sum of the lengths of the bases. (Trap. Midsegment Thm.)

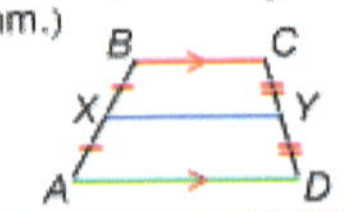

$\Rightarrow$ $\overline{XY} \parallel \overline{BC}, \overline{XY} \parallel \overline{AD}$

$XY = \frac{1}{2}(BC + AD)$

The midsegment of a trapezoid is parallel to each base and its length is one-half the sum of the lengths of the bases (the average of the base lengths).

Find *ED*.

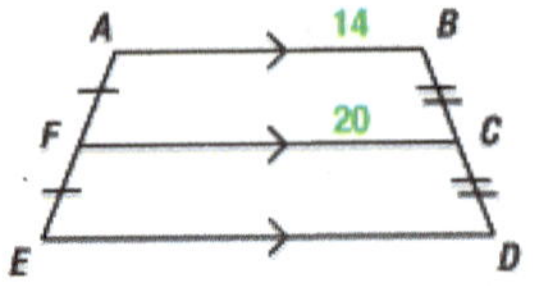

$FC = \frac{1}{2}(AB + ED)$	*Trap. Midsegment Thm.*
$20 = \frac{1}{2}(14 + ED)$	*Substitute the given values.*
$40 = 14 + ED$	*Multiply both sides by 2.*
$26 = ED$	*Subtract 14 from both sides.*

The length of a base of a trapezoid with a midsegment is determined in this example. The lengths of one base and the midsegment are given.

Use the formula for the length of the midsegment of a trapezoid to solve for the length of the unknown base. Substitute the given values for the length of the base and the length of the midsegment and solve for the length of the other base. The solution yields a length of 26 units.

7

7 Similarity

7.1 Similarity Relationships

Ratio and Proportion

Key Objectives

- Write and simplify ratios.
- Use proportions to solve problems.

Key Terms

- A **ratio** compares two numbers by division.
- A **proportion** is an equation stating that two ratios are equal.
- In the proportion $\frac{a}{b} = \frac{c}{d}$, the values a and d are the **extremes**. The values b and c are the **means**.
- The product of the extremes and the product of the means are called the **cross products**.

Theorems, Postulates, Corollaries, and Properties

- **Cross Products Property** In a proportion, if $\frac{a}{b} = \frac{c}{d}$ and b and $d \neq 0$, then $ad = bc$.
- **Properties of Proportions** The proportion $\frac{a}{b} = \frac{c}{d}$ is equivalent to the following:

$$ad = bc$$

$$\frac{b}{a} = \frac{d}{c}$$

$$\frac{a}{c} = \frac{b}{d}$$

Properties of Proportions

The proportion $\frac{a}{b} = \frac{c}{d}$ is equivalent to the following:

$$ad = bc$$

$$\frac{b}{a} = \frac{d}{c}$$

$$\frac{a}{c} = \frac{b}{d}$$

A proportion can be expressed in different ways that all produce the same cross products because they all express the same relationship between the factors. Substituting numbers for the variables will show that the different forms all express the same relationship.

Example 1 Writing Ratios

Write a ratio expressing the slope of ℓ.

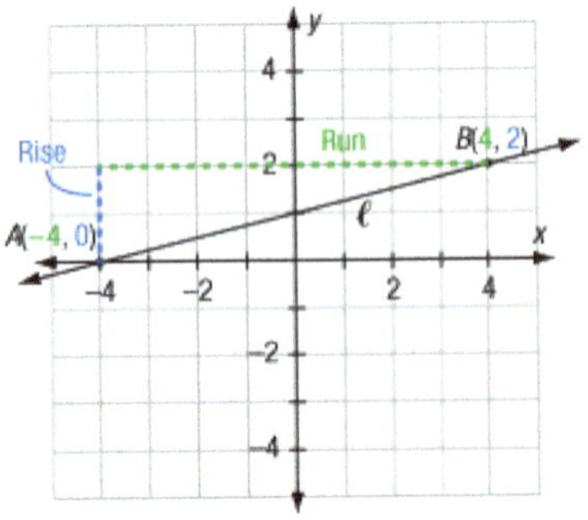

$$\text{slope} = \frac{y_2 - y_1}{x_2 - x_1}$$

$$= \frac{2 - 0}{4 - (-4)} \quad \textit{Substitute the given values.}$$

$$= \frac{2}{8} = \frac{1}{4} \quad \textit{Simplify.}$$

The slope of a line is expressed here as a ratio. The coordinates of the endpoints of the line are given.

The slope of a line is defined as the ratio of the rise of the line (change in y) to the run of the line (change in x). The slope is the difference in y values $(y_2 - y_1)$ over the difference in x values $(x_2 - x_1)$. To express the slope of the given line, substitute the appropriate coordinate values into the formula for slope and simplify.

The solution yields a slope of 1/4.

Example 2 Using Ratios

The ratio of the side lengths of a quadrilateral is 2:3:4:9, and its perimeter is 126 ft. What is the length of the longest side?

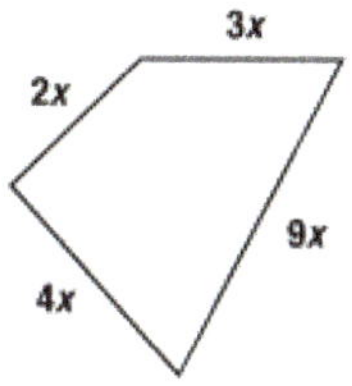

$s_1 + s_2 + s_3 + s_4 = 126$

$2x + 3x + 4x + 9x = 126$ *Let the side lengths be 2x, 3x, 4x, and 9x.*

$18x = 126$ *Combine like terms*

$x = 7$

Find the length of the longest side.

$9x = 9(7) = 63$

The length of the longest side is 63 feet.

The ratio of the lengths of the sides of a quadrilateral is used to solve for the length of the longest side in this example. It is given that the sides are in a ratio of 2:3:4:9 and that the perimeter is 126 ft.

To solve for the the actual length of the longest side, set x to equal the common factor of the side lengths. The actual side lengths can be expressed as the common factor times the ratio number. Write an equation that sets the sum of the side lengths equal to the perimeter and solve for the common factor x.

The solution yields $x = 7$ feet. This is not the answer to the example, which asked for the actual length of the longest side. The length of the longest side is $9x = 9(7) = 63$ feet.

Example 3 Solving Proportions

Solve the proportion.

Cross Products Property

In a proportion, if $\frac{a}{b} = \frac{c}{d}$ and b and $d \neq 0$, then $ad = bc$.

$\frac{3}{w} = \frac{60}{40}$

$\frac{3}{w} \times \frac{60}{40}$

$3(40) = (60)w$ *Cross Products Property*

$120 = 60w$ *Simplify.*

$w = 2$ *Divide both sides by 60.*

A proportion with an unknown is solved here for the value of the unknown. The given proportion is $\frac{3}{w} = \frac{60}{40}$.

To solve for the unknown, w, take the cross product of the proportion and solve for w. The solution yields $w = 2$, meaning that 3 is to 2 as 60 is to 40.

This is one of the most useful mathematical procedures to solve simple math questions in everyday life.

Solve the proportion.

Cross Products Property

In a proportion, if $\frac{a}{b} = \frac{c}{d}$ and b and $d \neq 0$, then $ad = bc$.

$\frac{x+3}{5} = \frac{20}{x+3}$

$(x + 3)^2 = 5(20)$ *Cross Products Property*

$(x + 3)^2 = 100$ *Simplify.*

$\sqrt{(x + 3)^2} = \sqrt{100}$

$x + 3 = \pm 10$ *Find the square root of both sides.*

$x + 3 = 10$ or $x + 3 = -10$ *Rewrite as two equations.*

$x = 7$ or $x = -13$ *Subtract 3 from both sides.*

A proportion with an unknown is solved for the value of the unknown in this example. The given proportion is

$\frac{x+3}{5} = \frac{20}{x+3}$.

To solve for the unknown, x, take the cross product of the proportion and solve for x. There are two possible solutions for the unknown: $x = 7$ or $x = -13$.

Example 4 Using Properties of Proportions

Given that $8x = 12y$, find the ratio of x to y in the simplest form.

$8x = 12y$

$\frac{8x}{y} = 12$ *Divide both sides by y.*

$\frac{x}{y} = \frac{12}{8}$ *Divide both sides by 8.*

$\frac{x}{y} = \frac{3}{2}$ *Simplify.*

The ratio of two unknowns in an algebraic equation is determined in this example.

To find the ratio of the two unknowns, manipulate the equation so that x and y are together in the form of a ratio. Simplifying yields the proportion $\frac{x}{y} = \frac{3}{2}$, which includes the ratio x to y.

Example 5 Problem-Solving Application

Jan made a model chair for a dollhouse with a height of 3 inches and a width of 1 inch. If the width of the full-size chair is 15 inches, what is its height?

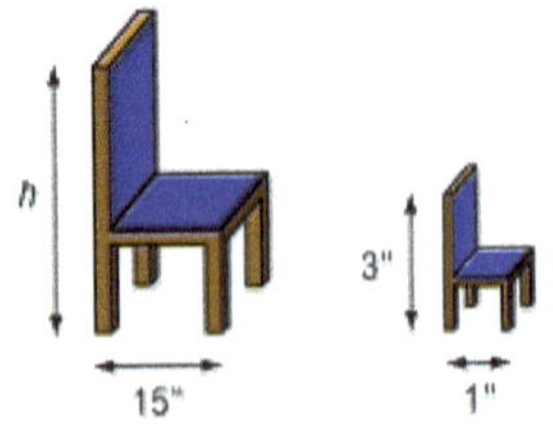

$$\frac{\text{height of doll house chair}}{\text{width of doll house chair}} = \frac{\text{height of full-size chair}}{\text{width of full-size chair}}$$

$$\frac{3}{1} = \frac{h}{15}$$

$\frac{3}{1} = \frac{h}{15}$

$h(1) = 3(15)$ *Cross Products Property*

$h = 45$ *Simplify.*

The height of the full-size chair is 45 inches.

A proportion with an unknown is solved for the value of the unknown in this application example. The given proportion is $\frac{3}{1} = \frac{h}{15}$.

The proportion sets the ratio of the height to the width of a model chair equal to the ratio of the unknown height to the width of a full-size chair.

To solve for the unknown height, h, take the cross product of the proportion and solve for h. The solution yields $h = 45$. The height of the full-size chair is 45 inches.

Ratios in Similar Polygons

Key Objectives

- Identify similar polygons.
- Apply properties of similar polygons to solve problems.

Key Terms

- Figures that are **similar** (~) have the same shape but not necessarily the same size.
- A **similarity ratio** is the ratio of the lengths of the corresponding sides of two similar polygons.
- Two polygons are **similar polygons** if and only if their corresponding angles are congruent and their corresponding sides are proportional.

Example 1 Describing Similar Polygons

Identify the pairs of congruent angles and corresponding sides.

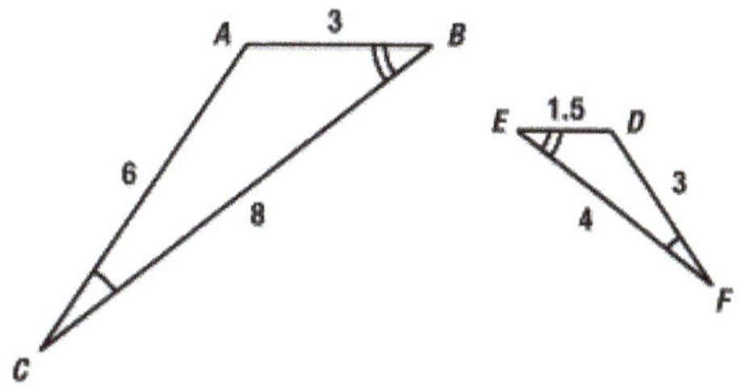

Congruent Angles

$\angle B \cong \angle E$ and $\angle C \cong \angle F$. By the Third Angles Theorem, $\angle A \cong \angle D$.

Congruent Sides

$\frac{AC}{DF} = \frac{6}{3} = 2$ $\frac{BC}{EF} = \frac{8}{4} = 2$ $\frac{AB}{ED} = \frac{3}{1.5} = 2$

The pairs of congruent sides and angles in two similar triangles are identified in this example. The lengths of the sides are given. Two pairs of angles are given as congruent.

According to the Third Angles Theorem, since $\angle B \cong \angle E$ and $\angle C \cong \angle F$, then the third angles, $\angle A$ and $\angle D$, are also congruent.

To determine whether the sides are congruent, calculate the ratios of the lengths of each hypothesized congruent pair. When setting up the ratios, make sure to put all the lengths from one triangle in the same position, on the top or on the bottom. These ratios are all equal to 2, so all three pairs of sides in the triangles are congruent.

Example 2 Identifying Similar Polygons

Determine whether rectangles *XYZW* and *ABCD* are similar. If so, write the similarity ratio and a similarity statement.

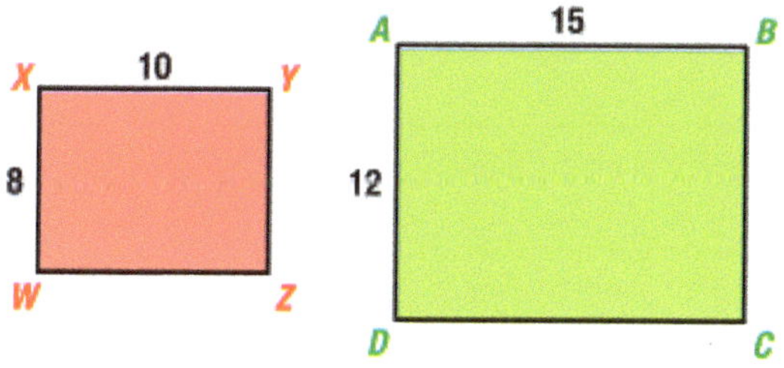

Since figures *XYZW* and *ABCD* are rectangles, all of the angles are right angles. So, corresponding angles are congruent.

Compare corresponding sides.

$\frac{AD}{XW} = \frac{12}{8} = \frac{3}{2}$

$\frac{AB}{XY} = \frac{15}{10} = \frac{3}{2}$

Thus, the similarity ratio is $\frac{3}{2}$, and rectangle $XYZW \sim$ rectangle $ABCD$.

In this example, two rectangles are compared to determine whether they are similar. The similarity ratio is calculated. It is given that both figures are rectangles. The lengths of two consecutive sides for each rectangle are given.

Since both figures are rectangles, all the angles are right angles and therefore congruent. To determine whether the sides are congruent, calculate the two ratios of side lengths. Both ratios of side lengths are 3/2, so the rectangles are similar (since these are rectangles, the other two sides are congruent pairs with the given sides). The similarity ratio for the two rectangles is the common ratio of the lengths of the sides, 3/2. Rectangle $XYZW \sim$ rectangle $ABCD$.

Determine whether $\triangle PQR$ and $\triangle STU$ are similar. If so, write the similarity ratio and a similarity statement.

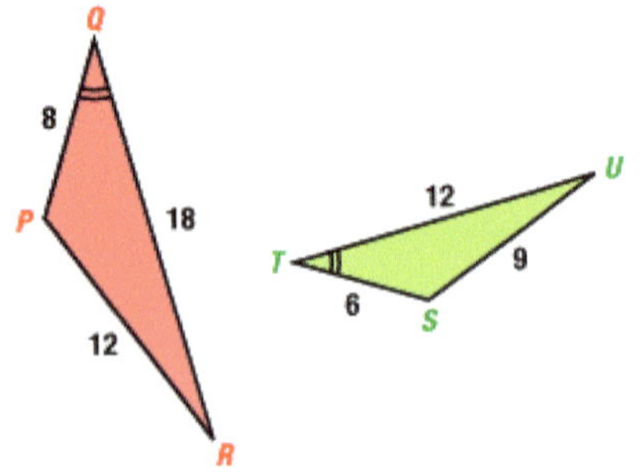

Compare the corresponding sides.

$\frac{PR}{SU} = \frac{12}{9} = \frac{4}{3}$ $\frac{PQ}{ST} = \frac{8}{6} = \frac{4}{3}$ $\frac{QR}{TU} = \frac{18}{12} = \frac{3}{2}$

Since corresponding sides are not proportional, the triangles are not similar.

In this example, two triangles are compared to determine whether they are similar. The similarity ratio is calculated. The lengths of all three sides for each rectangle are given. It is given that the triangles share one congruent angle.

To determine whether the triangles are similar, calculate the ratios of lengths of each hypothesized congruent pair of sides. Two of the sides have lengths in a ratio of 4/3. The third sides have lengths in a ratio of 3/2. Because the ratios are not the same, the sides are not all proportional, and these are not similar triangles.

Example 3 Architecture Application

The length of the kitchen in the blueprint is 5.5 inches and the width is 8.75 inches. The actual width of the kitchen is 12.25 feet. Find the actual length of the kitchen.

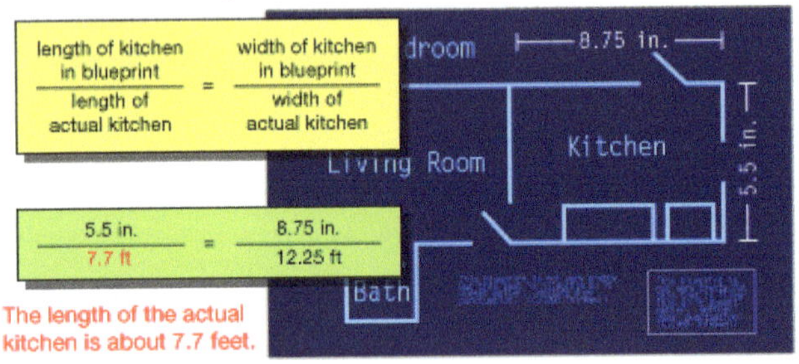

The length of the actual kitchen is about 7.7 feet.

$\frac{5.5}{x} = \frac{8.75}{12.25}$ *Let x be the length of the actual kitchen in feet.*

$8.75x = (5.5)(12.25)$ *Cross Products Property.*

$8.75x = 67.375$ *Simplify.*

$x \approx 7.7$ *Divide both sides by 8.75.*

A scale drawing of a rectangular room is compared with the actual rectangular room in this application example. The unknown length of a side of the actual room is determined. The lengths of two sides of the drawing of the rectangular room are given. The length of one side of the actual rectangular room is given.

To determine the unknown length of the side of the kitchen, set up the proportion using the ratios of the side lengths for the kitchen and the drawing. In this case, the proportion is: the length in the drawing is to the length of the kitchen (x) as the width in the drawing is to the width of the kitchen. Cross multiply the proportion and solve for the value of x, the length of the kitchen. The solution yields an approximate length of 7.7 feet for the length of the kitchen.

Triangle Similarity: AA, SSS, and SAS

Key Objectives

- Prove certain triangles are similar by using AA, SSS, and SAS.
- Use triangle similarity to solve problems.

Theorems, Postulates, Corollaries, and Properties

- **Angle-Angle Similarity Postulate** If two angles of one triangle are congruent to two angles of another triangle, then the triangles are similar.
- **Side-Side-Side Similarity Theorem** If the three sides of one triangle are proportional to the three corresponding sides of another triangle, then the triangles are similar.
- **Side-Angle-Side Similarity Theorem** If two sides of one triangle are proportional to two sides of another triangle and their included angles are congruent, then the triangles are similar.
- **Reflexive Property of Similarity** $\triangle ABC \sim \triangle ABC$
- **Symmetric Property of Similarity** If $\triangle ABC \sim \triangle DEF$, then $\triangle DEF \sim \triangle ABC$.
- **Transitive Property of Similarity** If $\triangle ABC \sim \triangle DEF$ and $\triangle DEF \sim \triangle XYZ$, then $\triangle ABC \sim \triangle XYZ$.

Example 1 Using the AA Similarity Postulate

Angle-Angle (AA) Similarity Postulate

If 2 $\angle$s of one $\triangle$ are $\cong$ to 2 $\angle$s of another $\triangle$, then the $\triangle$s are similar. (AA ~ Post.)

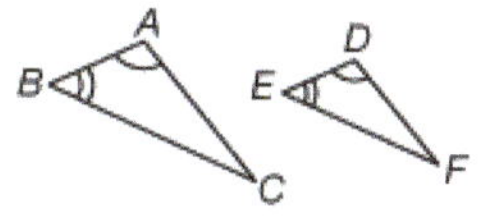

$\triangle ABC \sim \triangle DEF$

If two angles of one triangle are congruent to two angles of another triangle, then the triangles are similar.

Explain why the triangles are similar and write a similarity statement.

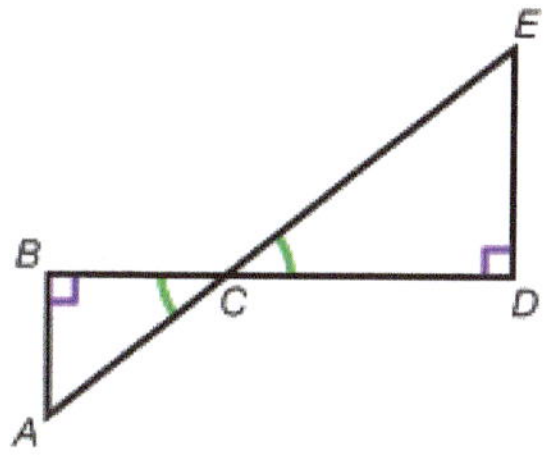

$\angle B$ and $\angle D$ are right angles, so $\angle B \cong \angle D$.

$\angle BCA \cong \angle DCE$ by the Vertical Angles Theorem.

Therefore, $\triangle ABC \sim \triangle EDC$ by AA~.

The reason why two triangles are similar is determined in this example. It is given that each triangle has a right angle.

Angles *B* and *D* are congruent because they are both right angles, as given. According to the Vertical Angles Theorem, $\angle BCA \cong \angle DCE$. Therefore, by Angle-Angle Similarity the two triangles are similar.

Example 2 Verifying Triangle Similarity

Side-Side-Side (SSS) Similarity Theorem

If 3 sides of one $\triangle$ are proportional to the 3 corresponding sides of another $\triangle$, then the $\triangle$s are similar. (SSS ~ Thm.)

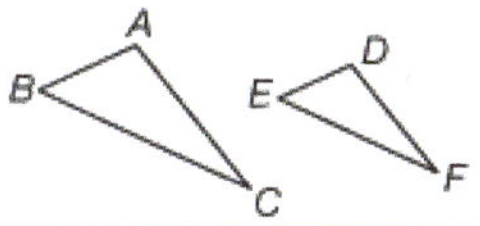

$$\frac{AB}{DE} = \frac{AC}{DF} = \frac{BC}{EF}$$

If three sides of one triangle are proportional to the three corresponding sides of another triangle, then the triangles are similar.

Verify that $\triangle PQS$ and $\triangle QRS$ are similar. 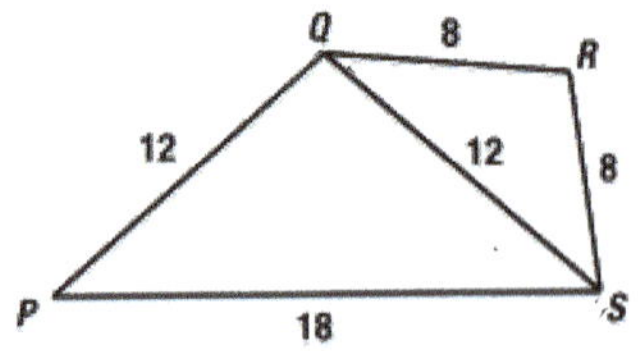 $\frac{PQ}{QR} = \frac{12}{8} = \frac{3}{2}$, $\frac{QS}{RS} = \frac{12}{8} = \frac{3}{2}$, $\frac{PS}{QS} = \frac{18}{12} = \frac{3}{2}$ $\triangle PQS \sim \triangle QRS$ by SSS~.	Two triangles are determined to be similar in this example. The lengths of the sides of the triangles are given. Calculate the ratios of the pairs of corresponding sides of the two triangles. Each of the ratios of the sides is 3/2, so the triangles are similar, according to the Side-Side-Side Similarity Theorem.
Side-Angle-Side (SAS) Similarity Theorem If 2 sides of one $\triangle$ are proportional to 2 sides of another $\triangle$ and their included angles are $\cong$, then the $\triangle$s are similar. (SAS ~ Thm.) 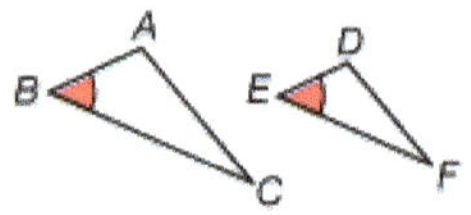 $\frac{AB}{DE} = \frac{BC}{EF}$ and $\angle B \cong \angle E \Rightarrow \triangle ABC \sim \triangle DEF$	If two sides of one triangle are proportional to two sides of another triangle and their included angles are congruent, then the triangles are similar.
Verify that $\triangle BDA$ and $\triangle CDE$ are similar. 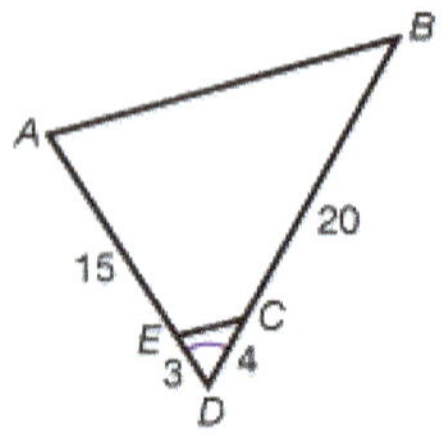 $\angle D \cong \angle D$ by the Reflexive Property of $\cong$. $\frac{AD}{ED} = \frac{18}{3} = 6$, $\frac{BD}{CD} = \frac{24}{4} = 6$ $\triangle BDA \sim \triangle CDE$ by SAS~.	Two triangles are determined to be similar in this example. The lengths of two corresponding sides of the triangles are given. The triangles share a common angle. By the Reflexive Property of Congruence, $\angle D \cong \angle D$. Calculate the ratios of the pairs of corresponding sides of the two triangles. Each of the two ratios of the sides is 6. $\angle D$ is the angle included between the two proportional sides, so the triangles are similar, according to the Side-Angle-Side Similarity Theorem.

Example 3 Finding Lengths in Similar Triangles

Explain why $\triangle PQR \sim \triangle TSR$ and then find SR. 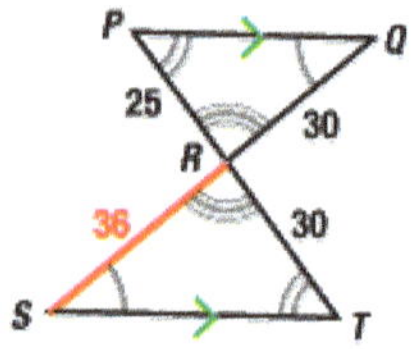 As shown, $PQ \parallel ST$. $\angle Q \cong \angle S$ by the Alternate Interior Angles Theorem, and similarly, $\angle P \cong \angle T$. By the Vertical Angles Theorem, $\angle PRQ \cong \angle TRS$. Therefore, $\triangle PQR \sim \triangle TSR$ by AA~ (using any two of the three congruent pairs of angles). $\triangle PQR \sim \triangle TSR$ by AA~. $\frac{SR}{QR} = \frac{TR}{PR}$ *Corr. sides are proportional.* $\frac{SR}{30} = \frac{30}{25}$ *Substitute 30 for QR, 30 for TR, and 25 for PR.* $\frac{SR}{30} = \frac{6}{5}$ *Simplify.* $5(SR) = 30(6)$ *Cross Products Prop.* $5(SR) = 180$ *Simplify.* $SR = 36$ *Divide both sides by 5.*	The length of an unknown side in a triangle is found in this example using the similarity ratio for two triangles. It is given that $\overline{PQ} \parallel \overline{ST}$, and the lengths PR, QR, and RT are given. According to the Alternate Interior Angles Theorem, $\angle Q \cong \angle S$ (because $\overline{PQ} \parallel \overline{ST}$). By the Vertical Angles Theorem, $\angle PRQ \cong \angle TRS$. Therefore, $\triangle PQR \sim \triangle TSR$ by Angle-Angle Similarity. Now, find the unknown side length SR by applying the similarity ratio for the triangles. Set up a proportion using the known corresponding lengths RT and PR and the corresponding lengths SR and QR. Substitute the known lengths of sides, cross multiply, and solve the resulting equation for the length of $\overline{SR}$. The solution yields $SR = 36$.

Example 4 Writing Proofs with Similar Triangles

Given:
Q is the midpoint of $\overline{PR}$.
T is the midpoint of $\overline{PS}$.

Prove: $\triangle PQT \sim \triangle PRS$

S: $PR = 2PQ$
A: $\angle P \cong \angle P$
S: $PS = 2PT$

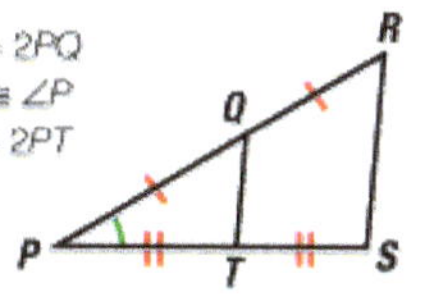

Statements:	Reasons:
1. Q is the midpoint of $\overline{PR}$. T is the midpoint of $\overline{PS}$.	1. Given
2. $\overline{PQ} \cong \overline{QR}$, $\overline{PT} \cong \overline{TS}$	2. Def. of mdpt.
3. $PQ = QR$, $PT = TS$	3. Def of $\cong$ seg.
4. $PR = PQ + QR$, $PS = PT + TS$	4. Seg. Add. Post.
5. $PR = PQ + PQ$, $PS = PT + PT$	5. Subst. Prop. *Steps 3, 4*
6. $PR = 2PQ$, $PS = 2PT$	6. Simplify.
7. $\frac{PR}{PQ} = 2$, $\frac{PS}{PT} = 2$	7. Div. Prop of =
8. $\frac{PR}{PQ} = \frac{PS}{PT}$	8. Trans. Prop. of =
9. $\angle P \cong \angle P$	9. Reflex. Prop. of $\cong$
10. $\triangle PQT \sim \triangle PRS$	10. SAS~ *Steps 8, 9*

Two triangles are proved to be similar in this example.

Since Q is the midpoint of $\overline{PR}$ and T is the midpoint of $\overline{PS}$ as given, then $PQ = QR$ and $PT = TS$. By the Segment Addition Postulate, $PR = PQ + QR$ and $PS = PT + TS$. Substituting gives $PR = PQ + PQ$ and $PS = PT + PT$. These equations are rearranged to find the ratios of two corresponding sides in the two triangles. Both these ratios are equal to 2. The angles between the proportional sides of the two triangles are congruent, because they are the same angle, P. Therefore, the triangles are similar by Side-Angle-Side Similarity.

Example 5 Problem-Solving Application

Here is a slice of cheese. $\overline{LC} \parallel \overline{SE}$, $SI = 10$ cm, $SE = 12$ cm, and $LC = 4.8$ cm. Use similar triangles to prove $\triangle SEI \sim \triangle LCI$, and then find LI.

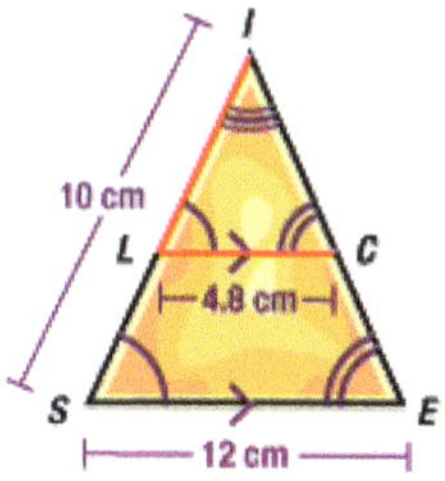

Statements:	Reasons:
1. $\overline{LC} \parallel \overline{SE}$	1. Given
2. $\angle ILC \cong \angle ISE$	2. Corr. $\angle$s Thm.
3. $\angle I \cong \angle I$	3. Reflexive Prop. of $\cong$
4. $\triangle SEI \cong \triangle LCI$	4. AA~ Thm.
5. $\frac{SI}{SE} = \frac{LI}{LC}$	5. Corr. sides are proportional.
6. $\frac{10}{12} = \frac{LI}{4.8}$	6. Substitute 10 for SI, 12 for SE, and 4.8 for LC.
7. $12(LI) = 4.8(10)$	7. Cross Products Prop.
8. $12(LI) = 48$	8. Simplify.
9. $LI = 4$ cm	9. Divide both sides by 12.

Two triangles are proved to be similar and the length of an unknown side is derived from the similarity ratio. It is given that $\overline{LC} \parallel \overline{SE}$. The lengths of $\overline{IS}$, $\overline{LC}$, and $\overline{SE}$ are given.

By the Corresponding Angles Theorem, $\angle ILC \cong \angle ISE$ (or $\angle ICL \cong \angle IES$). Because they are the same angle, $\angle I \cong \angle I$. Therefore, by Angle-Angle Similarity, $\triangle SEI \sim \triangle LCI$.

Find the unknown side length, LI, by applying the similarity ratio for the triangles. Set up a proportion using the known lengths SI and SE, and the corresponding lengths LI and LC. Substitute the known lengths of sides, cross multiply, and solve the resulting equation for the length of $\overline{LI}$. The solution yields $LI = 4$ cm.

8

8 Right Triangles and Trigonometry

8.1 Trigonometric Ratios

Similarity in Right Triangles

Key Objectives

- Use geometric mean to find segment lengths in right triangles.
- Apply similarity relationships in right triangles to solve problems.

Key Terms

- The **geometric mean** of two positive numbers is the positive square root of their product.

Theorems, Postulates, Corollaries, and Properties

- **Theorem** The altitude to the hypotenuse of a right triangle forms two triangles that are similar to each other and to the original triangle.
- **Corollary** The length of the altitude to the hypotenuse of a right triangle is the geometric mean of the lengths of the two segments of the hypotenuse.
- **Corollary** The length of a leg of a right triangle is the geometric mean of the lengths of the hypotenuse and the segment of the hypotenuse adjacent to that leg.

Example 1 Identifying Similar Right Triangles

Theorem

The altitude to the hypotenuse of a right triangle forms two triangles that are similar to each other and to the other triangle.

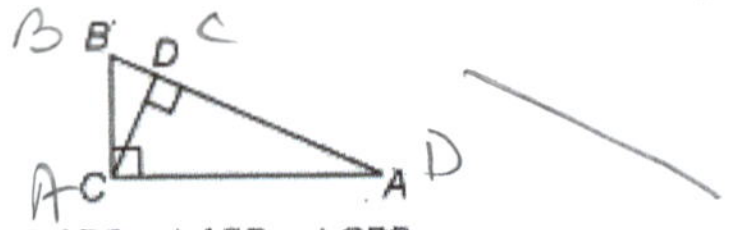

$\triangle ABC \sim \triangle ACD \sim \triangle CBD$

The altitude to the hypotenuse of a right triangle forms two triangles that are similar to each other and to the other triangle.

The altitude to $\triangle ABC$ forms a right angle with the hypotenuse.

Write a similarity statement comparing the three triangles.

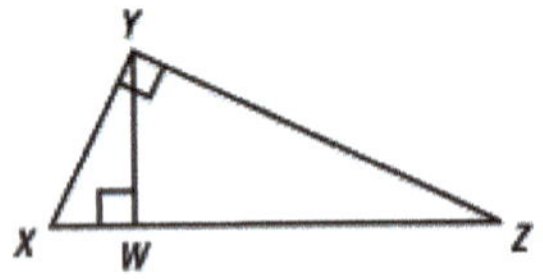

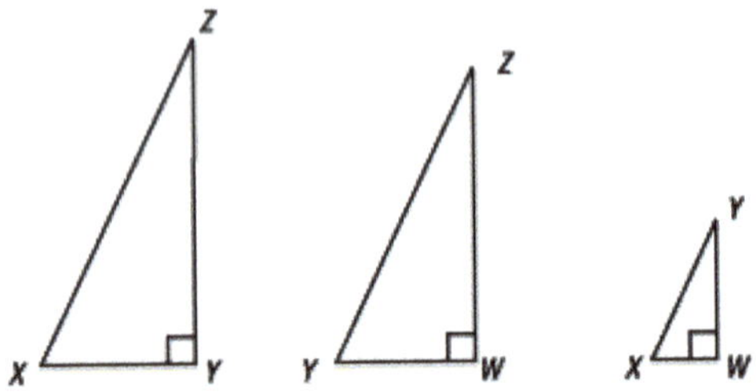

By the Similarity in Right Triangles Theorem,
$\triangle XYZ \sim \triangle YWZ \sim \triangle XWY$

The altitude $\overline{YW}$ forms two smaller triangles, $\triangle XWY$ and $\triangle YWZ$, from the orginal triangle, $\triangle XYZ$.

The three triangles are similar. This similarity is expressed as $\triangle XYZ \sim \triangle YWZ \sim \triangle XWY$.

Example 2 Finding Geometric Means

Find the geometric mean of each pair of numbers. If necessary, give the answer in the simplest radical form.

8 and 2

Let x be the geometric mean.

$x = \sqrt{8 \cdot 2}$	*Def. of geometric mean*
$= \sqrt{16} = 4$	*Simplify the square root.*

5 and 15

Let x be the geometric mean.

$x = \sqrt{5 \cdot 15}$	*Def. of geometric mean*
$= \sqrt{5 \cdot 5 \cdot 3}$	*Factor the terms inside the square root.*
$= \sqrt{5^2}\sqrt{3}$	
$= 5\sqrt{3}$	*Simplify the square root.*

The geometric means of pairs of numbers are calculated in these examples.

The geometric mean of two numbers is the positive square root of their product.

Example 3 Finding Side Lengths in Right Triangles

Geometric Means Corollaries

The length of the altitude to the hypotenuse of a right triangle is the geometric mean of the lengths of the two segments of the hypotenuse.

Example: $h^2 = xy$

The length of a leg of a right triangle is the geometric mean of the lengths of the hypotenuse and the segment of the hypotenuse adjacent to that leg.

Examples: $a^2 = xc$
$b^2 = yc$

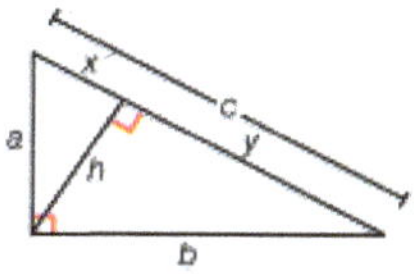

There are two corollaries that describe length relationships in triangles formed by dropping an altitude line in a right triangle.

The length of the altitude to the hypotenuse of a right triangle is the geometric mean of the lengths of the two segments of the hypotenuse.

The length of a leg of a right triangle is the geometric mean of the lengths of the hypotenuse and the segment of the hypotenuse adjacent to that leg.

Find *x*, *y*, and *z*.

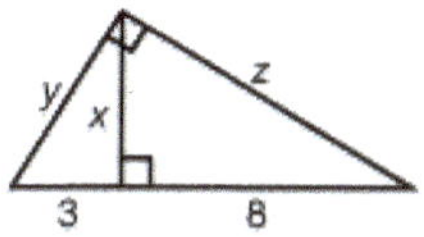

$\frac{y}{11} = \frac{3}{y}$	*Corr. sides of ~ △s are prop.*
$y^2 = 3(11)$	*y is the geometric mean of 3 and 11.*
$y = \sqrt{3(11)}$	*Find the positive square root.*
$= \sqrt{33}$	*Simplify.*
$\frac{x}{3} = \frac{8}{x}$	*Corr. sides of ~ △s are prop.*
$x^2 = 3(8)$	*x is the geometric mean of 3 and 8.*
$x = \sqrt{3(8)}$	*Find the positive square root.*
$= \sqrt{3 \cdot 2 \cdot 2 \cdot 2}$	*Factor the terms inside the square root.*
$= \sqrt{2^2}\sqrt{3 \cdot 2} = 2\sqrt{6}$	*Simplify.*
$\frac{\sqrt{33}}{3} = \frac{z}{2\sqrt{6}}$	*Corr. sides of ~ △s are prop.*
$3z = 2\sqrt{6}\sqrt{33}$	*Cross Prod. Prop.*
$3z = 2\sqrt{2 \cdot 3 \cdot 3 \cdot 11}$	*Multiply the square roots, and factor the terms inside.*
$3z = 6\sqrt{22}$	*Simplify.*
$z = 2\sqrt{22}$	*Divide both sides by 3.*

The lengths of the altitude and two sides of a right triangle are calculated using the geometric mean corollaries. The lengths of the two segments of the hypotenuse formed by the altitude are given.

To find the value of y, set up a proportion using ratios involving y and a known side in two of the similar triangles. For example, the ratio of y to the hypotenuse of the original triangle is $y/(8+3) = y/11$. Cross multiply and solve for the value of y. The solution yields the positive square root of 33 as the value of y.

Use similar proportions to find that $x = 2\sqrt{6}$ and $z = 2\sqrt{22}$.

Example 4 Measurement Application

To estimate the height of a statue, Holly steps away from the statue until her line of sight to the top of the statue and her line of sight to the bottom of the statue form a 90° angle. Her eyes are 5 ft 6 in. above the ground, and she is standing 25 ft from the statue. How tall is the statue to the nearest foot?

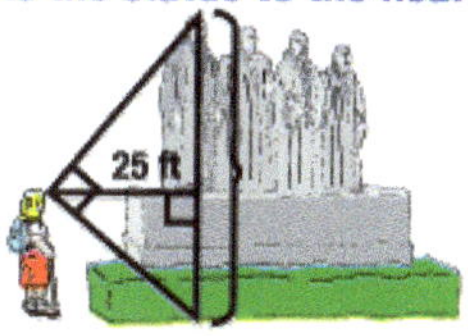

Let x be the height of the statue above eye level.

5 ft 6 in. = 5.5 ft	*Convert 6 in. to 0.5 ft.*
$\frac{25}{5.5} = \frac{x}{25}$	*Corr. sides of ~ △s are prop.*
$5.5x = (25)^2$	*25 is the geometric mean of 5.5 and x.*
$5.5x = 625$	*Simplify.*
$x = 113.\overline{63} \approx 113.6$ ft	*Solve for x and round.*

The statue is about 113.6 + 5.5 = 119.1 ft, or approximately 119 ft tall.

The height of an object is calculated from ground measurements in this application example. The distance from the observer to the statue is given, as is her height.

Recognize that the lines form a right triangle with an altitude. Let x be the height of the statue above eye-level, or the length of the upper line segment formed by the intersection with the altitude of the large right triangle. Set up a proportion using the ratios of corresponding sides of the similar triangles that includes the unknown length x. Cross multiply and solve for the value of x.

Notice that the equation resulting from cross multiplying is the equation for the geometric mean of the lengths of the two line segments formed by the altitude. Solving for x yields an approximate value of 113.6 ft. The statue height includes the height of the observer, so 113.6 + 5.5 = 119.1 feet.

Trigonometric Ratios

Key Objectives

- Find the sine, cosine, and tangent of an acute angle.
- Use trigonometric ratios to find side lengths in right triangles and to solve real-world problems.

Key Terms

- A **trigonometric ratio** is a ratio of two sides of a right triangle.
- The **sine** of an angle is the ratio of the length of the leg opposite the angle to the length of the hypotenuse.
- The **cosine** of an angle is the ratio of the length of the leg adjacent to the angle to the length of the hypotenuse.
- The **tangent** of an angle is the ratio of the length of the leg opposite the angle to the length of the leg adjacent to the angle.

Example 1 Finding Trigonometric Ratios

Trigonometric Ratios

Sine

$\sin A = \frac{\text{opposite leg}}{\text{hypotenuse}} = \frac{a}{c}$

$\sin B = \frac{\text{opposite leg}}{\text{hypotenuse}} = \frac{b}{c}$

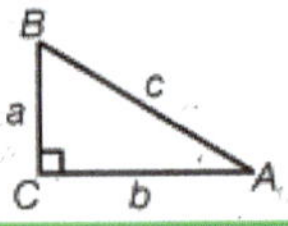

The sine of an angle is the ratio of the length of the leg opposite the angle to the length of the hypotenuse.

Cosine

$\cos A = \frac{\text{adjacent leg}}{\text{hypotenuse}} = \frac{b}{c}$

$\cos B = \frac{\text{adjacent leg}}{\text{hypotenuse}} = \frac{a}{c}$

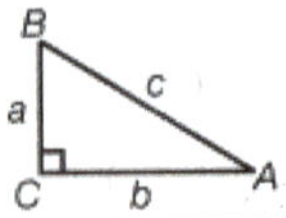

The cosine of an angle is the ratio of the length of the leg adjacent to the angle to the length of the hypotenuse.

Tangent

$\tan A = \frac{\text{opposite leg}}{\text{adjacent leg}} = \frac{a}{b}$

$\tan B = \frac{\text{opposite leg}}{\text{adjacent leg}} = \frac{b}{a}$

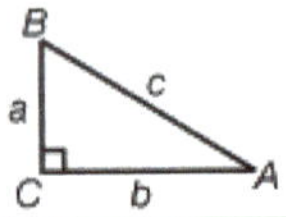

The tangent of an angle is the ratio of the length of the leg opposite the angle to the length of the leg adjacent to the angle.

Write each trigonometric ratio as a fraction and as a decimal rounded to the nearest hundredth.

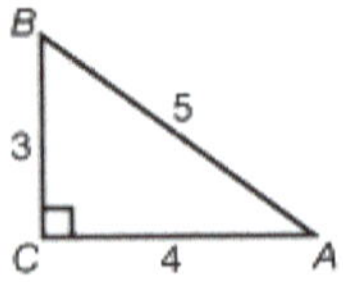

$\sin A = \frac{3}{5} = 0.6$ *The sine of an ∠ is* $\frac{\text{opp. leg}}{\text{hyp.}}$.

$\cos A = \frac{4}{5} = 0.8$ *The cosine of an ∠ is* $\frac{\text{adj. leg}}{\text{hyp.}}$.

$\tan B = \frac{4}{3} = 1.333... \approx 1.33$ *The tangent of an ∠ is* $\frac{\text{opp. leg}}{\text{adj. leg}}$.

Three trigonometric ratios for angles in a 3-4-5 right triangle are written in this example.

Sin A is the length of the opposite leg to $\angle A$ over the length of the hypotenuse of the triangle, or $3/5 = 0.6$.

Cos A is the length of the adjacent leg to $\angle A$ over the length of the hypotenuse of the triangle, or $4/5 = 0.8$.

Tan B is the length of the opposite leg to $\angle B$ over the length of the adjacent leg, or $4/3 \approx 1.33$.

Example 2 Finding Trigonometric Ratios in Special Right Triangles

Use a special right triangle to write tan 30° as a fraction.

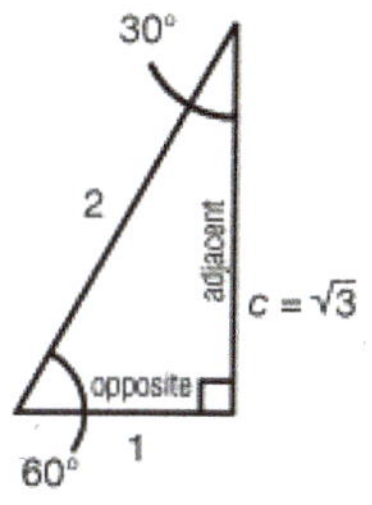

$1^2 + c^2 = 2^2$ *Pythagorean Thm.*

$1 + c^2 = 4$ *Simplify.*

$c^2 = 3$ *Subtract 1 from both sides.*

$c = \sqrt{3}$ *Take the positive square root.*

TIP

c is sometimes used to represent the length of the hypotenuse. In this case, it represents the length of a leg.

$\tan 30° = \frac{1}{\sqrt{3}} = \frac{\sqrt{3}}{3}$ *The tangent of an ∠ is* $\frac{\text{opp. leg}}{\text{adj. leg}}$.

Remember:

$\frac{1}{\sqrt{3}} = \frac{1}{\sqrt{3}} \cdot \frac{\sqrt{3}}{\sqrt{3}} = \frac{\sqrt{3}}{3}$

The tangent of the 30° angle of a special right triangle (30°-60°-90°) is written as a fraction in this example.

Recall that the lengths of the legs of a 30°-60°-90° right triangle are 1 and $\sqrt{3}$, and the length of the hypotenuse is 2.

The tangent of an angle is the ratio of the length of the opposite side over the length of the adjacent side.

$\text{Tan } 30° = \frac{1}{\sqrt{3}} = \frac{\sqrt{3}}{3}.$

Example 3 Calculating Trigonometric Ratios

Use your calculator to find each trigonometric ratio. Round to the nearest hundredth.

cos 51°	sin 29°	tan 78°
cos(51°) .629320391	sin(29°) .4848096202	tan(78°) 4.704630109
cos 51° ≈ 0.63	sin 29° ≈ 0.48	tan 78° ≈ 4.70

A calculator is used to calculate trigonometric ratios in this example.

Make sure the calculator is set to calculate ratios for angles expressed in degrees and not radians.

To calculate the cosine of 51°, enter 51 on the keypad and push the "cos" key. Round the result to the nearest hundredth as cos 51° ≈ 0.63.

To calculate the sine of 29°, enter 29 and push the "sin" key. Express the result to the nearest hundredth as sin 29° ≈ 0.48.

To calculate the tangent of 78°, enter 78 and push the "tan" key. Express the result to the nearest hundredth as tan 78° ≈ 4.70.

Example 4 Using Trigonometric Ratios to Find Lengths

Find each length. Round to the nearest hundredth.

PQ

$\overline{PQ}$ is adjacent to the given angle, ∠*P*. You are given the length of $\overline{QR}$, which is opposite ∠*P*.

Since the adjacent and opposite legs are involved, use a tangent ratio.

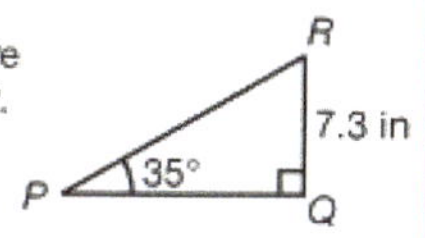

$\tan P = \frac{\text{opp. leg}}{\text{adj. leg}} = \frac{QR}{PQ}$ *Write a trigonometric ratio.*

$\tan 35° = \frac{7.3}{PQ}$ *Substitute the given values.*

$PQ = \frac{7.3}{\tan 35°}$ *Multiply both sides by PQ and divide by tan 35°.*

$PQ \approx 10.43$ in. *Simplify the expression.*

```
tan(35°)
    .7002075382
7.3/Ans
    10.42548045
```

The length of a side in a right triangle is obtained here using a trigonometric ratio. One angle is given as 35° and the length of its opposite side is given as 7.3 inches.

To find *PQ*, recognize that the known length is opposite the known angle, and that the unknown length is adjacent to the known angle. Thus, the tangent is the appropriate trigonometric ratio to use.

Write the trigonometric ratio for the tangent of 35° and substitute the given values and the unknown. Isolate the unknown, *PQ*, and use a calculator to find the value of tan 35° and solve for *PQ*. The value found for *PQ* is approximately 10.43 inches.

Find each length. Round to the nearest hundredth.

AB

$\overline{AB}$ is adjacent to the given angle, $\angle B$. You are given the length of $\overline{BC}$, which is the hypotenuse.

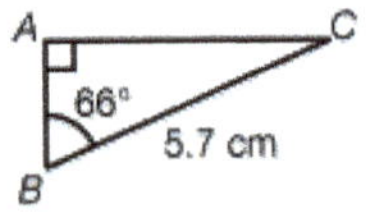

Since the adjacent side and hypotenuse are involved, use a cosine ratio.

$\cos B = \frac{\text{adj. leg}}{\text{hyp.}} = \frac{AB}{BC}$ *Write a trigonometric ratio.*

$\cos 66° = \frac{AB}{5.7}$ *Substitute the given values.*

$AB = 5.7(\cos 66°)$ *Multiply both sides by 5.7.*

$AB \approx 2.32$ cm *Simplify the expression.*

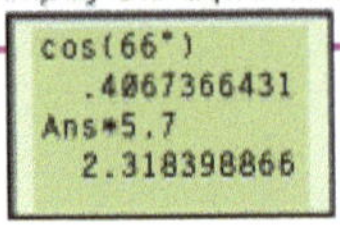

The length of a side in a right triangle is obtained here using a trigonometric ratio. One angle is given as 66° and the length of the hypotenuse is given as 5.7 cm.

To find *AB*, recognize that the known length is of the hypotenuse, and that the unknown length is of the side adjacent to the known angle. Thus, the cosine is the appropriate trigonometric ratio to use.

Write the trigonometric ratio for the cosine of 66° and substitute the given values and the unknown. Isolate the unknown, *AB*, and use a calculator to find the value of cos 66° and solve for *AB*. The value found for *AB* is approximately 2.32 cm.

Find each length. Round to the nearest hundredth.

YZ

$\overline{YZ}$ is the hypotenuse. You are given the length of $\overline{XZ}$, which is opposite the given angle, $\angle Y$.

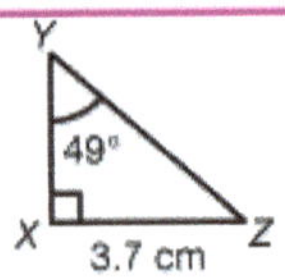

Since the opposite side and hypotenuse are involved, use a sine ratio.

$\sin Y = \frac{\text{opp. leg}}{\text{hyp.}} = \frac{XZ}{YZ}$ *Write a trigonometric ratio.*

$\sin 49° = \frac{3.7}{YZ}$ *Substitute the given values.*

$YZ = \frac{3.7}{\sin 49°}$ *Multiply both sides by YZ and divide by sin 49°.*

$YZ \approx 4.90$ cm *Simplify the expression.*

The length of a side in a right triangle is obtained here using a trigonometric ratio. One angle is given as 49° and the length of its opposite side is given as 3.7 cm.

To find *YZ*, recognize that the known length is of the side opposite the known angle, and that the unknown length is of the hypotenuse. Thus, the sine is the appropriate trigonometric ratio to use.

Write the trigonometric ratio for the sine of 49° and substitute the given values and the unknown. Isolate the unknown, *YZ*, and use a calculator to find the value of sin 49° and solve for *YZ*. The value found for *YZ* is approximately 4.90 cm.

Example 5 Problem-Solving Application

A contractor is building a wheelchair ramp for a doorway that is 1.6 ft above the ground. To meet ADA guidelines, the ramp will make an angle of 4.1° with the ground. To the nearest hundredth of a foot, what is the horizontal distance covered by the ramp?

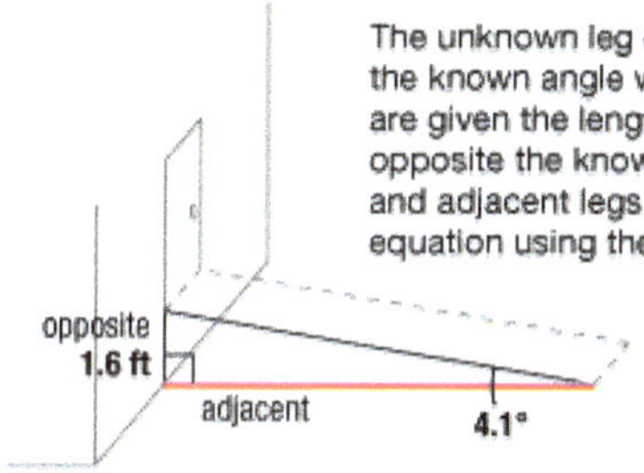

The unknown leg of length x is adjacent to the known angle with measure 4.1°. You are given the length of the other leg, which is opposite the known angle. Since the opposite and adjacent legs are involved, write an equation using the tangent ratio.

Let x be the length of the unknown leg.

$\tan 4.1° = \frac{\text{opp. leg}}{\text{adj. leg}}$ *Write a trigonometric ratio.*

$\tan 4.1° = \frac{1.6}{x}$ *Substitute the given values.*

$x = \frac{1.6}{\tan 4.1°}$ *Multiply both sides by x and divide by tan 4.1°.*

$x \approx 22.32$ ft *Simplify the expression.*

```
tan(4.1°)
   .0716808913
1.6/Ans
   22.32115102
```

The length of one leg of a right triangle is determined using a trigonometric ratio in this application example. The triangle represents a ramp and the angle with the ground is given as 4.1°, while the length of the rise is given as 1.6 ft.

The unknown length, x, is the distance along the ground of the ramp, which is the length of the side adjacent to the known angle. The side with known length is opposite the known angle. Thus, the tangent is the appropriate ratio to use.

Write the trigonometric ratio for the tangent of 4.1° and substitute the given values and the unknown, x. Isolate the unknown, and use a calculator to find the value of tan 4.1° and solve for x. The solution yields $x \approx 22.32$ ft.

Solving Right Triangles

Key Objectives

- Use trigonometric ratios to find angle measures in right triangles and to solve real-world problems.

Example 1 Identifying Angles from Trigonometric Ratios

Use the trigonometric ratio $\cos A = \frac{4}{5}$ to determine which angle of the triangle is $\angle A$.

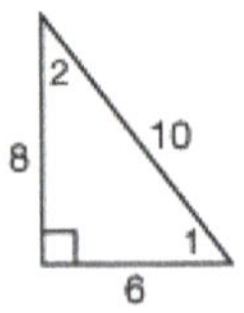

$\cos A = \frac{\text{adj. leg}}{\text{hyp.}}$ *Cosine is the ratio of the adjacent leg to the hypotenuse.*

$\cos \angle 1 = \frac{6}{10} = \frac{3}{5}$ $\frac{3}{5} \neq \frac{4}{5}$ ✗ *The leg adjacent to ∠1 is 6. The hypotenuse is 10.*

$\cos \angle 2 = \frac{8}{10} = \frac{4}{5}$ $\frac{4}{5} = \frac{4}{5}$ ✓ *The leg adjacent to ∠2 is 8. The hypotenuse is 10.*

Since $\cos A = \cos \angle 2$, $\angle 2$ is $\angle A$.

An angle is identified from its trigonometric ratio in this example. The lengths of the legs and hypotenuse of the triangle are given. The cosine of the angle is given as 4/5.

The cosine of $\angle 1$ is 6/10, or 3/5.

The cosine of $\angle 2$ is 8/10, or 4/5.

The calculated cosine of $\angle 2$ matches the cosine of $\angle A$, so $\angle A$ is $\angle 2$.

Example 2 Calculating Angle Measures from Trigonometric Ratios

Inverse Trigonometric Functions

If $\sin A = x$, then $\sin^{-1} x = m\angle A$.
If $\cos A = x$, then $\cos^{-1} x = m\angle A$.
If $\tan A = x$, then $\tan^{-1} x = m\angle A$.

Inverse trigonometric functions are equal to the measure of the angle that has a given trigonometric ratio.

For example, if $\sin 30° = 0.50$, then $\sin^{-1} 0.5 = 30°$.

Use your calculator to find each angle measure to the nearest hundredth of a degree.

$\sin^{-1}(0.5)$ $\cos^{-1}(0.25)$ $\tan^{-1}(5.1)$

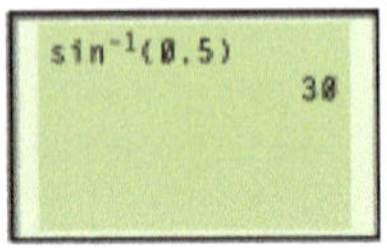

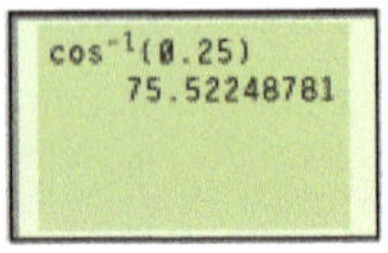

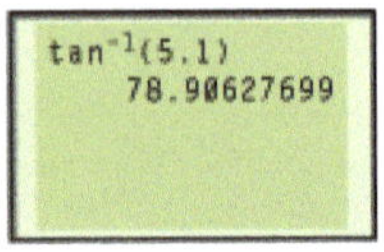

Tip #1: Make sure your calculator is set to show degrees.

Tip #2: When entering $\cos^{-1}$ on your calculator, do not use the x^{-1} key.

Tip #3: The answer you see is the measure of an angle, in degrees.

$\sin^{-1}(0.5) = 30°$ $\cos^{-1}(0.25) \approx 75.52°$ $\tan^{-1}(5.1) \approx 78.91°$

A calculator is used to calculate inverse trigonometric functions in this example.

Make sure the calculator is set to calculate with angles in degrees and not radians.

To calculate the inverse sine of 0.5, enter 0.5 on the keypad and push the "$\sin^{-1}$" key. This key may also be labeled "asin" or "arcsine," as this is another name for the inverse sine. "Arccosine" and "arctangent" are the other names for the other ratios. The calculator returns the value of 30, which is the answer in degrees. Thus, $\sin^{-1}(0.5) = 30°$.

The inverse cosine of 0.25 is the measure of the angle whose cosine is 0.25. Enter 0.25 on the keypad and push "$\cos^{-1}$." The result, rounded to hundredths, is 75.52°. Similarly, $\tan^{-1}(5.1) = 78.91°$, rounded to hundredths.

Example 3 Solving Right Triangles

Find the unknown measures. Round lengths to the nearest hundredth and angle measures to the nearest degree.

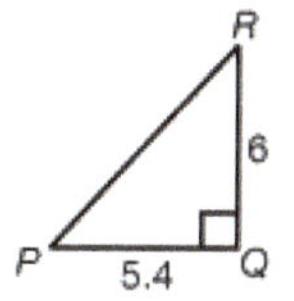

PR

Let q represent the length of side $\overline{PR}$.
By the Pythagorean Theorem,

$PR^2 = PQ^2 + QR^2$

$q^2 = 6^2 + (5.4)^2$ *Substitute the given values.*

$q^2 = 65.16$ *Simplify.*

$q \approx 8.07$ *Take the positive square root, and round the answer to the nearest hundredth.*

m∠P and m∠R

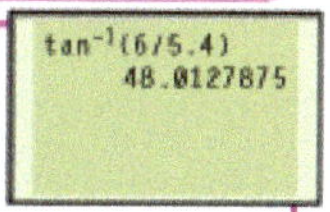

$\tan P = \frac{6}{5.4}$ *$\tan P = \frac{\text{opp.}}{\text{adj.}}$*

$m\angle P = \tan^{-1}\left(\frac{6}{5.4}\right) \approx 48°$ *Take the inverse tangent of both sides.*

Since the acute angles of a right triangle are complementary, $m\angle R \approx 90° - 48° = 42°$.

In this example, the length of a hypotenuse is calculated by the Pythagorean Theorem, and the measures of two unknown angles in a right triangle are calculated using inverse trigonometric ratios. The lengths of the two legs of the right triangle are given.

To find the length of the hypotenuse of the triangle, substitute the lengths of the legs as a and b into $a^2 + b^2 = c^2$ and solve for c, the length of the hypotenuse. The solution yields a length of approximately 8.07 units.

To find m∠P, recognize tan P = 6/5.4, and that m∠P = $\tan^{-1}(6/5.4) \approx 48°$. Since ∠$R$ is the supplement of ∠P in the right triangle, its measure is approximately $90° - 48° = 42°$.

Example 4 Solving a Right Triangle in the Coordinate Plane

The coordinates of the vertices of △JKL are J(2, 7), K(2, –5), and L(–3, –5). Find the side lengths to the nearest hundredth and the angle measures to the nearest degree.

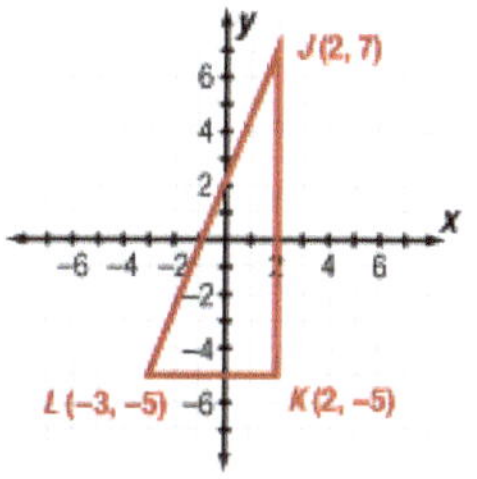

Find the side lengths.
$KL = 5$ $JK = 12$

By the Distance Formula,

$JL = \sqrt{(2-(-3))^2 + (7-(-5))^2}$
$= \sqrt{5^2 + 12^2}$
$= \sqrt{25 + 144}$
$= \sqrt{169} = 13$

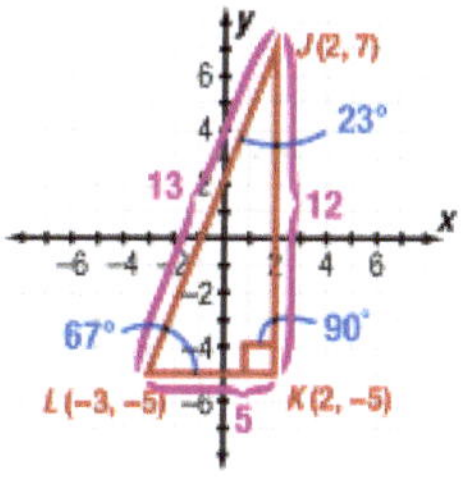

Find the angle measures.

$m\angle K = 90°$ *$\overline{JK}$ and $\overline{KL}$ are ⊥.*

$\tan L = \frac{12}{5}$ *Write a trigonometric ratio for L. $\overline{JK}$ is opp. ∠L, and $\overline{KL}$ is adj. to ∠L.*

$m\angle L = \tan^{-1}\left(\frac{12}{5}\right) \approx 67°$ *Use a calculator to solve for m∠L.*

$m\angle J \approx 90° - 67°$ *The acute ∠s of a rt. △ are comp.*

$m\angle J \approx 23°$

The side lengths and angle measures of a triangle in the coordinate plane are determined in this example. The coordinates of the vertices are given.

The lengths of the two legs of the triangle can be obtained by subtracting the x coordinates of the endpoints for the horizontal leg and subtracting the y coordinates of the endpoints for the vertical leg. The length of the hypotenuse is calculated using the Distance Formula. The lengths are $KL = 5$, $JK = 12$, and $JL = 13$; this is a Pythagorean triple.

Since $\overline{JK}$ and $\overline{KL}$ are perpendicular (they are each parallel to an axis of the coordinate plane), m∠K = 90°. To obtain the measure of ∠L, take the inverse tangent of the ratio of the opposite leg over the adjacent leg: m∠L = $\tan^{-1}(12/5) \approx 67°$. The final angle, ∠$J$, is complementary to ∠$L$ because they are the acute angles in a right triangle:
m∠J = $90° - 67° \approx 23°$.

Example 5 Problem-Solving Application

If a hill has a percent grade of 52%, what angle does the hill make with a horizontal line? Round to the nearest degree.

$52\% = \frac{52}{100}$ *Change the percent grade to a fraction.*

A 52% grade means the hill rises 52 ft for every 100 ft of horizontal distance.

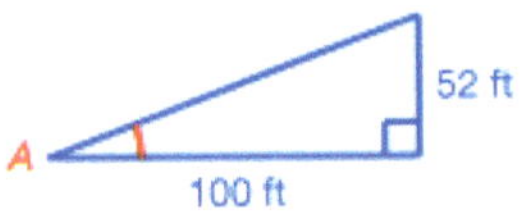

Draw a right triangle to represent the hill.

$\angle A$ is the angle the hill makes with a horizontal line.

$\tan A = \frac{52}{100}$

$m\angle A = \tan^{-1}\left(\frac{52}{100}\right) \approx 27°$

The angle associated with a given slope is derived in this application example. The grade of a hill is given as 52%.

The grade of 52% means a rise of 52 units for every 100 units of horizontal distance. In a representation of the hill as a right triangle, the legs of the triangle have lengths 52 and 100. The angle to be determined is the angle made by the slope of the hill with the horizontal line.

The measure of the angle is equal to the inverse tangent of 52 over 100 (opposite over adjacent). The solution, using a calculator, yields an angle of approximately 27°.

9

Developing Formulas for Triangles and Quadrilaterals

Key Objectives

- Develop and apply the formulas for the areas of triangles and special quadrilaterals.
- Solve problems involving perimeters and areas of triangles and special quadrilaterals.

Formulas

- The area of a parallelogram with base b and height h is $A = bh$.
- The area of a triangle with base b and height h is $A = \frac{1}{2}bh$.
- The area of a trapezoid with bases b_1 and b_2 and height h is $A = \frac{1}{2}(b_1 + b_2)h$, or $A = \frac{(b_1 + b_2)h}{2}$.
- The area of a rhombus or kite with diagonals d_1 and d_2 is $A = \frac{1}{2}d_1d_2$.

Theorems, Postulates, Corollaries, and Properties

- **Area Addition Postulate** The area of a region is equal to the sum of the areas of its nonoverlapping parts.

Example 1 Finding Measurements of Parallelograms

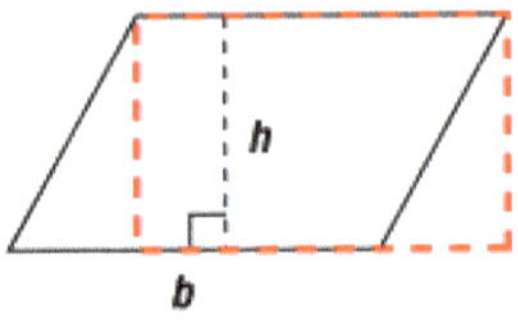

The area of a parallelogram with base *b* and height *h* is the same as the area of a rectangle with base *b* and height *h*: *A*=*bh*.

The area of a parallelogram with base b and height h is $A = bh$.

This is the same as the formula for the area of a rectangle. One way to think about why they are the same is to overlay the two figures and see that the rectangle can be constructed by removing a triangle from one end of the parallelogram and adding a congruent triangle to the other end of the parallelogram.

Find the measurement.

the area of the parallelogram

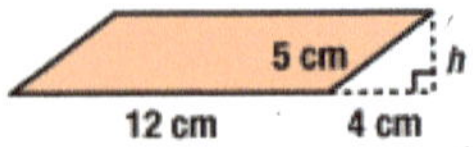

The area of the parallelogram is 36 cm².

$h^2 + 4^2 = 5^2$	*Pythagorean Thm.*
$h = 3$ cm	*Solve for h.*
$A = bh$	*Area of a parallelogram*
$A = 12(3) = 36$ cm²	*Substitute 12 for b and 3 for h.*
	Simplify.

The area of a parallelogram is determined in this example. The length of the base, the length of the sides, and the length of the overhang are given.

To calculate the height, use the Pythagorean Theorem with the parallelogram side as the hypotenuse and the overhang as a leg. The height of the parallelogram is 3 cm.

The area of the parallelogram is $A = bh = 12(3) = 36$ cm^2.

Find the measurement.

the height of the rectangle

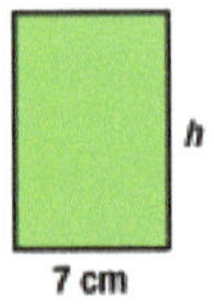

The height of the rectangle is $2x + x^3$ cm.

$A = bh$	*Area of a rectangle*
$14x + 7x^3 = 7h$	*Substitute 14x + 7x³ for A and 7 for b.*
$2x + x^3 = h$	*Divide both sides by 7.*

The height of a rectangle is determined in this example. The length of the base (or one side) of the rectangle is given. The area of the rectangle is given as an algebraic expression.

Substitute the known values into the formula for area and solve for the unknown h. The height is found to be $2x + x^3$ cm.

Find the measurement.

the perimeter of the rectangle

The perimeter of the rectangle is $4x^2 + 34$ in.

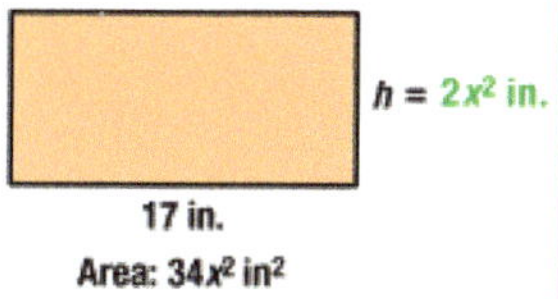

$A = bh$	*Area of a rectangle*
$34x^2 = 17h$	*Substitute $34x^2$ for A and 17 for b.*
$2x^2 = h$	*Divide both sides by 17.*
$P = 2h + 2b$	*Perimeter of a rectangle*
$P = 2h + 34$	*Substitute 17 for b, and simplify.*
$P = 2(2x^2) + 34$	*Substitute $2x^2$ for h.*
$= 4x^2 + 34$ in.	*Simplify.*

The perimeter of a rectangle is determined in this example. The length of the base of the rectangle is given. The area of the rectangle is given as an algebraic expression.

Begin by determining the height. The formula for area relates the area to the height and base (or side) length of the rectangle. Substitute the known values into the formula for area and solve for the unknown h. The height is found to be $2x^2$ inches.

The formula for the perimeter of a rectangle is $P = 2h + 2b$. Substitute the values for b and h into the formula. Simplifying yields $P = 4x^2 + 34$ inches.

Example 2 Finding Measurements in Triangles and Trapezoids

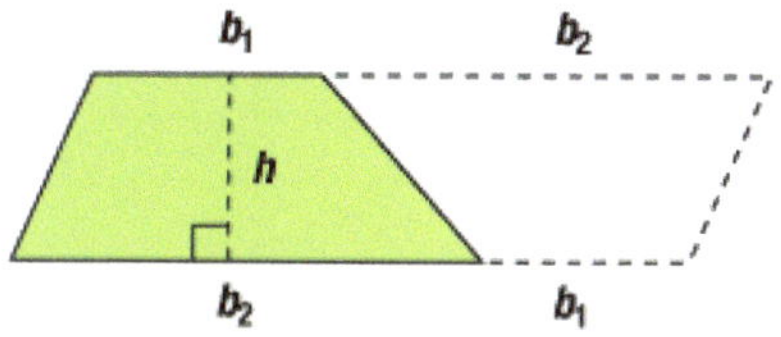

The area of a trapezoid with bases b_1 and b_2 and height h is half the area of the parallelogram formed by two of the trapezoids, with base $b_1 + b_2$ and height h:

$A_{parallelogram} = (b_1 + b_2)h \qquad A_{trapezoid} = \frac{(b_1 + b_2)}{2} \cdot h$

The area of a trapezoid with bases b_1 and b_2 and height h is $A = \frac{1}{2}(b_1 + b_2)h$. This formula is the same as the formula for half the area of a parallelogram formed by two congruent trapezoids.

To visualize why this is the case, add an inverted congruent trapezoid to one end of a trapezoid and see what the area of the parallelogram formed would be.

Find each measurement.

the area of a trapezoid in which $b_1 = 8$ cm, $b_2 = 10$ cm, and $h = 5$ cm

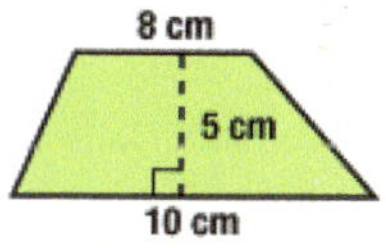

The area of the trapezoid is 45 cm².

$A = \frac{(b_1 + b_2)}{2} \cdot h$	*Area of a trapezoid*
$A = \frac{(8 + 10)}{2} \cdot 5$	*Substitute 8 for b_1, 10 for b_2, and 5 for h.*
$= 9 \cdot 5 = 45$ cm²	*Simplify.*

The area of a trapezoid is determined in this example. The lengths of the bases and the height are given.

To calculate the area, substitute the given values into the formula for the area of a trapezoid. Simplify and solve for area. Remember that the units for area are square units, in this case square centimeters (cm²).

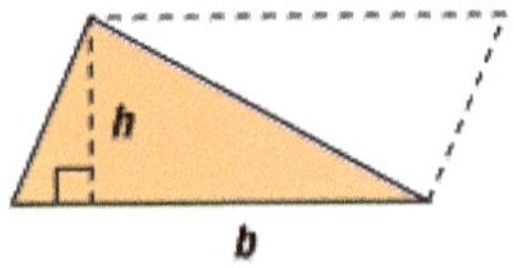

The area of a triangle with base b and height h is half the area of the parallelogram formed by two of the triangles, with base b and height h.

$A_{parallelogram} = bh$ $\quad A_{triangle} = \frac{1}{2}bh$

The area of a triangle with base b and height h is $A = \frac{1}{2}bh$.

This formula would be the formula for half the area of a parallelogram formed by two congruent triangles.

To visualize why this is the case, add an inverted congruent triangle to one end of a triangle and see that the area of the parallelogram formed would be $A = bh$, or twice the area for one of the triangles.

Find each measurement.

the base of the triangle

The base of the triangle measures $2x + 2$ ft.

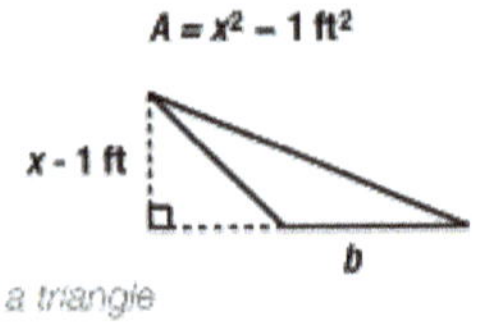

$A = \frac{1}{2}bh$	*Area of a triangle*
$x^2 - 1 = \frac{1}{2}b(x - 1)$	*Substitute $x^2 - 1$ for A, and $x - 1$ for h.*
$2(x^2 - 1) = b(x - 1)$	*Multiply both sides by 2.*
$\frac{2(x^2 - 1)}{x - 1} = b$	*Divide both sides by $x - 1$.*
$\frac{2(x + 1)(x - 1)}{x - 1} = b$	*Factor the numerator.*
$2(x + 1) = b$	*Simplify.*
$2x + 2 = b$	*Distribute the 2.*

The length of the base of a triangle is determined in this example. The height and the area of the triangle are given as algebraic expressions.

To calculate the base length, substitute the given values into the formula for the area of a triangle. Isolate the term for the base, simplify, and solve for the base length. The length is also an expression with the unknown x. The units are feet.

Find each measurement.

b_1 of the trapezoid, in which $A = 22$ in²

b_1 of the trapezoid is 5 in.

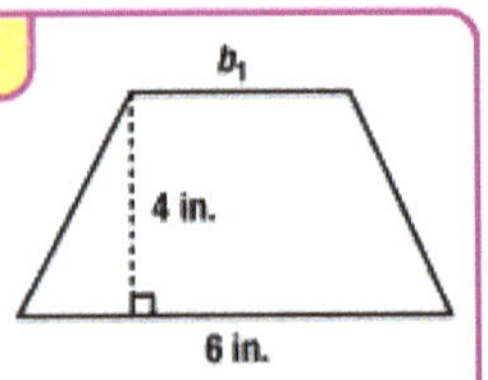

$A = \frac{b_1 + b_2}{2} \cdot h$	*Area of a trapezoid*
$22 = \frac{b_1 + 6}{2} \cdot 4$	*Substitute 22 for A, 6 for b, and 4 for h.*
$22 = (b_1 + 6)2$	*Simplify.*
$11 = b_1 + 6$	*Divide both sides by 2.*
$5 = b_1$	*Subtract 6 from both sides.*

The length of one base of a trapezoid is determined in this example. The lengths of one base, the height, and the area are given.

To calculate the length of the other base, substitute the given values into the formula for the area of a trapezoid. Simplify and solve for b_1. The value of b_1 is found to be 5 inches.

Example 3 Finding Measurements of Rhombuses and Kites

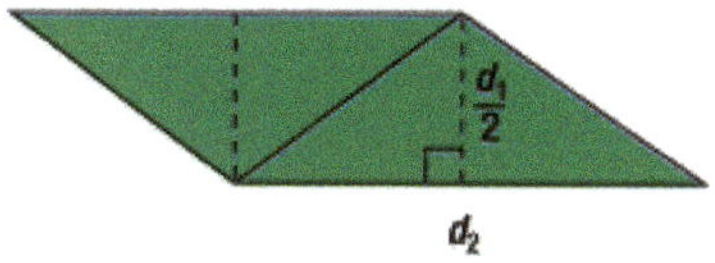

The area of a rhombus or kite with diagonals, d_1 and d_2, is the same as the area of the parallelogram obtained by cutting the rhombus or kite along the diagonal d_2, and arranging the two parts to form a parallelogram with base d_2 and height $\frac{1}{2}d_1$.

$A_{parallelogram} = bh = \frac{1}{2}d_1d_2 \quad A_{rhombus} = \frac{1}{2}d_1d_2 \quad A_{kite} = \frac{1}{2}d_1d_2$

The area of a rhombus or kite with diagonals d_1 and d_2 is

$$A = \frac{1}{2}d_1d_2.$$

This is the same area as a parallelogram. Imagine taking the rhombus, cutting it along one diagonal and moving the halves to form the parallelogram shown. Now the parallelogram base equals the length of d_2, and the parallelogram height equals one-half of d_1.

Find each measurement.

d_2 of a kite in which $d_1 = 10$ m and $A = 55$ m^2

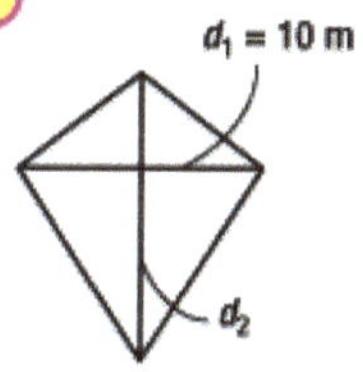

d_2 of the kite is 11 m.

$A = \frac{1}{2}d_1d_2$	*Area of a kite*
$55 = \frac{1}{2}(10)d_2$	*Substitute 55 for A and 10 for d_1.*
$55 = 5d_2$	*Simplify.*
$11 = d_2$	*Divide both sides by 5.*

The length of one diagonal of a kite is determined in this example. The length of the other diagonal and the area are given.

To calculate the length of the diagonal, substitute the given values into the formula for the area of a kite. Simplify and solve for d_2. The value of d_2 is found to be 11 meters.

Find each measurement.

the area of the rhombus

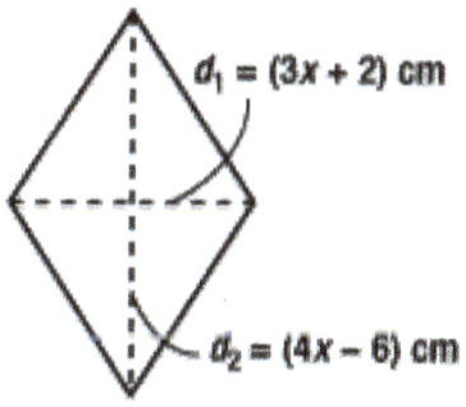

The area of the rhombus is $6x^2 - 5x - 6$ cm^2.

$A = \frac{1}{2}d_1d_2$	*Area of a rhombus*
$A = \frac{1}{2}(3x + 2)(4x - 6)$	*Substitute 3x + 2 for d_1 and 4x − 6 for d_2.*
$A = \frac{1}{2}(3x + 2)2(2x - 3)$	*Factor out a 2 from 4x − 6.*
$A = (3x + 2)(2x - 3)$	*Simplify.*
$A = 6x^2 - 5x - 6$	*Multiply the binomials (FOIL).*

The area of a rhombus is determined in this example. The length of the diagonals are given as algebraic expressions.

To calculate the area, substitute the given values into the formula for the area of a rhombus. Simplify the expression for area. The area is also an expression with the unknown x.

Find each measurement.

the area of the kite

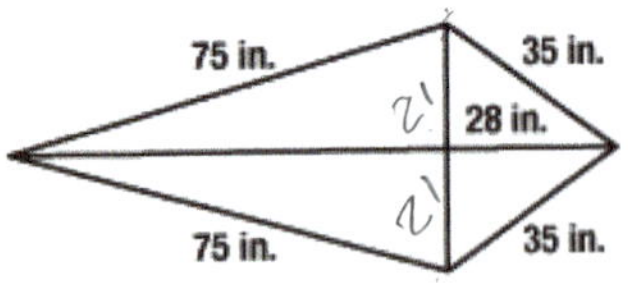

The area of the kite is 2100 in^2.

$?^2 + 28^2 = 35^2$	*Pythagorean Thm.*
$? = 21$	*Solve for the unknown quantity.*
$d_1 = 2(21) = 42$ in.	*Multiply.*
$?^2 + 21^2 = 75^2$	*Pythagorean Thm.*
$? = 72$	*Solve for the unknown quantity.*
$d_2 = 72 + 28 = 100$ in.	*Add.*
$A = \frac{1}{2} d_1 d_2$	*Area of a kite*
$A = \frac{1}{2}(42)(100)$	*Substitute 42 for d_1 and 100 for d_2.*
$= 21(100) = 2100$ in^2	*Simplify.*

The area of a kite is determined in this example. The length of the sides of the kite and the length of one segment of a diagonal are given.

Begin by determining the length of d_1, the shorter diagonal. Use the Pythagorean Theorem to determine the length of one-half of the diagonal, which is one leg of a right triangle. Substitute the known length of the other leg and the length of the hypotenuse into $a^2 + b^2 = c^2$. The length of the leg is found to be 21 inches, so the length of d_1 is 42 inches.

The 21-inch leg is also a leg in the larger triangle that has the other half of d_2 as a leg. Use the Pythagorean Theorem to calculate the length of that leg and add it to 28 inches to find the length of d_2. The diagonal d_2 is found to be $72 + 28 = 100$ inches.

The area of the kite is $1/2(d_1 d_2) = 1/2(42)(100) = 2100$ in^2.

Example 4 Games Application

Some pieces of a tangram are arranged in a rectangle with dimensions 2 cm by 3 cm. Use the grid to find the perimeter and area of the red trapezoid.

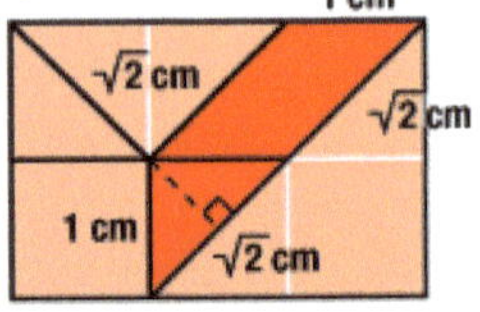

The perimeter of the trapezoid is $P = 2\sqrt{2} + \sqrt{2} + 1 + 1 = 3\sqrt{2} + 2$ cm.

The area of the trapezoid can be found by using the formula for the area:

$A = \frac{b_1 + b_2}{2} \cdot h$	*Area of a trapezoid*
$A = \frac{3\sqrt{2}}{2} \cdot \frac{\sqrt{2}}{2}$	*Substitute $\sqrt{2}$ for b_1, $2\sqrt{2}$ for b_2, and $\frac{\sqrt{2}}{2}$ for h.*
$= \frac{3}{2} \cdot \frac{1}{2} \cdot 2 = \frac{3}{2}$ cm^2	*Simplify.*

The area of the trapezoid is $\frac{3}{2}$ cm^2.

Another way to find the area is to count the half-squares. How many are there?

The perimeter and the area of a trapezoid within a tangram is calculated in this example. The squares of the tangram are given as 1 by 1 cm.

To determine the perimeter, the lengths of the bases and sides of the trapezoid must be determined. The length of a diagonal of one of the squares of the tangram is $\sqrt{2}$, according to the Pythagorean Theorem. The lengths of the sides of the trapezoid are equal to the length of the side of the tangram square (1 cm). Thus, $P = 2\sqrt{2} + \sqrt{2} + 1 + 1 = 3\sqrt{2} + 2$ cm.

To determine the area of the trapezoid, substitute the lengths of the bases and the height (one-half the length of a diagonal of a tangram square) into the formula for the area of a trapezoid. The area is found to be 3/2 cm^2.

Developing Formulas for Circles and Regular Polygons

Key Objectives

- Develop and apply the formulas for the area and circumference of a circle.
- Develop and apply the formula for the area of a regular polygon.

Key Terms

- A **circle** is the locus of points in a plane that are a fixed distance from a point called the **center of the circle**.
- The **center of a regular polygon** is equidistant from the vertices.
- The **apothem** is the distance from the center to the side of a regular polygon.
- A **central angle of a regular polygon** has its vertex at the center, and its sides pass through consecutive vertices.

Formulas

- A circle with diameter d and radius r has circumference $C = \pi d$ or $C = 2\pi r$ and area $A = \pi r^2$.
- The area of a regular polygon with apothem a and perimeter P is $A = \frac{1}{2}aP$.

Example 1 Finding Measurements of Circles

Find each measurement.

the area of $\odot B$ in terms of π

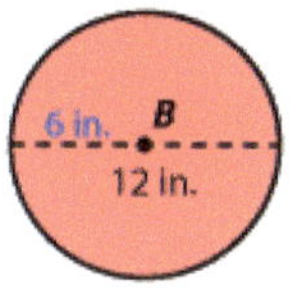

$r = \frac{d}{2}$ *The radius is half of the diameter.*

$r = \frac{12}{2} = 6$ in. *Substitute 12 for d and simplify.*

$A = \pi r^2$ *Area of a circle*

$A = \pi(6)^2 = 36\pi$ in^2 *Substitute 6 for r and simplify.*

The area of a circle is determined in this example. The diameter of the circle is given.

The radius of a circle is equal to one-half the diameter. The radius of the circle is 12/2 = 6 inches. Substitute the value for the length of the radius into the formula for the area of a circle. The solution is $A = \pi r^2 = \pi(6)^2 = 36\pi$ in^2.

Find each measurement.

the radius of $\odot Y$ in which $C = 32\pi$ cm

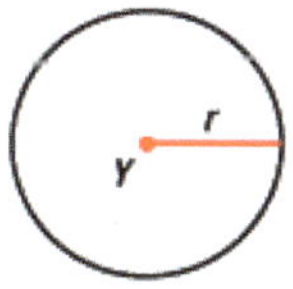

$C = 2\pi r$ *Circumference of a circle*

$32\pi = 2\pi r$ *Substitute 32π for C.*

$16 = r$ *Divide both sides by 2π.*

$r = 16$ cm

The length of the radius of a circle is determined in this example. The circumference of the circle is given.

Use the formula for the circumference of a circle, $C = 2\pi r$. Substitute the value for the circumference into the formula and solve for the length of the radius. The solution yields $r = 16$ cm.

Find each measurement.

the circumference of ⊙M in which $A = 4x^2\pi$ m²

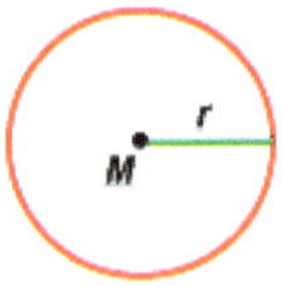

$A = \pi r^2$	*Area of a circle*
$4x^2\pi = \pi r^2$	*Substitute $4x^2\pi$ for A.*
$4x^2 = r^2$	*Divide both sides by π.*
$2x = r$	*Take the positive square root of both sides. (Assume $x \geq 0$.)*
$C = 2\pi r$	*Circumference of a circle*
$C = 2\pi(2x)$	*Substitute 2x for r.*
$= 4\pi x$ m	*Simplify.*

The circumference of a circle is determined in this example. The area of the circle is given as an algebraic expression.

Begin by determining the length of the radius of the circle from the expression for the area. Substitute the expression for the area into the formula for the area of a circle and isolate r. The value of r is found to be $2x$.

Substitute the value for r into the formula for the circumference of a circle. The solution yields $C = 4\pi x$ meters for the circumference.

Example 2 Cooking Application

A plastic food storage container set consists of three circular containers with diameters of 4 in., 8 in., and 10 in. Find the area of the lid of each container. Round to the nearest tenth.

$r = \frac{d}{2} = \frac{4}{2} = 2$ in.

$A = \pi r^2$

$A = \pi(2^2)$

$= 4\pi \text{ in}^2 \approx 12.6 \text{ in}^2$

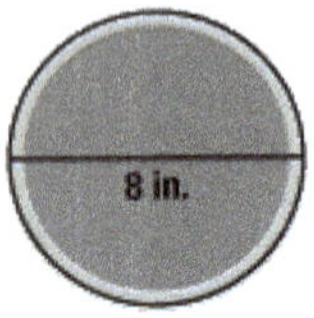

$r = \frac{d}{2} = \frac{8}{2} = 4$ in.

$A = \pi r^2$

$A = \pi(4^2)$

$= 16\pi \text{ in}^2 \approx 50.3 \text{ in}^2$

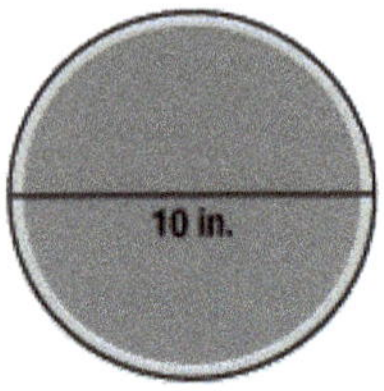

$r = \frac{d}{2} = \frac{10}{2} = 5$ in.

$A = \pi r^2$

$A = \pi(5^2)$

$= 25\pi \text{ in}^2 \approx 78.5 \text{ in}^2$

The areas of three circles are determined in this application example. The diameters of each of the circles are given.

Divide each diameter by 2 to find the length of the radius of each of the circles. Then substitute the length of the radius into the formula for the area of a circle. Use a calculator to find the actual area in square inches. The three circles have areas 12.6 in² ($r = 2$ in.), 50.3 in² ($r = 4$ in.), and 78.5 in² ($r = 5$ in.).

Example 3 Finding the Area of a Regular Polygon

Find the area of each regular polygon. Round to the nearest tenth.

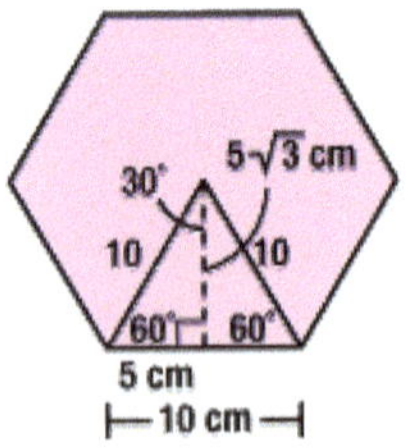

By the 30° - 60° - 90° Triangle Theorem, $a = 5\sqrt{3}$ cm.

$A = \frac{1}{2}aP$	*Area of a regular polygon*
$A = \frac{1}{2}(5\sqrt{3})(60)$	*Substitute $5\sqrt{3}$ for a, and 60 for P.*
$= 150\sqrt{3} \text{ cm}^2$	*Simplify.*
$\approx 259.8 \text{ cm}^2$	*Round to the nearest tenth.*

The area of a regular hexagon is determined in this example. The length of a side of the hexagon is given.

The area of a regular polygon is one-half times the length of the apothem times the perimeter. Notice that the formula combines the areas for the six triangles that fit inside the hexagon. The perimeter is equal to the sum of the six triangle bases and the apothem is the height of each triangle.

The interior angles of the hexagon have a measure of 120°, so the angles at the base of the triangle (and the third angle) have measure 60°. The height forms a 30°-60°-90° triangle, which has one leg equal to half of the hexagon side, or 5 cm, and the other leg equal to $5\sqrt{3}$.

Substitute the value for the height into the apothem value in the formula and 60 cm for the perimeter. Using a calculator the area of the hexagon is found to be 259.8 cm².

Find the area of each regular polygon. Round to the nearest tenth.

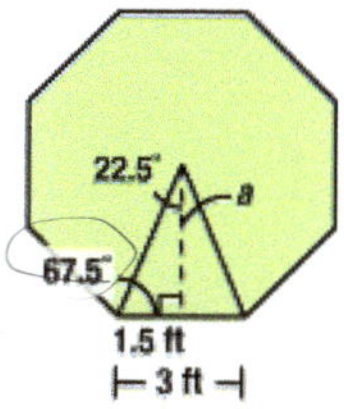

$\tan 22.5° = \frac{1.5}{a}$	*The tangent of an angle is* $\frac{\text{opp. leg}}{\text{adj. leg}}$.
$a = \frac{1.5}{\tan 22.5°}$	*Solve for a.*
$A = \frac{1}{2}aP$	*Area of a regular polygon*
$A = \frac{1}{2}\left(\frac{1.5}{\tan 22.5°}\right)(24)$	*Substitute* $\frac{1.5}{\tan 22.5°}$ *for a, and 24 for P.*
$= \frac{18}{\tan 22.5°}\ \text{ft}^2$	*Simplify.*
$\approx 43.5\ \text{ft}^2$	*Round to the nearest tenth.*

The area of a regular octagon is determined in this example. The length of a side of the octagon is given.

The area of a regular polygon is one-half times the length of the apothem times the perimeter.

The interior angles of the octagon have a measure of 135°, so the angles at the base of the triangle shown have measure 67.5°. The height of the triangle forms a triangle with a third angle equal to 22.5°. The tangent of the 22.5° angle can be used to find the length of the height (the apothem of the octagon).

Substitute the value for the height into the apothem value in the formula and (3)(8) = 24 for the perimeter. Using a calculator, the area of the hexagon is found to be 43.5 ft^2.

10

10 Spatial Reasoning

10.1 Surface Area and Volume

Surface Area of Prisms and Cylinders

Key Objectives

- Learn and apply the formula for the surface area of a prism.
- Learn and apply the formula for the surface area of a cylinder.

Key Terms

- A **lateral face** is not a base.
- A **lateral edge** is not an edge of a base.
- The lateral faces of a **right prism** are all rectangles.
- An **oblique prism** has at least one nonrectangular lateral face.
- An **altitude** of a prism or cylinder is a perpendicular segment joining the planes of the bases.
- **Surface area** is the total area of all faces and curved surfaces of a three-dimensional figure.
- The **lateral surface** of a cylinder is the curved surface that connects the two bases.
- The **axis of a cylinder** is the segment with endpoints at the centers of the bases.
- The axis of a **right cylinder** is perpendicular to its bases.
- The axis of an **oblique cylinder** is not perpendicular to its bases.

Formulas

- **Lateral Area of a Right Prism** The lateral area of a right prism with base perimeter P and height h is $L = Ph$.
- **Surface Area of a Right Prism** The surface area of a right prism with lateral area L and base area B is $S = L + 2B$, or $S = Ph + 2B$.
- **Surface Area of a Cube** The surface area of a cube with edge length s is $S = 6s^2$.
- **Lateral Area of a Right Cylinder** The lateral area of a right cylinder with radius r and height h is $L = 2\pi rh$.
- **Surface Area of a Right Cylinder** The surface area of a right cylinder with lateral area L and base area B is $S = L + 2B$, or $S = 2\pi rh + 2\pi r^2$.

Example 1 Finding Lateral Areas and Surface Areas of Prisms

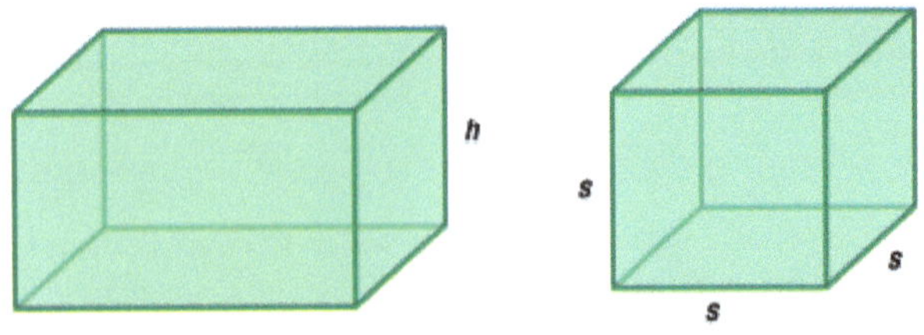

Lateral area of a right prism	$L = Ph$
Surface area of a right prism	$S = L + 2B$ $S = Ph + 2B$
Surface area of a cube	$S = 6s^2$

P - base perimeter
B - base area
L - lateral area

The lateral area of a right prism is equal to the product of the perimeter of the base and the height.

The surface area of a right prism is equal to the lateral area plus the area of the two bases.

The surface area of a cube is equal to the length of an edge cubed.

Find the lateral area and surface area of each right prism.

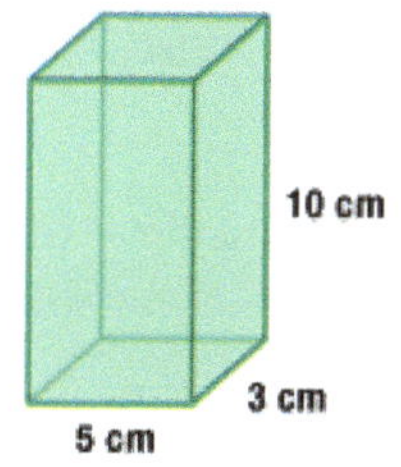

$P = 5 + 3 + 5 + 3$	*Find the perimeter.*
$= 16$	*Simplify.*
$L = Ph$	*Lateral area of a right prism*
$= 16 \cdot 10 = 160 \text{ cm}^2$	*Substitute 16 for P and 10 for h, and simplify.*
$S = L + 2B$	*Surface area of a right prism*
$= 160 + 2(15)$	*$B = 5 \cdot 3 = 15 \text{ cm}^2$. Substitute 160 for L and 15 for B, and simplify.*
$= 190 \text{ cm}^2$	

The lateral and surface area of a right prism are determined in this example. The length and width of the base and the height of the prism are given.

To find the lateral area of the right prism, first calculate the perimeter of the base using the given values for length and width. The perimeter of the base is equal to 16 cm. Substitute the perimeter of the base and the height of the prism into the formula for the lateral area. The lateral area is found to be 160 cm^2.

To find the surface area of the right prism, calculate the area of the base by multiplying the given length and width. The area of the base is equal to 15 cm^2. To find the surface area of the prism, substitute the lateral area and base area into the formula for surface area. The surface area is 190 cm^2.

Find the lateral area and surface area of each right prism.

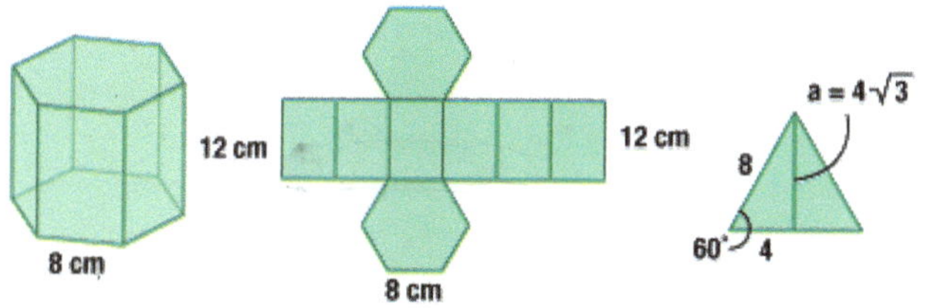

$L = Ph$	*Lateral area of a right prism*
$= 48 \cdot 12$	*$P = 8 \cdot 6 = 48$ cm. Substitute 48 for P and 12 for h and simplify.*
$= 576 \text{ cm}^2$	
$B = \frac{1}{2}aP$	*Find the area of one of the bases.*
$= \frac{1}{2}(4\sqrt{3})(48) = 96\sqrt{3}$	*Substitute and simplify.*
$S = L + 2B$	*Surface area of a right prism*
$S = 576 + 2(96\sqrt{3})$	*Substitute 576 for L and $96\sqrt{3}$ for B.*
$= 576 + 192\sqrt{3} \approx 908.6 \text{ cm}^2$	*Simplify.*

The lateral and surface area of a right prism are determined in this example. The length of an edge of the base and the height of the prism are given.

To find the lateral area of the right prism, calculate the perimeter of the base using the given value for the length of an edge of the base. The perimeter of the base is 48 cm. Substitute the perimeter of the base and the height of the prism into the formula for the lateral area of the prism. The lateral area is 576 cm^2.

To find the surface area of the right prism, calculate the area of the base. The formula for the area of the base is $B = (1/2)aP$, where a is the length of the apothem. The base is a hexagon, so the perimeter is 8(6) = 48 cm long. The apothem is the longer leg of a 30°-60°-90° right triangle with short leg length of 4.

Substitute the values of the base area and the lateral area into the formula for surface area to find the surface area approximately equal to 908.6 cm^2.

Example 2 Finding Lateral Areas and Surface Areas of Right Cylinders

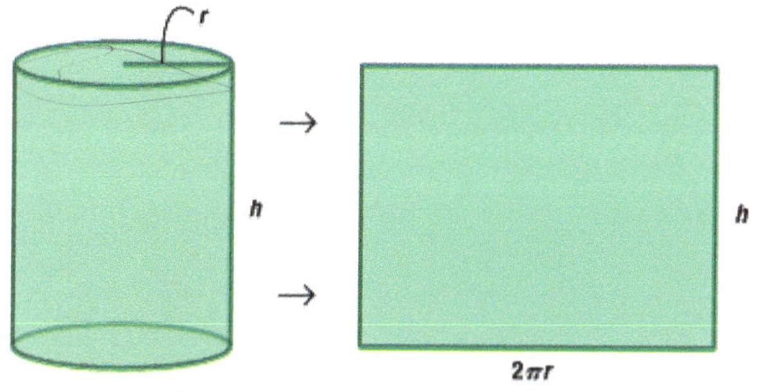

Lateral area of a right cylinder	$L = 2\pi rh$
Surface area of a right cylinder	$S = L + 2B$ $S = 2\pi rh + 2\pi r^2$

B - base area
L - lateral area

The lateral area of a cylinder is equal to the circumference of the base times the height of the cylinder.

The surface area of a right cylinder is equal to the lateral area plus the areas of the two bases.

Find the lateral area and surface area of the right cylinder. Give Your answers in terms of π.

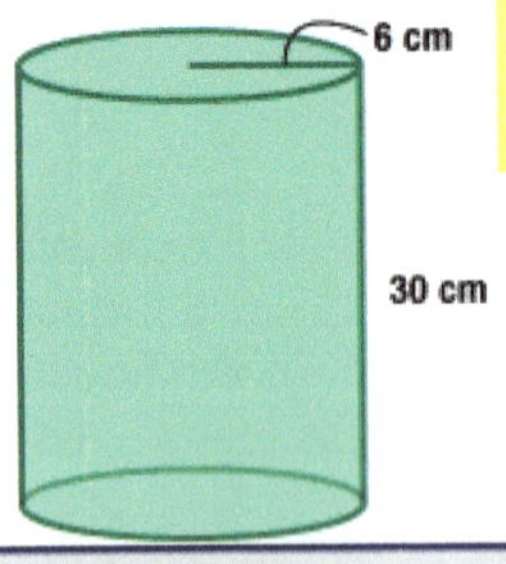

A cylinder with a circumference of 12π cm and a height equal to 5 times the radius.

$C = 12\pi$ cm
$r = 6$ cm
$h = 30$ cm

$2\pi r = 12\pi \Rightarrow r = 6$ cm
$h = 5(6) = 30$ cm

$B = \pi r^2$
$= \pi \cdot 6^2$
$= 36\pi$ cm^2

$L = 2\pi rh$	Lateral area of a right cylinder
$= 2\pi \cdot 6 \cdot 30$	Substitute 6 for r and 30 for h.
$= 360\pi$ cm^2	Simplify.
$S = L + 2B$	Surface area of a right cylinder
$= 360\pi + 2(36\pi)$	Substitute 360π for L and 36π for B.
$= 360\pi + 72\pi$	Simplify.
$= 432\pi$ cm^2	Simplify.

The lateral and surface areas of a right cylinder are determined in this example. The circumference is given as 12π cm and the height is given as $5r$.

To calculate the lateral area, begin by determining r by substituting the value for the circumference into the formula for circumference: $C = 2\pi r$, $12\pi = 2\pi r$, therefore $r = 6$ cm. Since $r = 6$, the height equals 30 cm. Substituting the values for circumference and height into the formula for lateral area gives 360π cm^2.

To calculate the surface area, find the area of the base. The area of the base is πr^2, or 36π. Substituting the values for the area of the base and the lateral area into the formula for surface area gives a value of 432π cm^2.

Example 3 Finding Surface Areas of Composite Three-Dimensional Figures

Find the surface area of the composite figure.

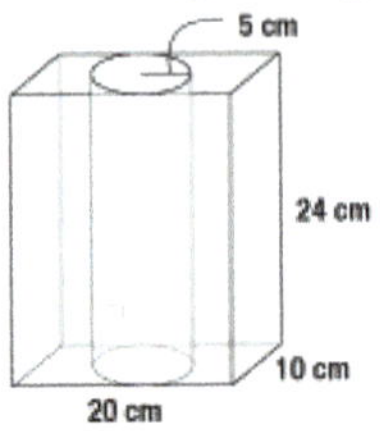

lateral area of right rectangular prism	lateral area of cylinder	area of each base of composite figure
$L = Ph$	$L = 2\pi rh$	$B = B_{prism} - B_{cylinder}$
$= 60 \cdot 24$	$= 2\pi(5)(24)$	$= \ell w - \pi r^2$
$= 1440$ cm^2	$= (10\pi)(24)$	$= 20(10) - \pi(5)^2$
	$= 240\pi$ cm^2	$= 200 - 25\pi$ cm^2

L = (lateral area of prism) + (lateral area of cylinder) + 2(area of each base)
$= 1440\pi \quad + 240\pi \quad + 2(200 - 25\pi)$
$= 1440 + 400 + 240\pi - 50\pi$
$= 1840 + 190\pi$ cm^2

The surface area of a composite figure, a rectangular prism with a missing volume the shape of a right cylinder, is determined in this example. The radius of the cylinder, the height of the prism (and cylinder), and the length and width of the prism base are given.

The lateral area of the rectangular prism is obtained by substituting the known values for base dimensions and height into the formula. The area is 1440 cm^2. The lateral area of the cylinder is found by substituting the known values for the radius and height into the formula. The area is 240π cm^2.

The area of the bases of the figure are equal to the area of the prism base minus the area of the cylinder base. These are obtained by substituting the known values into the formulas for each. The area is $200 - 25\pi$ cm^2.

The surface area is the sum of the lateral areas plus twice the base area. Using a calculator, the area is found to be approximately 2437 cm^2.

Example 4 Exploring Effects of Changing Dimensions

The length, width, and height of the right rectangular prism are doubled. Describe the effect on the surface area.

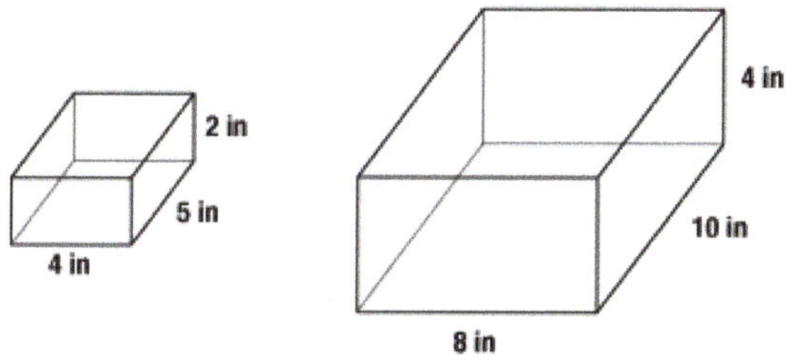

original dimensions:

$S = Ph + 2B$
$= (18)(2) + 2(5)(4)$
$= 76 \text{ in}^2$

length, width and height doubled:

$S = Ph + 2B$
$= (36)(4) + 2(10)(8)$
$= 304 \text{ in}^2$

Notice that 304 = 4(76). If the length, width and height are doubled, the surface area is multipled by 2^2, or 4.

The effect of a proportional doubling of the dimensions of a rectangular prism on the surface area of the prism are demonstrated in this example.

The original dimensions of the rectangular prism are 2, 4, and 5 inches. The surface area of this prism is obtained by substituting the given values into the formula for surface area. The area is 76 in^2.

Doubling the dimensions proportionately gives dimensions of a new rectangular prism of 4, 8, and 10 inches. The new area of the prism is 304 in^2.

The new surface area is 2^2 times the surface area of the original prism, so the surface area changes by a factor of the proportional increase squared.

Example 5 Problem-Solving Application

If two pieces of ice have the same volume, the one with the greater surface area will melt faster because more of it is exposed to the air. One piece of ice shown is a rectangular prism, and the other is half a cylinder. Given that the volumes are approximately equal, which will melt faster?

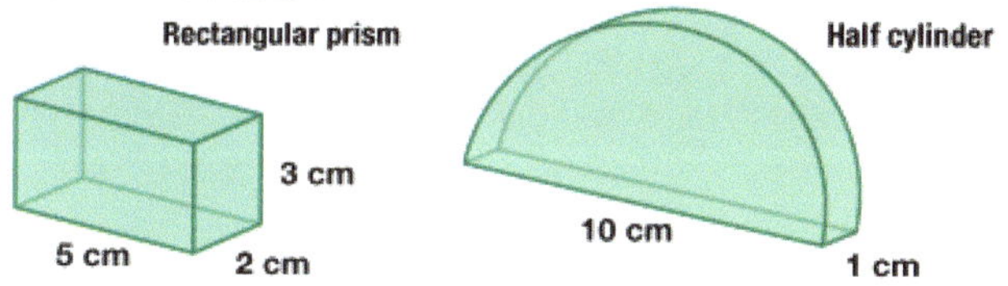

rectangular prism:

$S = Ph + 2B$
$= 14(3) + 2(10)$
$= 62 \text{ cm}^2$

half cylinder:

$S = \pi rh + \pi r^2 + 2rh$
$= \pi(5)(1) + \pi(5)^2 + (10)(1)$
$= 10 + 30\pi$
$\approx 104.25 \text{ cm}^2$

The half cylinder of ice has a greater surface area, so it will melt faster.

The surface areas of two solid figures are determined to answer a question about the physical behavior of the two objects in this application example. The base length and width and height of a rectangular prism are given. The diameter and height of a half cylinder are given.

The surface area of the rectangular prism is obtained by substituting the given values into the formula for surface area. The area is 62 cm^2.

The surface area for the half cylinder is obtained by substituting half the diameter into a formula for the area of a right cylinder that only counts one base and half the lateral area. The area is approximately 104.25 cm^2.

The half cylinder area is far greater than the rectangular prism area, so this block of ice will melt much faster.

Surface Area of Pyramids and Cones

Key Objectives

- Learn and apply the formula for the surface area of a pyramid.
- Learn and apply the formula for the surface area of a cone.

Key Terms

- The **vertex of a pyramid** is the point opposite the base of the pyramid.
- The base of a **regular pyramid** is a regular polygon, and the lateral faces are congruent isosceles triangles.
- The **slant height of a regular pyramid** is the distance from the vertex to the midpoint of an edge of the base.
- The **altitude of a pyramid** is the perpendicular segment from the vertex to the plane of the base.
- The **vertex of a cone** is the point opposite the base.
- The **axis of a cone** is the segment with endpoints at the vertex and the center of the base.
- The axis of a **right cone** is perpendicular to the base.
- The axis of an **oblique cone** is not perpendicular to the base.
- The **slant height of a right cone** is the distance from the vertex of a right cone to a point on the edge of the base.
- The **altitude of a cone** is a perpendicular segment from the vertex of the cone to the plane of the base.

Formulas

- **Lateral Area of a Regular Pyramid** The lateral area of a regular pyramid with perimeter P and slant height ℓ is $L = \frac{1}{2}P\ell$.
- **Surface Area of a Regular Pyramid** The surface area of a regular pyramid with lateral area L and base area B is $S = L + B$, or $S = \frac{1}{2}P\ell + B$.
- **Lateral Area of a Right Cone** The lateral area of a right cone with radius r and slant height ℓ is $L = \pi r \ell$.
- **Surface Area of a Right Cone** The surface area of a right cone with lateral area L and base area B is $S = L + B$, or $S = \pi r \ell + \pi r^2$.

Example 1 Finding Lateral Area and Surface Area of Pyramids

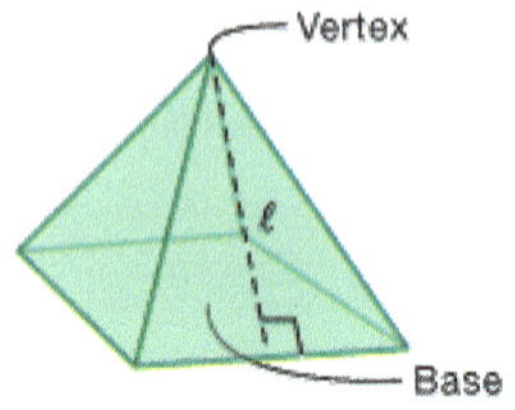

Regular Pyramid	
Lateral Area	$L = \frac{1}{2}P\ell$
Surface Area	$S = \frac{1}{2}P\ell + B$

P - perimeter
B - base area
ℓ - slant height

The lateral area of a right pyramid is equal to one half the product of the perimeter of the base and the slant height.

The surface area of a right pyramid is equal to the lateral area plus the base area.

Find the lateral area and surface area of each pyramid.

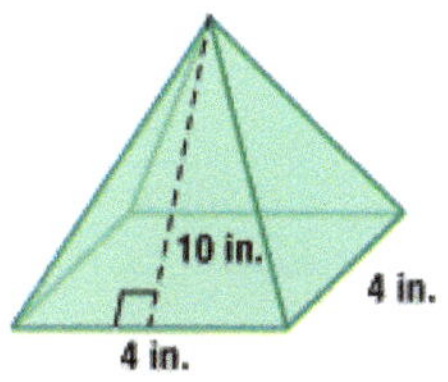

A regular square pyramid with base edge length 4 in. and slant height 10 in.

$L = \frac{1}{2}P\ell$	*Lateral area of a regular pyramid*
$= \frac{1}{2} \cdot 16 \cdot 10 = 80 \text{ in}^2$	*$P = 4(4) = 16$ in. Substitute 16 for P and 10 for ℓ.*
$S = L + B$	*Surface area of a regular pyramid*
$= 80 + 16 = 96 \text{ in}^2$	*$B = 4^2 = 16 \text{ in}^2$*

The lateral and surface area of a right pyramid are determined in this example. The figure is given as a square pyramid with base edge length 4 inches and slant height 10 inches.

To calculate the lateral surface area, substitute the known values into the formula for lateral surface area. The lateral surface area is 80 in^2.

To calculate the surface area of the pyramid, substitute the known values into the formula for surface area. The area of the base is the square of the edge length because the base is a square. The surface area is 96 in^2.

Find the lateral area and surface area of each pyramid.

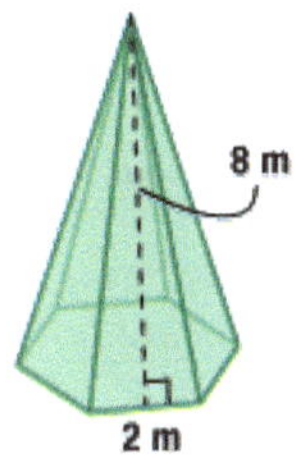

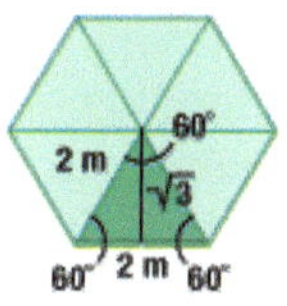

$L = \frac{1}{2}P\ell$	*Lateral area of a regular pyramid*
$= \frac{1}{2}(12)(8) = 48 \text{ m}^2$	*$P = 2 \cdot 6 = 12$ m and $\ell = 8$ m.*
$S = L + B$	*Surface area of a regular pyramid*
$= 48 + 6\sqrt{3} \text{ m}^2$	*Base perimeter: $6(2) = 12$ m, apothem: $\sqrt{3}$ m* *$B = \frac{1}{2}aP = \frac{1}{2}(\sqrt{3}) \cdot 12 = 6\sqrt{3} \text{ m}^2$* *Substitute $6\sqrt{3}$ for B.*

The lateral and surface area of a regular right pyramid are determined in this example. The figure is given as a hexagonal pyramid with base edge length 2 meters and slant height 8 meters.

To calculate the lateral surface area, substitute the known values into the formula for lateral surface area. The lateral surface area is 48 m^2.

To calculate the surface area of the pyramid, substitute the known values into the formula for surface area. The area of the base is one half the product of the apothem and the perimeter. The apothem is determined by recognizing that it is the long leg of a 30°-60°-90° right triangle whose short leg has length 1 m. The length of the apothem is $\sqrt{3}$. The surface area is $48 + 6\sqrt{3}$ m^2.

Example 2 Finding Lateral Area and Surface Area of Right Cones

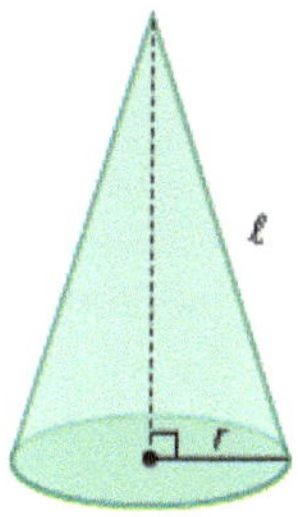

Right Cone	
Lateral Area	$L = \pi r\ell$
Surface Area	$S = \pi r\ell + B$

B - base area
ℓ - slant height

The lateral area of a right cone is equal to $\pi r \ell$, where ℓ is the slant height. This is equivalent to one half the base circumference times the slant height.

The surface area of a right cone is equal to the lateral area plus the base area. The base area is equal to the area of a circle with radius r.

Find the lateral area and surface area of each cone.

a right cone with a radius of 4 m and a slant height of 7 m

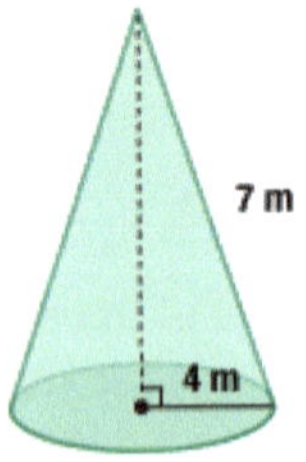

$L = \pi r \ell$ — *Lateral area of a cone*

$= \pi(4)(7) = 28\pi \text{ m}^2$ — *Substitute 4 for r, and 7 for ℓ.*

$S = L + \pi r^2$ — *Surface area of a cone*

$= 28\pi + \pi(4)^2 = 44\pi \text{ m}^2$ — *Substitute 28π for L, and 4 for r.*

The lateral and surface area of a right cone are determined in this example. The figure is given as a right cone with radius 4 meters and slant height 7 meters.

To calculate the lateral surface area, substitute the known values into the formula for lateral surface area. The lateral surface area is 28π m^2.

To calculate the surface area of the cone, substitute the known values into the formula for surface area. The area of the base is πr^2. The surface area is 44π m^2.

Find the lateral area and surface area of each cone.

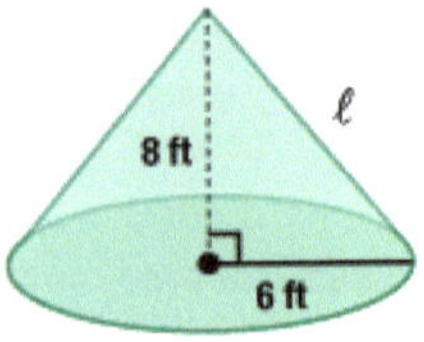

$\ell = \sqrt{6^2 + 8^2} = 10$ ft — *Use the Pythagorean Theorem to find ℓ*

$L = \pi r \ell$ — *Lateral area of a right cone*

$= \pi(6)(10) = 60\pi \text{ ft}^2$ — *Substitute 6 for r, and 10 for ℓ.*

$S = L + \pi r^2$ — *Surface area of a right cone*

$= 60\pi + \pi(6)^2 = 96\pi \text{ ft}^2$ — *Substitute 60π for L, and 6 for r.*

The lateral and surface area of a right cone are determined in this example. The figure is given as a right cone with radius 6 feet and altitude 8 feet.

To calculate the lateral surface area, substitute the known values into the formula for lateral surface area. Use the Pythagorean Theorem to calculate the slant length, or recognize that it is the hypotenuse of a 6, 8, 10 Pythagorean triple. The lateral surface area is 60π ft^2.

To calculate the surface area of the cone, substitute the known values into the formula for surface area. The area of the base is πr^2. The surface area is 96π ft^2.

Example 3 Exploring Effects of Changing Dimensions

The radius and slant height of the right cone are doubled. Describe the effect on the surface area.

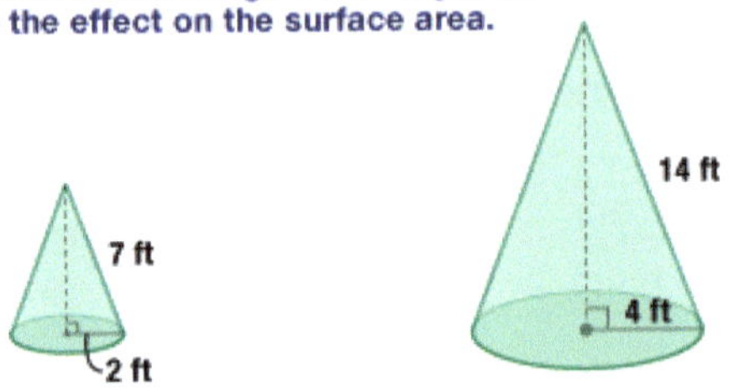

original dimensions:

$S = \pi r \ell + \pi r^2$

$= \pi(2)(7) + \pi(2)^2$

$= 18\pi \text{ ft}^2$

radius and slant height doubled:

$S = \pi r \ell + \pi r^2$

$= \pi(4)(14) + \pi(4)^2$

$= 56\pi + 16\pi = 72\pi \text{ ft}^2$

Notice that $72\pi = 4(18\pi)$. If the radius and slant height are doubled, the surface area is multiplied by 2^2, or 4.

The effect on surface area of a proportional increase in the dimensions of a right cone are examined in this example. The dimensions of the cone are given as a radius of 2 feet and a slant height of 7 feet.

To find the surface area of the right cone, substitute the given values into the formula. The surface area of the cone is 18π ft^2.

In the new cone, the radius and slant height are doubled and have the values 4 ft and 14 ft. To find the surface area of the new cone, substitute these values into the formula. The larger cone has surface area 72π ft^2.

The larger cone has a surface area that is $2^2 = 4$ times the surface area of the original cone, which is the square of the factor used to increase the dimensions.

Example 4 Finding Surface Area of Composite Three-Dimensional Figures

Find the surface area of the composite figure.

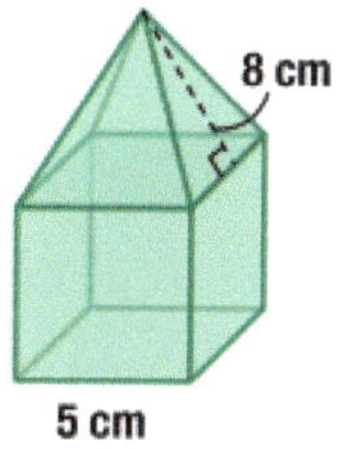

Lateral area of pyramid:

$$L = \frac{1}{2}P\ell$$

$$= \frac{1}{2}(20)(8) = 80 \text{ cm}^2$$

Area of five sides of cube:

$$A = 5(25) = 125 \text{ cm}^2$$

Surface area of the composite figure:

$$S = (\text{area of five sides of cube}) + (\text{lateral area of pyramid})$$

$$= 125 + 80 = 205 \text{ cm}^2$$

The surface area of a composite solid is calculated in this example. The solid is given as a cube with a square pyramid on top of it. The length of an edge of the cube is given as 5 cm and the slant height of the pyramid is given as 8 cm.

First, calculate the lateral area of the pyramid. The length of the edge of the cube is the same as the length of the edge of the pyramid base, so the perimeter of the base is $4(5) = 20$ cm. Substitute the slant height and base perimeter into the formula for lateral area. The lateral area of the pyramid is 80 cm^2.

Second, calculate the area of five sides of the cube, since the sixth is not on the surface of the composite object. The surface area of one side is $5^2 = 25 \text{ cm}^2$, so the surface area of five sides is 125 cm^2.

The total area of the object is $125 + 80 = 205 \text{ cm}^2$.

Example 5 Problem-Solving Application

Find the diameter of the circular base created by folding the figure on the left into a right cone.

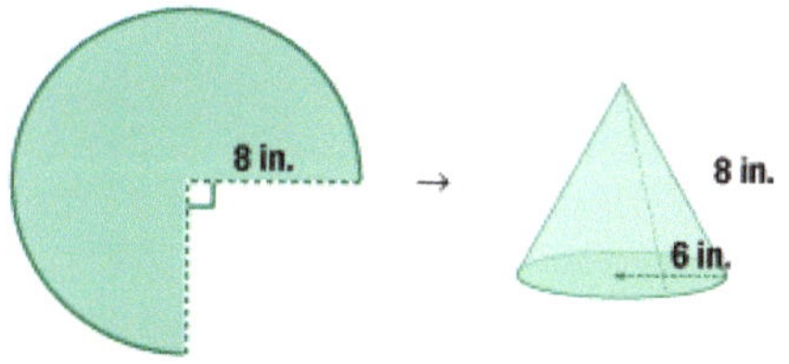

Professor Burger's Method:

$$\frac{3}{4}\begin{pmatrix}\text{Circumference} \\ \text{of big circle}\end{pmatrix} = \begin{pmatrix}\text{Circumference} \\ \text{of base of cone}\end{pmatrix}$$

$\frac{3}{4}(2\pi \cdot 8) = 2\pi r$	*8 is the radius of the big circle.*
$12\pi = 2\pi r$	*Simplify.*
$r = 6$ in.	*Solve for r.*
$d = 2 \cdot 6 = 12$ in.	*$d = 2r$*

The diameter of the base of a right cone made by folding a two-dimensional shape is determined in this example.

The circumference of the base of the cone is equal to three quarters of the length of the circumference of the two-dimensional circle. The circumference of the two-dimensional circle is $2\pi 8 = 16\pi$. Three quarters of this value is 12π. This is the circumference of the base of the cone, or $2\pi r$. Set these two values equal to each other and solve for r. The radius of the cone is 6 inches. The diameter of the cone is 12 inches.

Volume of Prisms and Cylinders

Key Objectives

- Learn and apply the formula for the volume of a prism.
- Learn and apply the formula for the volume of a cylinder.

Key Terms

- The **volume** of a three-dimensional figure is the number of nonoverlapping unit cubes of a given size that will exactly fill the interior.

Formulas

- **Volume of a Prism** The volume of a prism with base area B and height h is $V = Bh$.
- **Volume of a Right Rectangular Prism** The volume of a right rectangular prism with length ℓ, width w, and height h is $V = \ell wh$.
- **Volume of a Cube** The volume of a cube with edge length s is $V = s^3$.
- **Volume of a Cylinder** The volume of a cylinder with base area B, radius r, and height h is $V = Bh$, or $V = \pi r^2 h$.

Example 1 Finding Volumes of Prisms

Find the volume of a rectangular prism with length 5 cm, width 2 cm, and height 3 cm. Round to the nearest tenth.

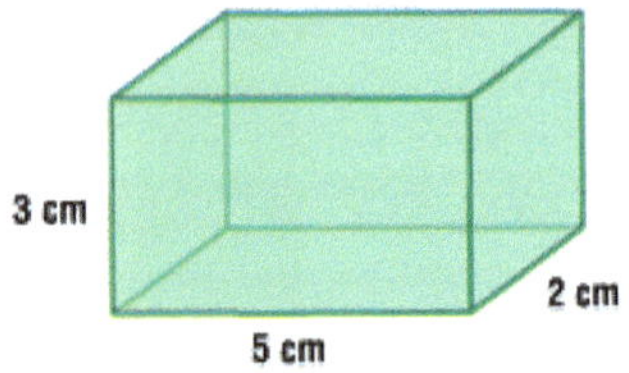

$V = \ell wh$ *Volume of a rectangular prism*

$= 5 \cdot 2 \cdot 3$ *Substitute.*

$= 30\ \text{cm}^3$ *Simplify.*

The volume of a rectangular prism is determined in this example. The length, width, and height of the prism are given.

To find the volume of the prism, substitute the given values into the formula for the volume of a rectangular prism. The volume is equal to 30 cm^3. Notice that the units of volume are cubed.

Find the volume of a cube with edge length 5 cm. Round to the nearest tenth.

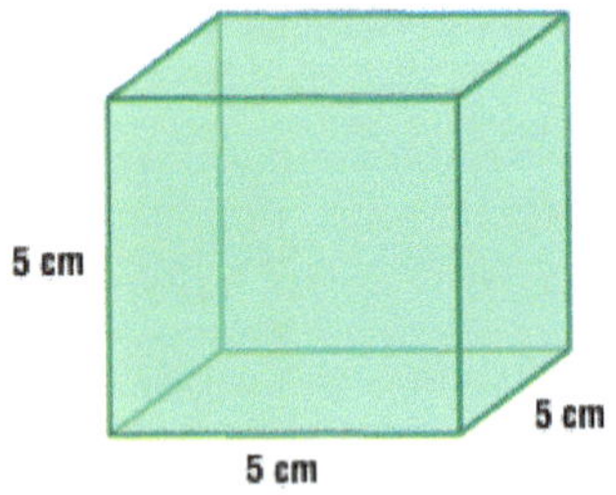

$V = s^3$ *Volume of a cube*

$= 5^3$ *Substitute 5 for s.*

$= 125\ \text{cm}^3$ *Simplify.*

The volume of a cube is determined in this example. The length of the edges of the cube are given.

To find the volume of the cube, substitute the given value into the formula for the volume of a cube. The volume is 125 cm^3.

Find the volume of a right regular pentagonal prism with base edge length 4 m and height 10 m.

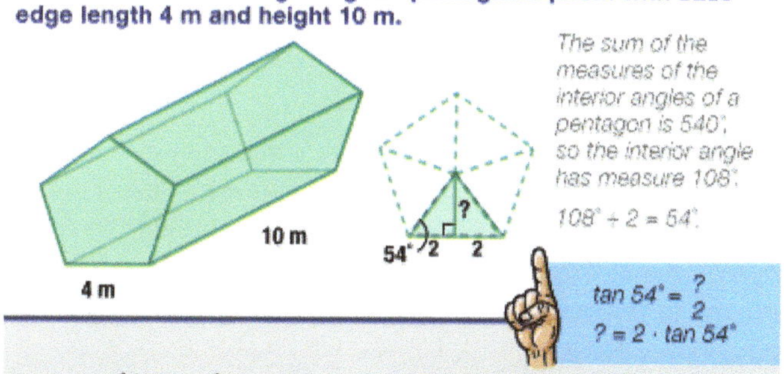

$B = 5 \cdot \left(\frac{1}{2} \cdot 4 \cdot ?\right)$ *Area of a pentagonal base*

$B = 5 \cdot \left(\frac{1}{2} \cdot 4 \cdot 2 \cdot \tan 54°\right)$ *Substitute 2 tan 54° for ?.*

$= 20 \tan 54° \text{ m}^2$ *Simplify.*

$V = Bh$ *Volume of a prism*

$V = (20 \tan 54°) \cdot 10$ *Substitute 20 tan 54° for B and 10 for h.*

$= 200 \tan 54° \approx 275.3 \text{ m}^3$ *Simplify.*

The volume of a regular pentagonal prism is determined in this example. The base edge length and height of the prism are given.

To find the volume of the prism, first calculate the area of the base. To find the apothem of the pentagonal base, recognize that the sum of the measures of the interior angles in a pentagon is 540°. The base angle of the triangles with the pentagon's sides as their base is one-half of 108°, or 54°. The apothem bisects these triangles to form two right triangles with a short leg length of 2 m. The tangent of 54° equals the apothem length over 2 m. The apothem length equals 2tan 54°. Therefore, the base area equals 20tan 54°.

Substituting the calculated and given values into the formula for the volume of a prism yields a volume of approximately 275.3 m^3.

Example 2 Marine Biology Application

Estimate the volume of water in an aquarium with width 100 feet, length 50 feet, and height 6 feet. The density of water is about 8.33 pounds per gallon. Estimate the weight of the water in pounds.

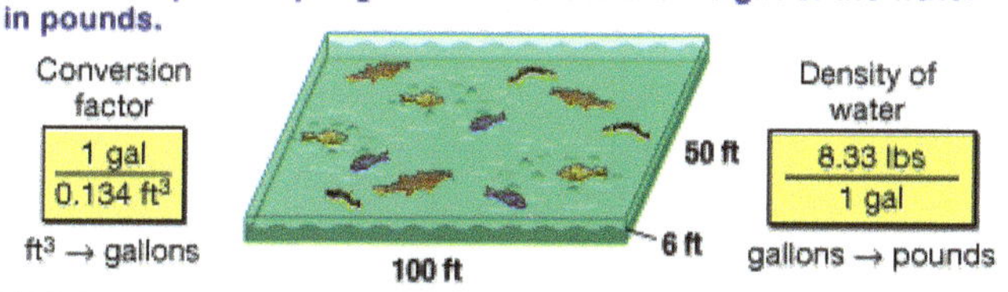

$V = \ell wh$ *Volume of right rectangular prism*

$= 100 \cdot 50 \cdot 6$ *Substitute 100 for ℓ, 50 for w, and 6 for h.*

$= 30{,}000 \text{ ft}^3$ *Simplify.*

$G = \frac{30{,}000 \text{ ft}^3}{1} \cdot \frac{1 \text{ gal}}{0.134 \text{ ft}^3}$ *Convert the volume to gallons.*

$= \frac{30{,}000 \text{ gal}}{0.134}$ *Simplify.*

$\approx 223{,}881 \text{ gal}$ *Simplify.*

$W = \frac{223{,}881 \text{ gal}}{1} \cdot \frac{8.33 \text{ lbs}}{1 \text{ gal}}$ *Convert the gallons to pounds.*

$\approx 1{,}864{,}929 \text{ lbs}$ *Simplify.*

The weight of a volume of water contained in a pool in the shape of a rectangular prism is calculated in this application example. The dimensions of the rectangular prism and the density of water are given.

First, calculate the volume of the pool by substituting the given dimensions into the formula for the volume of a rectangular prism. The pool contains 30,000 ft^3 of water.

Second, convert the volume to gallons using the conversion factor 1 gallon per 0.134 ft^3. The pool contains 223,881 gallons of water.

Finally, convert the gallons of water to pounds of water using the density of water, 8.33 lbs per gallon. The weight of the water is approximately 1,864,929 lbs.

Example 3 Finding Volumes of Cylinders

Find the volume of the cylinder.

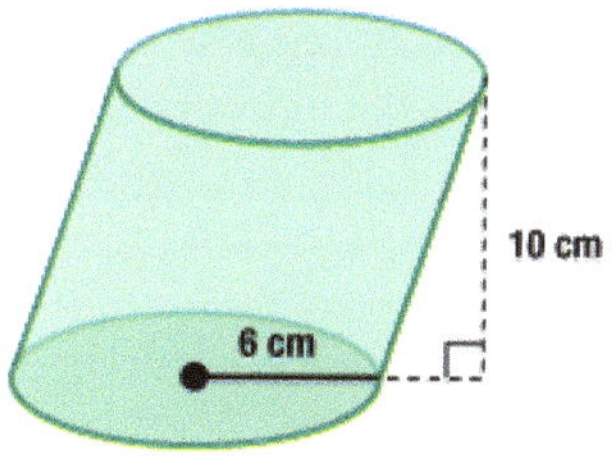

$V = \pi r^2 h$ *Volume of a cylinder*

$= \pi \cdot 6^2 \cdot 10$ *Substitute 6 for r and 10 for h.*

$= 360\pi \text{ cm}^3$ *Simplify.*

The volume of an oblique cylinder is determined in this example. The radius of the base and height of the cylinder are given.

To find the volume of the cylinder, substitute the given values into the formula for the volume of a cylinder. The volume of the cylinder is 360π cm^3. Notice that the formula works for both a right cylinder and an oblique cylinder.

Find the volume of a cylinder with a base area of 25π in^2 and a height equal to the radius.

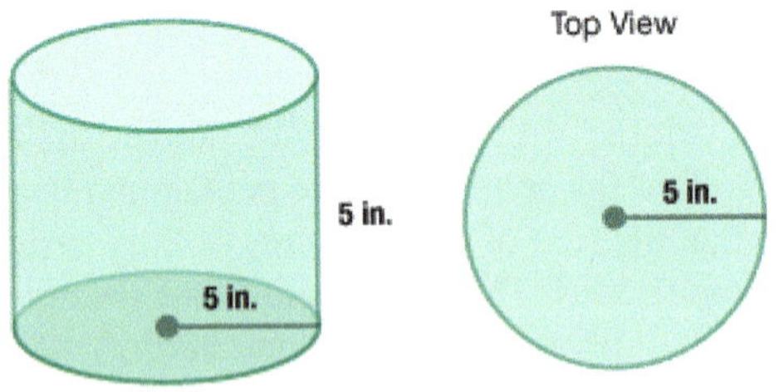

$A = \pi r^2$	*Area of a circle*
$25\pi = \pi r^2$	*Substitute 25π for A.*
$25 = r^2$	*Divide both sides by π.*
$5 \text{ in.} = r$	*Take the positive square root of both sides.*
$V = \pi r^2 h$	*Volume of a cylinder*
$= \pi \cdot 5^2 \cdot 5$	*Substitute 5 for r and 5 for h.*
$= 125\pi \text{ in}^3$	*Simplify.*

The volume of a right cylinder is determined in this example. The base area is given and the height is given as equal to the radius.

Use the formula for the area of a circle and the given area of the base to solve for the radius of the cylinder. The radius is equal to 5 inches, which is also equal to the height. To find the volume of the cylinder, substitute the radius and height into the formula for the volume of a cylinder. The volume of the cylinder is 125π in^3.

Example 4 Exploring Effects of Changing Dimension

The radius and the height are multiplied by $\frac{1}{2}$. Describe the effects on the volume.

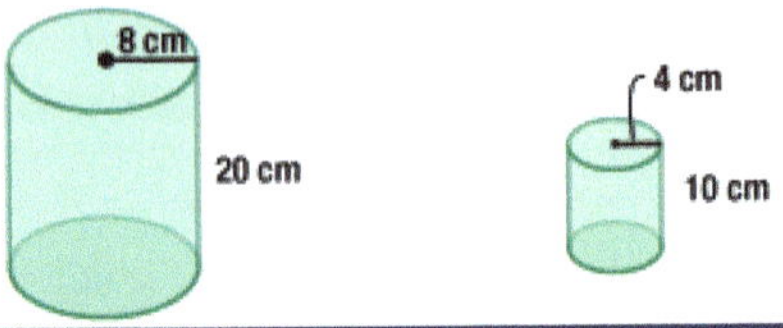

original dimensions:

$V = \pi r^2 h$	*Volume of a cylinder*
$= \pi \cdot 64 \cdot 20$	*Substitute 8 for r and 20 for h.*
$= 1{,}280\pi \text{ cm}^3$	*Simplify.*

radius and height halved:

$V = \pi r^2 h$

$= \pi\left(\frac{8}{2}\right)^2\left(\frac{20}{2}\right)$

$= \pi\left(\frac{64}{4}\right)\left(\frac{20}{2}\right)$

$= \frac{1}{8}(1280\pi) \text{ cm}^3$

$= 160\pi \text{ cm}^3$

If the radius and the height are multiplied by $\frac{1}{2}$, the volume is multiplied by $\left(\frac{1}{2}\right)^3$, or $\frac{1}{8}$.

The effects of a proportional increase in dimensions on the volume of a cylinder is examined in this example. The radius of a right cylinder is given as 8 cm and its height is 20 cm.

To calculate the volume of the cylinder, substitute the values for the radius and height into the formula for volume. The volume of the cylinder is 1280π cm^3. The new cylinder has dimensions that are half those of the original cylinder: the radius is 4 cm and the height is 10 cm. To calculate the volume, substitute the values into the formula for volume. The new volume is 160π cm^3.

The new volume is one-eighth the original volume, or $\left(\frac{1}{2}\right)^3$. The change in volume is a cube of the factor by which the dimensions were changed.

Example 5 Finding Volumes of Composite Three-Dimensional Figures

Find the volume of the composite three-dimensional figure.

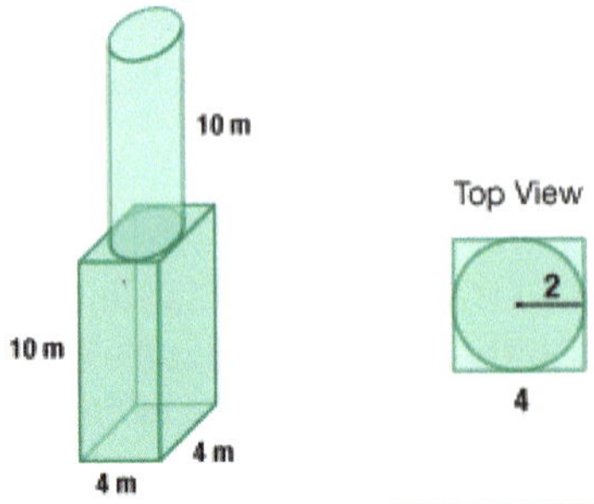

$V_{prism} = \ell \cdot w \cdot h$	*Volume of a rectangular prism*
$V_{prism} = 4 \cdot 4 \cdot 10$	*Substitute 4 for ℓ, 4 for w, and 10 for h.*
$= 160 \text{ m}^3$	*Simplify.*
$V_{cylinder} = \pi r^2 h$	*Volume of a cylinder*
$V_{cylinder} = \pi \cdot 2^2 \cdot 10$	*Substitute 2 for r and 10 for h.*
$= 40\pi \text{ m}^3$	*Simplify.*

$V_{composite} = V_{prism} + V_{cylinder} = 160 + 40\pi \text{ m}^3$

The volume of a composite solid is determined in this example. The solid is a rectangular prism with a cylinder on top of it. The length, width, and height of the rectangular prism and the height of the cylinder are given.

To calculate the volume of the prism, substitute the length, width, and height into the formula for the volume of a rectangular prism. The volume is 160 m^3.

To calculate the volume of the cylinder, first notice that the base of the cylinder is as wide as the base of the rectangular prism, meaning it has a diameter of 4 m and a radius of 2 m. Substitute the radius and the given height into the formula for the volume of a cylinder. The volume of the cylinder is 40π m^3.

The volume of the composite solid is $160 + 40\pi$ m^3.

11

11 Circles

Lines That Intersect Circles

Key Objectives

- Identify tangents, secants, and chords.
- Use properties of tangents to solve problems.

Key Terms

- The **interior of a circle** is the set of all points inside the circle.
- The **exterior of a circle** is the set of all points outside the circle.
- A **chord** is a segment whose endpoints lie on a circle.
- A **secant** is a line that intersects a circle at two points.
- A **tangent** is a line in the same plane as a circle that intersects it at exactly one point.
- The point where the tangent and a circle intersect is called the **point of tangency**.
- Two circles are **congruent circles** if and only if they have congruent radii.
- **Concentric circles** are coplanar circles with the same center.
- Two coplanar circles that intersect at exactly one point are called **tangent circles**.
- A **common tangent** is a line that is tangent to two circles.

Theorems, Postulates, Corollaries, and Properties

- **Theorem** If a line is tangent to a circle, then it is perpendicular to the radius drawn to the point of tangency.
- **Theorem** If a line is perpendicular to a radius of a circle at a point on the circle, then the line is tangent to the circle.
- **Theorem** If two segments are tangent to a circle from the same external point, then the segments are congruent.

Example 1 Identifying Lines and Segments that Intersect Circles

Identify each line or segment that intersects $\odot A$.

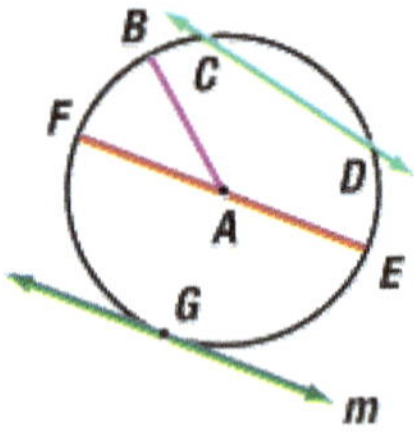

chord: $\overline{CD}$, $\overline{EF}$
secant: $\overleftrightarrow{CD}$
tangent: m
diameter: $\overline{EF}$
radius: $\overline{AB}$, $\overline{AE}$, $\overline{AF}$

TERM	DIAGRAM
A **chord** is a segment whose endpoints lie on a circle.	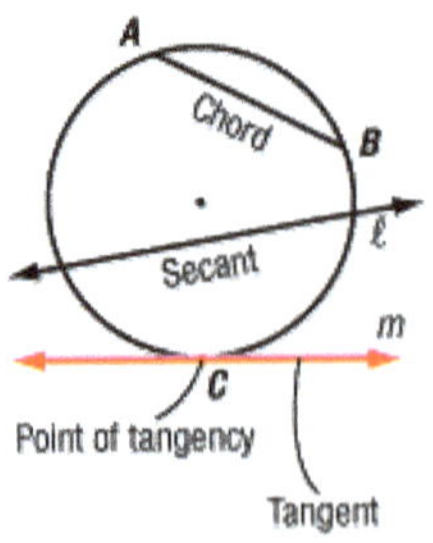
A **secant** is a line that intersects a circle at two points.	
A **tangent** is a line in the same plane as a circle that intersects it at exactly one point.	
The point where the tangent and circle intersect is called the **point of tangency**.	

A chord is a line segment that has its endpoints on a circle. $\overline{EF}$ is an example of a chord, and the segment $\overline{CD}$ is also a chord.

A secant is a line that intersects a circle at two points. Line CD is an example of a secant.

A tangent is a line that intersects a circle at one point. A tangent is in the plane of a circle. Line m is an example of a tangent. The point of intersection with the circle of a tangent is the point of tangency. The point of tangency for line m is G.

A radius is a segment that extends from the center of the circle to a point on the circle.

A diameter is a chord that passes through the center of a circle.

Example 2 Identifying Tangents of Circles

Find the length of each radius. Identify the point of tangency and write the equation of the tangent line at this point.

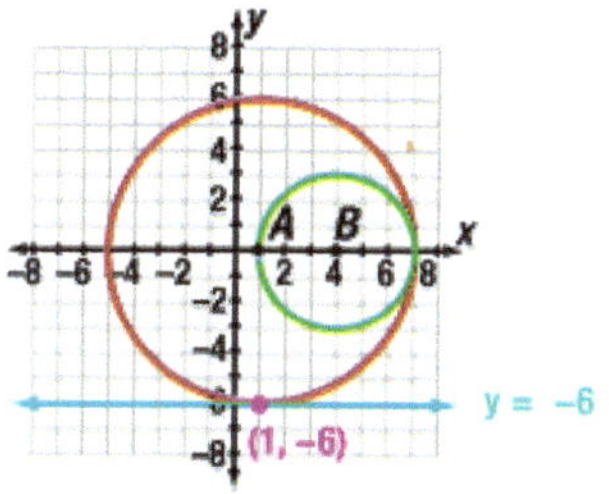

radius ⊙*A*: 6

radius ⊙*B*: 3

point of tangency: (1, −6)

tangent line: $y = -6$

Attributes of two circles are identified in this example. The circles are drawn in the coordinate plane.

The length of the radius of circle *A* can be determined by counting along the *x*-axis, since the center of circle *A* is on the *x*-axis at *A*(2, 0). The length of the radius of circle *A* is 6 units.

The length of the radius of circle *B* can also be determined by counting along the *x*-axis, since it is centered at *B*(4, 0). The length of the radius of circle *B* is 3 units.

The point of tangency of the line intersecting with circle *A* appears to be (1, −6).

The equation of the tangent line to circle *A* is $y = -6$.

Example 3 Problem-Solving Application

Theorem

If a line is tangent to a circle, then it is perpendicular to the radius drawn to the point of tangency.
(Line tangent to ⊙ → line ⊥ to radius)

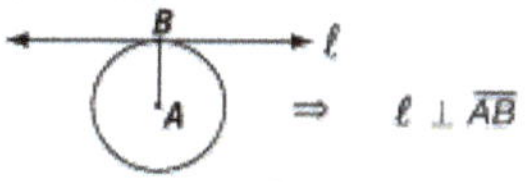

ℓ is tangent to ⊙*A*.

A line tangent to a circle is perpendicular to the radius that extends from the center of the circle to the point of tangency.

Theorem

If a line is perpendicular to a radius of a circle at a point on the circle, then the line is tangent to the circle.
(Line ⊥ to radius → line tangent to ⊙)

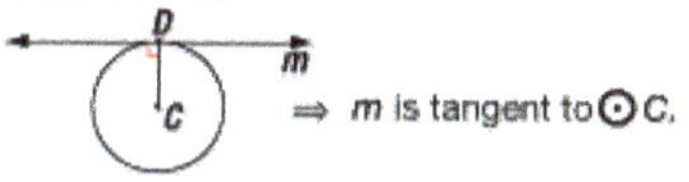

m is ⊥ to $\overline{CD}$ at *D*.

A line that is perpendicular to a radius of a circle at a point on the circle is tangent to the circle.

The summit of Mount McKinley is approximately 20,300 ft above sea level. What is the distance from the summit to the horizon to the nearest mile?

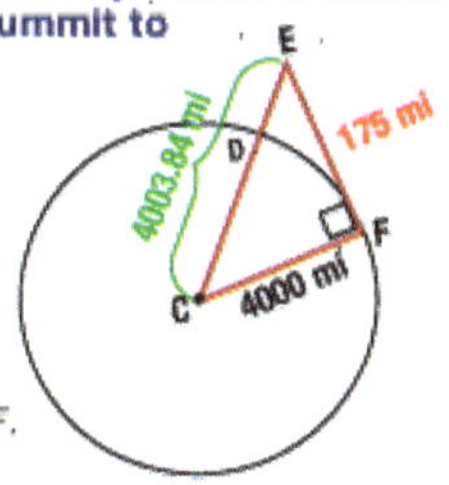

C - Center of the earth
E - Summit of Mount McKinley
F - Horizon

The distance from the summit to the horizon is *EF*.

$\overline{EF}$ is tangent to ⊙*C* at *F*.	*Given*
$\overline{EF} \perp \overline{CF}$	*line tangent to ⊙ → line ⊥ to radius*
△*CFE* is a right triangle.	*Def of right △*
ED = 20,300 ft	
$ED \approx 3.84$ mi	*Divide by 5,280, since there are 5,280 feet in a mile.*
$EC = 4000 + 3.84 = 4003.84$ mi	
$4000^2 + EF^2 = 4003.84^2$	*Pythagorean Theorem*
$EF \approx \sqrt{4003.84^2 - 4000^2} \approx 175$ mi	

The distance from the summit to the horizon is approximately 175 miles.

The length of the leg of a right triangle that is tangent to a circle is determined in this application example. The length of one leg of the triangle is given.

To solve this example, recognize that the line of sight from the mountain top to the horizon is a tangent line with the point of tangency as the endpoint to be measured to. The radius of Earth forms a right angle with the point on the horizon. The hypotenuse that forms a right triangle with the radius and line of sight is a line from the center of Earth to the point of view on top of the mountain. The length of the hypotenuse is the radius of Earth plus the height of the mountain, or 4003.84 mi.

Use the Pythagorean Theorem to find the length of the unknown leg, the distance to the horizon. The distance is approximately 175 mi.

Example 4 Using Properties of Tangents

Theorem

If two segments are tangent to a circle from the same external point, then the segments are congruent.
(2 segs. tangent to ⊙ from same ext. pt. → segs. ≅)

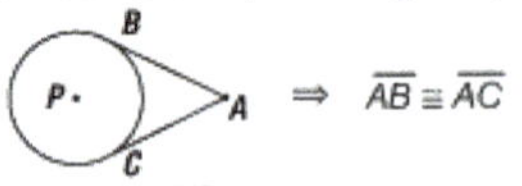

$\overline{AB}$ and $\overline{AC}$ are tangent to ⊙P.

Two line segments drawn from one point external to a circle to points of tangency with the circle are congruent.

$\overline{BC}$ and $\overline{DC}$ are tangent to ⊙A. Find BC.

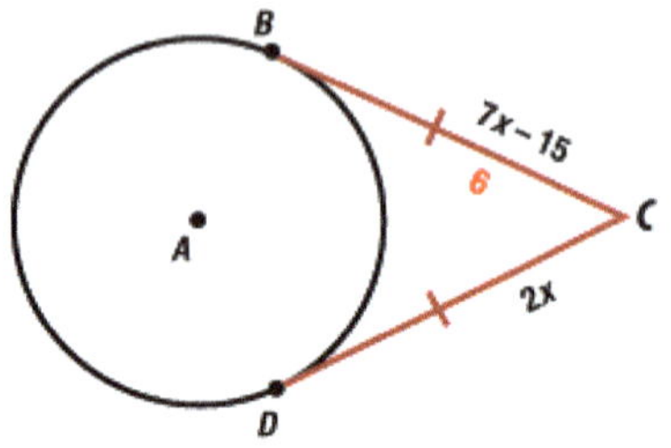

$BC = DC$	*2 segs. tangent to ⊙ from same ext. pt. → segs. ≅*
$7x - 15 = 2x$	*Substitute 7x – 15 for BC and 2x for DC.*
$5x = 15$	*Subtract 2x from both sides and add 15 to both sides.*
$x = 3$	*Divide both sides by 5.*
$BC = 7(3) - 15 = 6$	*Substitute 3 for x and simplify.*

The length of a line tangent to a circle is determined in this example. The length of two lines tangent to a circle and drawn from the same external point are given as algebraic expressions.

Since they extend from one point external to a circle and are both tangent to the circle, $\overline{BC} \cong \overline{DC}$ and $BC = DC$. Set up an equation that equates the given expressions for the lengths of the two line segments and solve for x. To find BC, substitute $x = 3$ into the expression for the length. The solution yields $BC = 6$.

Arcs and Chords

Key Objectives

- Apply properties of arcs.
- Apply properties of chords.

Key Terms

- A **central angle** is an angle whose vertex is the center of a circle.
- An **arc** is an unbroken part of a circle consisting of two points called the endpoints and all the points on the circle between them.
- A **minor arc** is an arc whose points are on or in the interior of a central angle.
- A **major arc** is an arc whose points are on or in the exterior of a central angle.
- If the endpoints of an arc lie on a diameter, the arc is a **semicircle**.
- **Adjacent arcs** are the arcs of the same circle that intersect at exactly one point.
- Within a circle or congruent circles, **congruent arcs** are two arcs that have the same measure.

Theorems, Postulates, Corollaries, and Properties

- **Arc Addition Postulate** The measure of an arc formed by two adjacent arcs is the sum of the measure of the two arcs.
- **Theorem** In a circle or congruent circles:
  (1) Congruent central angles have congruent chords.
  (2) Congruent chords have congruent arcs.
  (3) Congruent arcs have congruent central angles.
- **Theorem** In a circle, if a radius (or diameter) is perpendicular to a chord, then it bisects the chord and its arc.
- **Theorem** In a circle, the perpendicular bisector of a chord is a radius (or diameter).

Example 1 Data Application

The circle graph shows the type of shoes worn by students in a mall. Find $m\overset{\frown}{ABC}$.

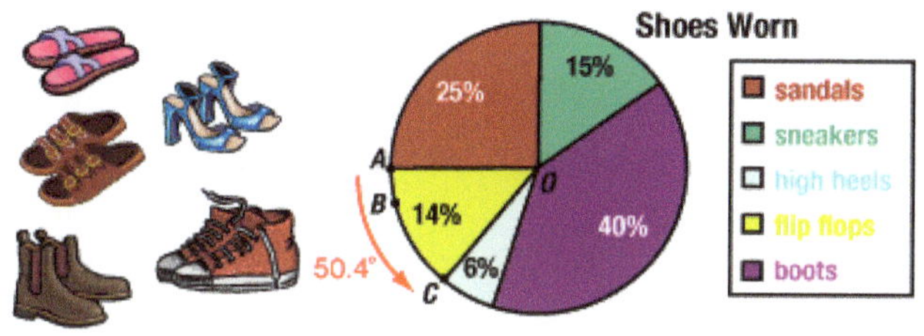

$m\overset{\frown}{ABC} = m\angle AOC$ *m of arc = m of central ∠.*

$m\overset{\frown}{ABC} = 0.14(360°)$ *Central ∠ is 14% of the ⊙.*

$= 50.4°$ *Simplify.*

The measure of an arc is determined in this application example. The percent value of the sectors in a pie chart are given.

To find $m\overset{\frown}{ABC}$, recognize that the percents given for the pie chart can represent the percent area of the sector of the circle, the percent measure of the central angle that forms the sector, or the percent measure of the length of the arc.

The measure of an arc is equal to the measure of the central angle between the radii that form its endpoints. The measure of the central angle that forms $\overset{\frown}{ABC}$ is 14 percent of the whole circle, or (0.14)360°. The measure of the arc is 50.4°.

Example 2 Using the Arc Addition Postulate

Arc Addition Postulate

The measure of an arc formed by two adjacent arcs is the sum of the measures of the two arcs.
(Arc Add. Post.)

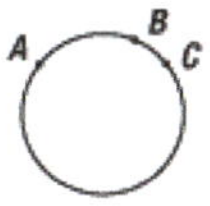

$m\overset{\frown}{ABC} = m\overset{\frown}{AB} + m\overset{\frown}{BC}$

The measure of an arc that is formed by two adjacent arcs (arcs that share an endpoint) is the sum of the measure of the two arcs.

Find $m\overset{\frown}{CDE}$.

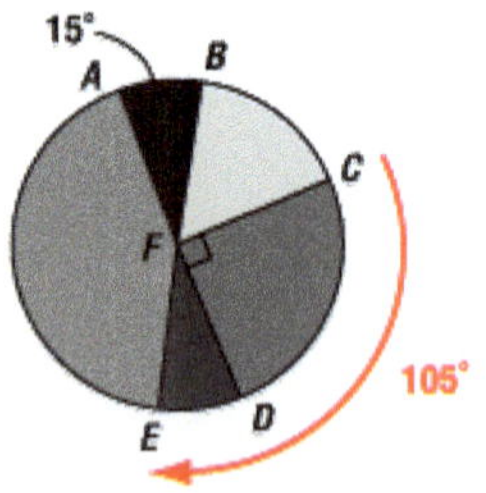

$m\overset{\frown}{CD} = 90°$	*$m\angle CFD = 90°$*
$m\angle DFE = 15°$	*Vert. ∠s Thm.*
$m\overset{\frown}{DE} = 15°$	*$m\angle DFE = 15°$*
$m\overset{\frown}{CE} = m\overset{\frown}{CD} + m\overset{\frown}{DE}$	*Arc Add. Post.*
$m\overset{\frown}{CE} = 90 + 15 = 105°$	*Substitute and simplify.*

The measure of an arc is determined in this example. The measures of two central angles in the circle are given. Segment AD and segment BE are given as diameters.

$m\overset{\frown}{CD}$ is 90°, since this is the measure of its central angle.

m AFB = m EFD by the Vertical Angles Theorem.

So $m\overset{\frown}{CDE}$ is 90° + 15° = 105° by the Arc Addition Postulate.

Example 3 Applying Congruent Angles, Arcs, and Chords

Theorem

In a circle or congruent circles, congruent central angles have congruent chords. Congruent chords have congruent arcs. Congruent arcs have congruent central angles.
(≅ arcs have ≅ central ∠s have ≅ chords)

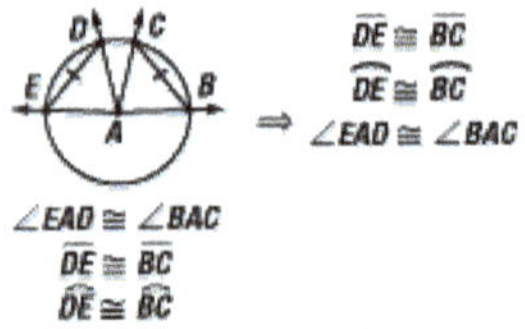

In a circle, or in congruent circles:

(1) Congruent central angles have congruent chords.
(2) Congruent chords have congruent arcs.
(3) Congruent arcs have congruent central angles.

Find each measure.

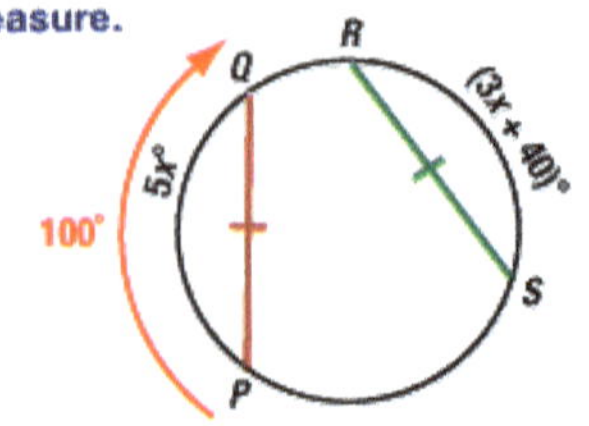

$m\overset{\frown}{PQ}$

$\overline{PQ} \cong \overline{RS}$	*Given*
$\overset{\frown}{PQ} \cong \overset{\frown}{RS}$	*≅ chords have ≅ arcs.*
$m\overset{\frown}{PQ} = m\overset{\frown}{RS}$	*Def. of ≅ arcs*
$5x = 3x + 40$	*Substitute the given measures.*
$2x = 40$	*Subtract 3x from both sides.*
$x = 20$	*Divide both sides by 2.*
$m\overset{\frown}{PQ} = 5(20) = 100°$	*Substitute 20 for x and simplify.*

The measure of $\overset{\frown}{PQ}$ is determined in this example. It is given that $\overline{PQ} \cong \overline{RS}$. The lengths of $\overset{\frown}{PQ}$ and $\overset{\frown}{RS}$ are given as algebraic expressions.

Since congruent chords have congruent arcs, $m\overset{\frown}{PQ} = m\overset{\frown}{RS}$. Substitute the algebraic expressions given for the lengths of each arc into the equation. Solve for x. Substitute $x = 20$ back into the expression for the length of $\overset{\frown}{PQ}$. Solving yields $m\overset{\frown}{PQ} = 100°$.

Find each measure.

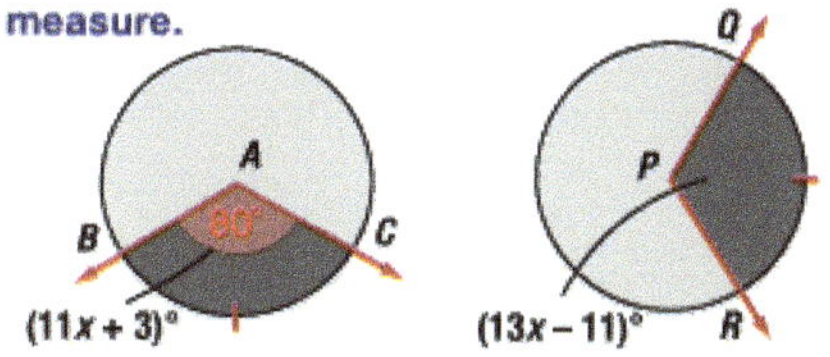

Given: $\odot A \cong \odot P$, and $\overset{\frown}{BC} \cong \overset{\frown}{QR}$.

m∠BAC

$\overset{\frown}{BC} \cong \overset{\frown}{QR}$	Given
$\angle BAC \cong \angle QPR$	≅ arcs have ≅ central ∠s.
$m\angle BAC = m\angle QPR$	Def. of ≅ ∠s
$11x + 3 = 13x - 11$	Substitute the given measures.
$3 = 2x - 11$	Subtract 11x from both sides.
$14 = 2x$	Add 11 to both sides.
$7 = x$	Divide both sides by 2.
$m\angle BAC = 11(7) + 3 = 80°$	Substitute 7 for x and simplify.

The measure of a central angle is determined in this example. The two circles are given to be congruent and $\overset{\frown}{BC} \cong \overset{\frown}{QR}$.

Since congruent arcs have congruent central angles, $m\angle BAC = m\angle QPR$. Substitute the algebraic expressions into the equation. Solve for x. Substitute $x = 7$ into the expression for $m\angle BAC$. The solution yields $m\angle BAC = 80°$.

Example 4 Using Radii and Chords

Theorem

In a circle, if a radius (or diameter) is perpendicular to a chord, then it bisects the chord and its arc.
(Diam. ⊥ chord → diam. bisects chord and arc)

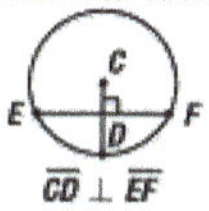

$\Rightarrow$ $\overline{CD}$ bisects $\overline{EF}$ and $\overset{\frown}{EF}$.

$\overline{CD} \perp \overline{EF}$

If a radius or diameter of a circle is perpendicular to a chord, then it bisects the chord and its arc.

Theorem

In a circle, the perpendicular bisector of a chord is a radius (or diameter).
(⊥ bisector of chord is diam.)

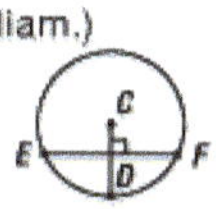

$\Rightarrow$ $\overline{CD}$ is a diameter of $\odot C$.

$\overline{CD}$ is ⊥ bisector of $\overline{EF}$.

In a circle, the perpendicular bisector of a chord is a radius or diameter.

Find RS.

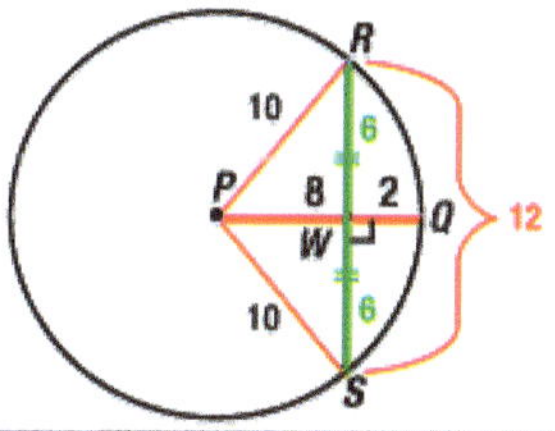

$PR = PS = 10$	Radii of a ⊙ are ≅.
$\overline{RW} \cong \overline{SW}$	$\overline{PQ} \perp \overline{RS}$, so $\overline{PQ}$ bisects $\overline{RS}$.
$RW^2 + PW^2 = PR^2$	Pythagorean Thm.
$RW^2 + 8^2 = 10^2$	Substitute 8 for +PW and 10 for PR.
$RW^2 = 36$	Subtract 8^2 from both sides.
$RW = 6$	Take the positive square root of both sides.
$RS = 2(RW)$	$\overline{PQ} \perp \overline{RS}$, so $\overline{PQ}$ bisects $\overline{RS}$.
$RS = 2(6) = 12$	Substitute 6 for RW and simplify.

The measure of a chord is determined in this example. The length of the radius of the circle is given. It is given that $\overline{PQ}$ is perpendicular to $\overline{RS}$. It is also given that $\overline{PQ}$ is divided into lengths of 8 and 2 units by the intersection of $\overline{RS}$.

The trick to solving this example is to draw radii, the lengths of which are known, to R and S. Because $\overline{PQ}$ and $\overline{RS}$ are perpendicular, $\overline{PQ}$ bisects $\overline{RS}$, and RW and SW are equal. Use the Pythagorean Theorem to determine RW (or SW). Substituting the known lengths gives $RW = 6$. Since RW and SW are equal, $RS = 2(6) = 12$.

Inscribed Angles

Key Objectives

- Find the measure of an inscribed angle.
- Use inscribed angles and their properties to solve problems.

Key Terms

- An **inscribed angle** is an angle whose vertex is on a circle and whose sides contain chords of the circle.
- An **intercepted arc** consists of endpoints that lie on the sides of an inscribed angle and all the points of the circle between them.
- A chord or arc **subtends** an angle if its endpoints lie on the sides of the angle.

Theorems, Postulates, Corollaries, and Properties

- **Inscribed Angle Theorem** The measure of an inscribed angle is half the measure of its intercepted arc.

$$m\angle ABC = \frac{1}{2} m\overset{\frown}{AC}$$

- **Corollary** If inscribed angles of a circle intercept the same arc or are subtended by the same chord or arc, then the angles are congruent.
- **Theorem** An inscribed angle subtends a semicircle if and only if the angle is a right angle.
- **Theorem** If a quadrilateral is inscribed in a circle, then its opposite angles are supplementary.

Example 1 Finding Measures of Arcs and Inscribed Angles

Inscribed Angle Theorem

The measure of an inscribed angle is half the measure of its intercepted arc.
(Inscribed ∠ Thm.)

$m\angle ABC = \frac{1}{2} m\overset{\frown}{AC}$

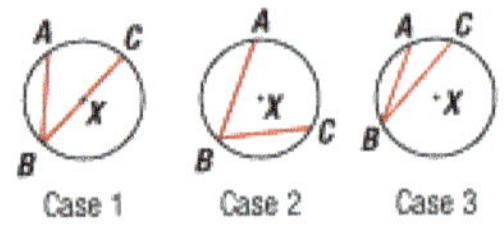

According to the Inscribed Angle Theorem, the measure of an inscribed angle is half the measure of its intercepted arc.

Notice how this theorem covers three different cases of inscribed angles. Also, think about how this differs from the relationship of a central angle and its arc.

Find each measure.

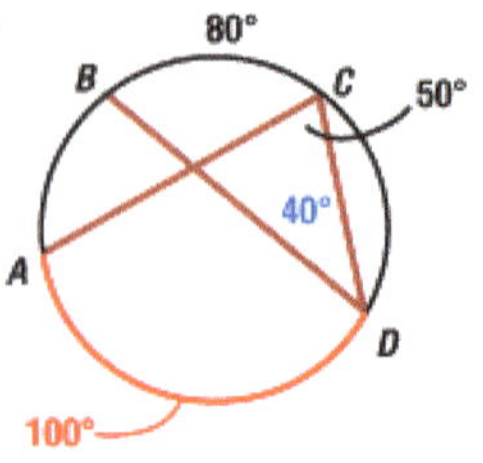

m∠CDB

$m\angle CDB = \frac{1}{2} m\overset{\frown}{CB}$ *Inscribed ∠ Thm.*

$= \frac{1}{2}(80°) = 40°$ *Substitute 80° for $m\overset{\frown}{CB}$.*

$m\overset{\frown}{AD}$

$m\overset{\frown}{AD} = 2m\angle ACB$ *Inscribed ∠ Thm.*

$= 2(50°) = 100°$ *Substitute 50° for m∠ACD.*

The measure of an inscribed angle and the measure of an arc intercepted by an inscribed angle are determined in these examples.

In the first example, the measure of the arc intercepted by the inscribed angle *CDB* is given. According to the Inscribed Angle Theorem, the measure of the inscribed angle is equal to one-half the measure of the arc,

$m\angle CDB = (1/2)m\overset{\frown}{CB} = (1/2)80° = 40°$.

In the second example, the measure of the inscribed angle that intercepts the arc is given. According to the Inscribed Angle Theorem, the measure of the arc is twice the measure of the inscribed angle that intercepts it,

$m\overset{\frown}{AD} = 2m\angle ACB = 2(50°) = 100°$.

Example 2 Recreation Application

Corollary

If inscribed angles of a circle intercept the same arc or are subtended by the same chord or arc, then the angles are congruent.

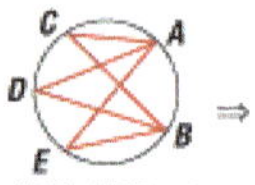

⇒ ∠ACB ≅ ∠ADB ≅ ∠AEB
(and ∠CAE ≅ ∠CBE)

∠ACB, ∠ADB, and ∠AEB intercept $\overset{\frown}{AB}$

If inscribed angles of a circle intercept the same arc or are subtended by the same chord, then the angles are congruent.

Match up the inscribed angles that intercept the same arcs in the diagram with those angles listed as congruent.

Find m∠QZR.

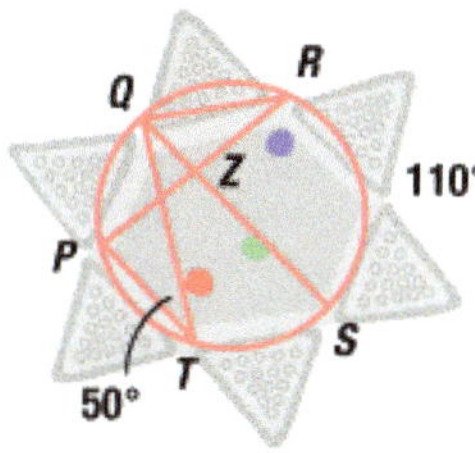

$m\angle PTQ = 50^\circ$	*Given*
$m\overset{\frown}{PQ} = 100^\circ$	*Inscribed ∠ Thm.*
$m\angle QRZ = \frac{1}{2}(100^\circ) = 50^\circ$	*Inscribed ∠ Thm.*
$m\overset{\frown}{RS} = 110^\circ$	*Given*
$m\angle RQZ = \frac{1}{2}(110^\circ) = 55^\circ$	*Inscribed ∠ Thm.*
$m\angle QZR + m\angle QRZ + m\angle RQZ = 180^\circ$	*△ Sum Thm.*
$m\angle QZR + 50^\circ + 55^\circ = 180^\circ$	*Substitute 50° for m∠QRZ and 55° for m∠RQZ.*
$m\angle QZR + 105^\circ = 180^\circ$	*Simplify.*
$m\angle QZR = 75^\circ$	*Subtract 105° from both sides.*

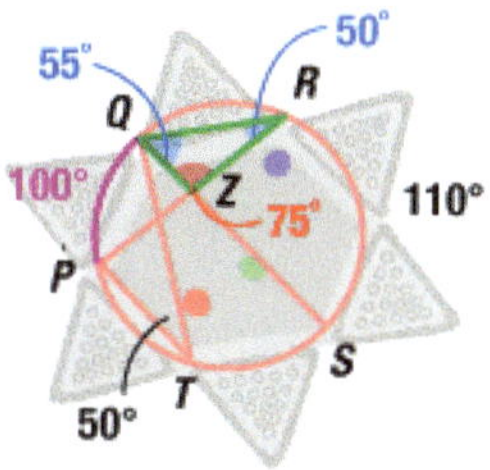

The measure of an inscribed angle is determined in this application example. $m\angle PTQ$ is given as 50° and $m\overset{\frown}{RS}$ is given as 110°.

It is possible to use the given information to find the measures of angles in the triangle that contains $\angle QZR$. According to the Inscribed Angle Theorem, $m\overset{\frown}{PQ} = 100^\circ$ and therefore $m\angle QRZ = 50^\circ$. Also according to the Inscribed Angle Theorem, $m\angle RQZ = 55^\circ$. Knowing two of the three angles in $\triangle QRZ$, the Triangle Sum Theorem can be used to find the measure of the unknown angle. Thus, $m\angle QZR = 75^\circ$.

Example 3 Finding Angle Measures in Inscribed Triangles

Theorem

An inscribed angle subtends a semicircle if and only if the angle is a right angle.

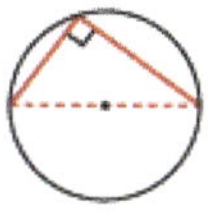

An inscribed angle subtends a semicircle if and only if the angle is a right angle.

To subtend something means to be opposite it and to extend from one endpoint to the other endpoint.

Think about the measure of the right angle and the measure of the arc, a semicircle, formed by the right angle. Is this theorem a special case of the Inscribed Angles Theorem?

Find each value.

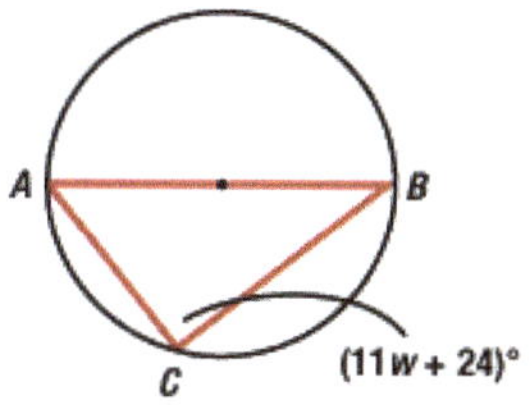

w

∠ACB is a right angle.	*∠ACB is inscribed in a semicircle.*
$m\angle ACB = 90^\circ$	*Def. of right ∠*
$11w + 24 = 90$	*Substitute 11w + 24 for m∠ACB.*
$11w = 66$	*Subtract 24 from both sides.*
$w = 6$	*Divide both sides by 11.*

An unknown in the expression for the measure of an inscribed angle is determined in this example. The expression for the angle measure is given. It is given that $\overline{AB}$ is a diameter.

Because $\overline{AB}$ is a diameter, $\angle ACB$ is inscribed in a semicircle and $m\angle ACB$ is 90°. Equate $m\angle ACB$ with the given expression and solve for w. The solution yields $w = 6$.

Find each value.

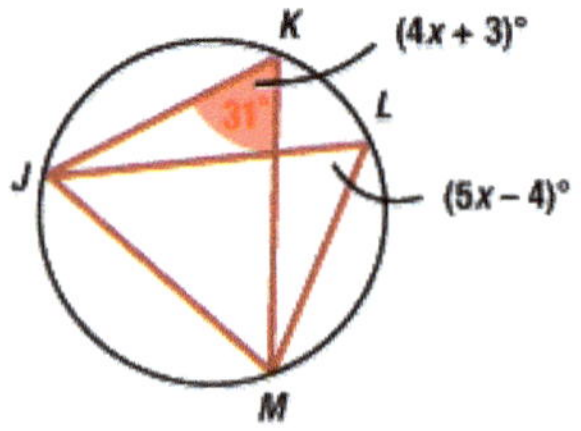

m∠JKM

m∠JKM = m∠JLM	*Inscribed ∠ Thm.*
4x + 3 = 5x − 4	*Substitute 4x + 3 for m∠JKM and 5x − 4 for m∠JLM.*
3 = x − 4	*Subtract 4x from both sides.*
7 = x	*Add 4 to both sides.*
m∠JKM = 4(7) + 3 = 31°	*Substitute 7 for x.*

The measure of an inscribed angle is determined in this example. The expression for the angle measure is given. The expression for the measure of an inscribed angle that subtends the same arc is also given.

According to the Inscribed Angle Theorem, m∠*JKM* = m∠*JLM*. Substitute the two expressions for the measures of the angle into the equation and solve for x. The solution yields $x = 7$. Substitute this value back into the expression for m∠*JKM* and solve. The solution yields m∠*JKM* = 31°.

Example 4 Finding Measures in Inscribed Quadrilaterals

Theorem

If a quadrilateral is inscribed in a circle, then its opposite angles are supplementary.
(Quad. inscribed in circle → opp. ∠s supp.)

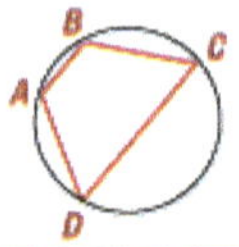

⇒ ∠*A* and ∠*C* are supplementary. ∠*B* and ∠*D* are supplementary.

If a quadrilateral is inscribed in a circle, then both pairs of opposite angles are supplementary.

Look at the two intercepted arcs of opposite angles to understand why this theorem is true. How are the two arcs related?

Find the angle measures of *PQRS*.

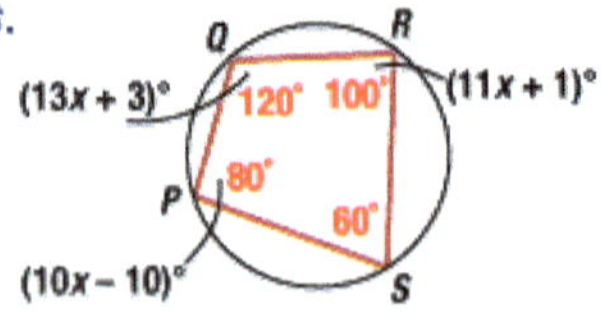

m∠P + m∠R = 180°	*PQRS is inscribed in a ⊙.*
10x − 10 + 11x + 1 = 180	*Substitute 10x − 10 for m∠P and 11x + 1 for m∠R.*
21x − 9 = 180	*Combine like terms.*
21x = 189	*Add 9 to both sides.*
x = 9	*Divide both sides by 21.*
m∠P = 10(9) − 10 = 80°	*Substitute 9 for x in each expression.*
m∠R = 11(9) + 1 = 100°	
m∠Q = 13(9) + 3 = 120°	
m∠Q + m∠S = 180°	*∠Q and ∠S are supp.*
120° + m∠S = 180°	*Substitute 120° for ∠Q.*
m∠S = 60°	*Subtract 120° from both sides.*

The angle measures of a quadrilateral inscribed in a circle are determined in this example. The measures of three of the angles are given as algebraic expressions in one variable, x.

Since ∠*P* and ∠*R* are opposite angles and therefore supplementary, the sum of their measures is 180°. Substitute the expressions for the measures of the angles into the equation and solve for x. The solution yields $x = 9$. Substitute this value of x back into the expressions for the angles to find their measures. The solutions yield m∠*P* = 80°, m∠*R* = 100°, and m∠*Q* = 120°. The sum of the angles in a quadrilateral is 360°, so m∠*S* = 360° − (80° + 100° + 120°) = 60°. Or, since ∠*S* is opposite ∠*Q*, m∠*S* = 180° − 120° = 60°.

Angle Relationships in Circles

Key Objectives

- Find the measures of angles formed by lines that intersect circles.
- Use angle measures to solve problems.

Theorems, Postulates, Corollaries, and Properties

- **Theorem** If a tangent and a secant (or chord) intersect on a circle at the point of tangency, then the measure of the angle formed is half the measure of its intercepted arc.
- **Theorem** If two secants or chords intersect in the interior of a circle, then the measure of each angle formed is half the sum of the measures of its intercepted arcs.
- **Theorem** If a tangent and a secant, two tangents, or two secants intersect in the exterior of a circle, then the measure of the angle formed is half the difference of the measures of its intercepted arcs.

Example 1 Using Tangent-Secant and Tangent-Chord Angles

Theorem

If a tangent and a secant (or chord) intersect on a circle at the point of tangency, then the measure of the angle formed is half the measure of its intercepted arc.

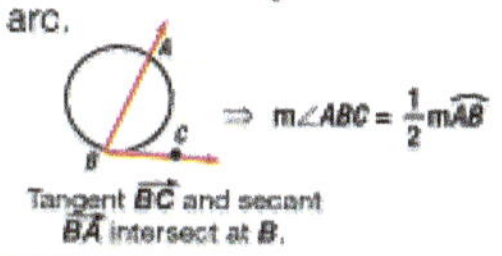

If a tangent and a secant (or chord) intersect on a circle at the point of tangency, then the measure of the angle formed is half the measure of its intercepted arc.

Compare this theorem to the Inscribed Angles Theorem.

Find each measure.

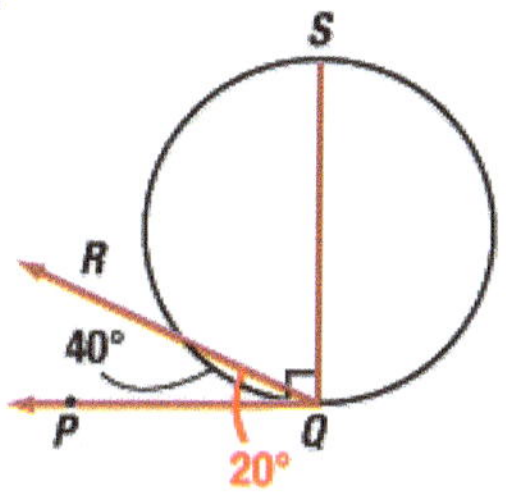

$m\angle PQR$

$m\angle PQR = \frac{1}{2}m\overset{\frown}{QR}$ — *Tangent $\overrightarrow{QP}$ and secant $\overrightarrow{QR}$ intersect at Q.*

$m\angle PQR = \frac{1}{2}(40°)$ — *Substitute 40° for $m\overset{\frown}{QR}$.*

$= 20°$ — *Simplify.*

The measure of an angle between a tangent and a secant to a circle is determined in this example. The measure of the intercepted arc is given.

The tangent and the secant intersect the circle at the point of tangency, so the measure of $\angle PQR$ is equal to one-half the measure of the arc it intercepts. Substituting the given value and simplifying gives $m\angle PQR = 20°$.

Find each measure.

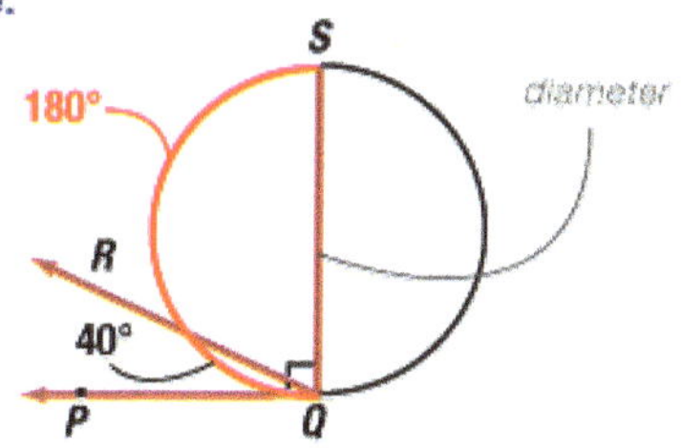

$m\overset{\frown}{SRQ}$

$m\angle SQP = \frac{1}{2}m\overset{\frown}{SRQ}$ — *Tangent $\overleftrightarrow{QP}$ and chord $\overline{QS}$ intersect at Q.*

$90° = \frac{1}{2}m\overset{\frown}{SRQ}$ — *$\angle SQP$ is a right $\angle$.*

$180° = m\overset{\frown}{SRQ}$ — *Multiply both sides by 2.*

The measure of an arc intercepted by a tangent and a chord to a circle is determined in this example. It is given that the chord is a diameter.

The tangent and the chord intersect the circle at the point of tangency. Since the chord is a diameter the angle at the intersection is a right angle. $m\overset{\frown}{SRQ}$ is equal to twice the measure of the angle that intercepts it. Substituting the given value and simplifying gives $m\overset{\frown}{SRQ} = 180°$.

Example 2 Finding Angle Measures Inside a Circle

Theorem

If two secants or chords intersect in the interior of a circle, then the measure of each angle formed is half the sum of the measures of its intercepted arcs.

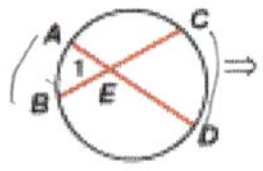

$\Rightarrow m\angle 1 = \frac{1}{2}(m\widehat{AB} + m\widehat{CD})$

If two secants or chords intersect in the interior of a circle, then the measure of each angle formed is half the sum of the measures of its intercepted arcs.

Another way to think of this is that the measure of the angle is equal to the average of the measures of the arcs it intercepts.

Find each measure.

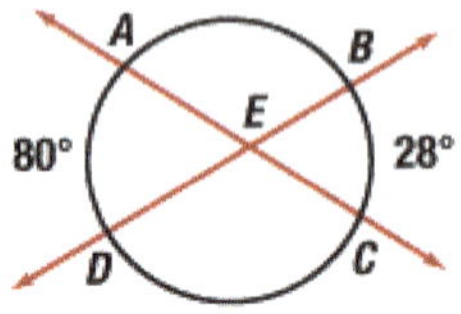

m∠BEC

$m\angle BEC = \frac{1}{2}(m\widehat{AD} + m\widehat{BC})$ *Secants $\overline{AC}$ and $\overline{BD}$ intersect at E.*

$= \frac{1}{2}(80^\circ + 28^\circ)$ *Substitute 80° for $m\widehat{AD}$ and 28° for $m\widehat{BC}$.*

$= \frac{1}{2}(108^\circ)$ *Simplify.*

$= 54^\circ$ *Simplify.*

The measure of an angle formed by the intersection of two secants inside a circle is determined in this example. The measures of the two intercepted arcs are given.

The measure of $\angle BEC$ equals one-half the sum of $m\widehat{AD}$ and $m\widehat{BC}$. Substituting the given values yields $m\angle BEC = 54^\circ$.

Example 3 Finding Measures Using Tangents and Secants

Theorem

If a tangent and a secant, two tangents, or two secants intersect in the exterior of a circle, then the measure of the angle formed is half the difference of the measures of its intercepted arcs.

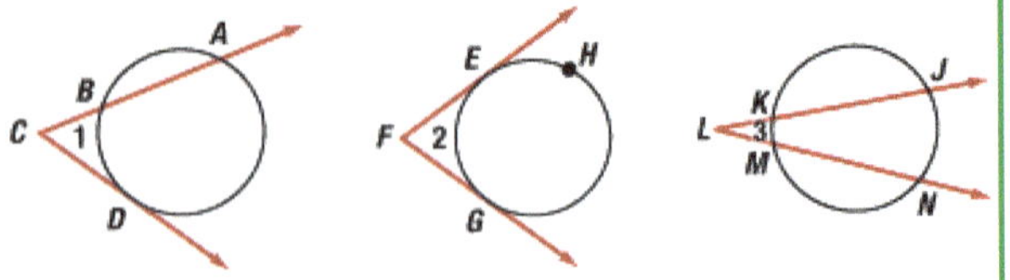

$m\angle 1 = \frac{1}{2}(m\widehat{AD} - m\widehat{BD})$ $m\angle 2 = \frac{1}{2}(m\widehat{EHG} - m\widehat{EG})$ $m\angle 3 = \frac{1}{2}(m\widehat{JN} - m\widehat{KM})$

If a tangent and a secant, two tangents, or two secants intersect in the exterior of a circle, then the measure of the angle formed is half the difference of the measures of its intercepted arcs.

Find the value of *x*.

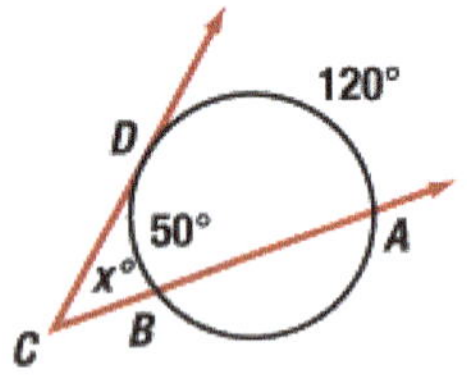

$x = \frac{1}{2}(m\widehat{AD} - m\widehat{BD})$ *Write the formula.*

$= \frac{1}{2}(120^\circ - 50^\circ)$ *Substitute 120° for $m\widehat{AD}$ and 50° for $m\widehat{BD}$.*

$= \frac{1}{2}(70^\circ) = 35^\circ$ *Simplify.*

The measure of an angle formed by a tangent and a secant to a circle is determined in this example. The measures of the arcs intercepted by the angle are given.

The measure of the angle is equal to twice the difference between the arcs intercepted by the tangent and secant. Substituting the given values into the formula gives $x = (1/2)(120^\circ - 50^\circ) = 35^\circ$.

Find the value of y.

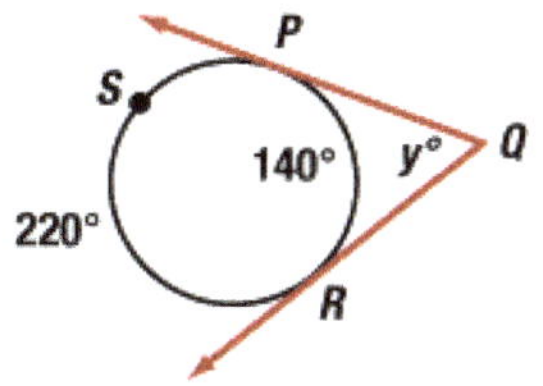

$y = \frac{1}{2}(m\overparen{RSP} - m\overparen{RP})$ — *Write the formula.*

$= \frac{1}{2}(m\overparen{RSP} - 140°)$ — *Substitute 140° for $m\overparen{RP}$.*

$= \frac{1}{2}(220° - 140°)$ — *$m\overparen{RSP} = 360° - 140° = 220°$*

$= \frac{1}{2}(80°) = 40°$ — *Simplify.*

The measure of an angle formed by two tangents to a circle is determined in this example. The measures of the arcs intercepted by the angle are given.

The measure of the angle is equal to twice the difference between the arcs intercepted by the tangents. Substituting the given values into the formula gives
$y = (1/2)(220° - 140°) = 40°$.

Example 4 Cycling Application

In the bicycle frame shown, if $m\overparen{AE} = 50°$ and $m\overparen{BD} = 10°$, what is $m\angle ACE$?

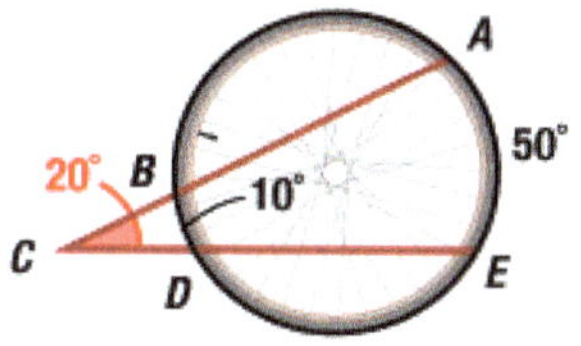

$$m\angle ACE = \frac{1}{2}(m\overparen{AE} - m\overparen{BD})$$
$$= \frac{1}{2}(50° - 10°)$$
$$= \frac{1}{2}(40°) = 20°$$

The measure of an angle formed by two secants to a circle is determined in this application example. The measures of the arcs intercepted by the angle are given.

The measure of the angle is equal to twice the difference between the arcs intercepted by the secants. Substituting the given values into the formula gives
$m\angle ACE = (1/2)(50° - 10°) = 20°$.

Example 5 Finding Arc Measures

Find $m\overparen{QT}$.

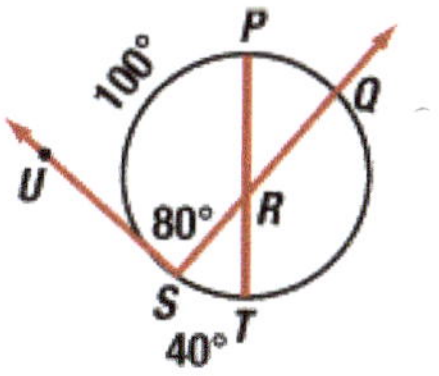

$m\overparen{SPQ} = 2m\angle USQ$
$= 2(80°) = 160°$

$m\overparen{STQ} = 360° - m\overparen{SPQ}$
$= 360° - 160° = 200°$

$m\overparen{PQ} = m\overparen{SPQ} - m\overparen{SP}$
$= 160° - 100° = 60°$

$m\overparen{QT} = m\overparen{STQ} - m\overparen{ST}$
$= 200° - 40° = 160°$

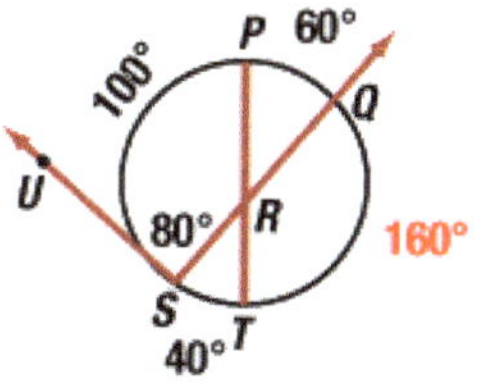

The measure of an arc is determined in this example. The measures of two arcs and an angle formed by the two chords and a tangent intersecting the circle are given.

The following approach is different, and simpler, than the one illustrated. The measure of $\overparen{SPQ}$ is twice the measure of $\angle USQ$ according to the Inscribed Angles Theorem, or 160°. And $m\overparen{PQ} = 160° - 100° = 60°$. Since the measure of the circumference of the circle is 360°,
$m\overparen{QT} = 360° - (m\overparen{ST} + m\overparen{SP} + m\overparen{PQ}) = 160°$.

Segment Relationships in Circles

Key Objectives

- Find the lengths of segments formed by lines that intersect circles.
- Use the lengths of segments in circles to solve problems.

Key Terms

- A **secant segment** is a segment of a secant with at least one endpoint on the circle.
- An **external secant segment** is a secant segment that lies in the exterior of the circle with one endpoint on the circle.
- A **tangent segment** is a segment of a tangent with one endpoint on the circle.

Theorems, Postulates, Corollaries, and Properties

- **Chord-Chord Product Theorem** If two chords intersect in the interior of a circle, then the products of the lengths of the segments of the chords are equal.
- **Secant-Secant Product Theorem** If two secants intersect in the exterior of a circle, then the product of the lengths of one secant segment and its external segment equals the product of the lengths of the other secant segment and its external segment. (whole · outside = whole · outside)
- **Secant-Tangent Product Theorem** If a secant and a tangent intersect in the exterior of a circle, then the product of the lengths of the secant segment and its external segment equals the length of the tangent segment squared. (whole · outside = tangent^2)

Example 1 Applying the Chord-Chord Product Theorem

Chord-Chord Product Theorem

If two chords intersect in the interior of a circle, then the products of the lengths of the segments of the chords are equal. (Chord-Chord Product Thm.)

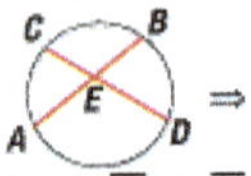

$\Rightarrow AE \cdot EB = CE \cdot ED$

Chords $\overline{AB}$ and $\overline{CD}$ intersect at E.

If two chords intersect in the interior of a circle, then the products of the lengths of the segments of the chords are equal.

Find the value of x and the length of each chord.

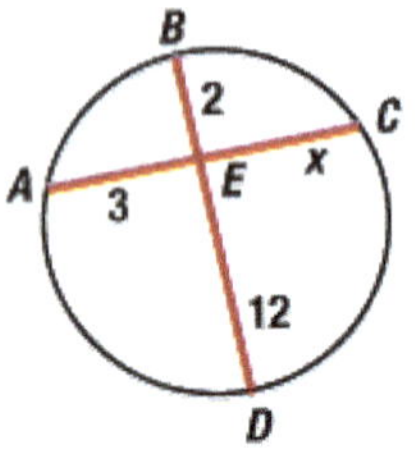

$AE \cdot EC = BE \cdot ED$	*Chord-Chord Product Thm.*
$3x = 2 \cdot 12$	*Substitute the given values.*
$3x = 24$	*Simplify.*
$x = 8$	*Divide both sides by 3.*

Chord Lengths:

$BD = 12 + 2 = 14$

$AC = 3 + 8 = 11$

The Chord-Chord Product Theorem is used in this example to determine the length of two chords. The lengths of three of the segments formed by two chords intersecting inside a circle are given.

Substitute the given values into the equation for the Chord-Chord Product Theorem and solve for x. The solution yields $x = 8$. The lengths of the chords are $BD = 14$ and $AC = 11$.

Example 2 Environmental Application

Chord $\overline{PR}$ is a perpendicular bisector of $\overline{QS}$. Find the pond's diameter.

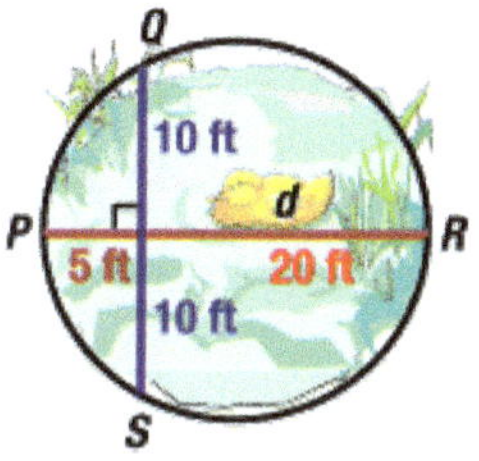

Since $\overline{PR}$ is the perpendicular bisector of a chord, $\overline{PR}$ is a diameter of the pond.

$5d = 10 \cdot 10$	*Chord-Chord Product Thm.*
$5d = 100$	*Simplify.*
$d = 20$	*Divide both sides by 5.*

$PR = 20 + 5 = 25$ ft

The diameter of the pond is 25 ft.

The Chord-Chord Product Theorem is used in this application example to determine the diameter of a circle. It is given that one chord is the perpendicular bisector of a second chord. One segment length from each of the chords, 5 ft and 10 ft, are also given.

$\overline{PR}$ is a diameter because it is perpendicular to and bisects $\overline{QS}$. Substitute the given values into the equation for the Chord-Chord Product Theorem and solve for d. The solution yields $d = 20$. The diameter of the pond is the sum of the lengths of the two segments that make up the diameter, which is 25 ft.

Example 3 Applying the Secant-Secant Product Theorem

Secant-Secant Product Theorem

If two secants intersect in the exterior of a circle, then the product of the lengths of one secant segment and its external segment equals the product of the lengths of the other secant segment and its external segment.
(Secant-Secant Product Thm.)

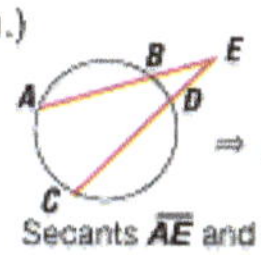

$\Rightarrow AE \cdot BE = CE \cdot DE$

Secants $\overline{AE}$ and $\overline{CE}$ intersect at E.

If two secants intersect in the exterior of a circle, then the product of the lengths of one secant segment and its external segment equals the product of the lengths of the other secant segment and its external segment.

(whole · outside = whole · outside)

Find the value of x and the length of each secant segment.

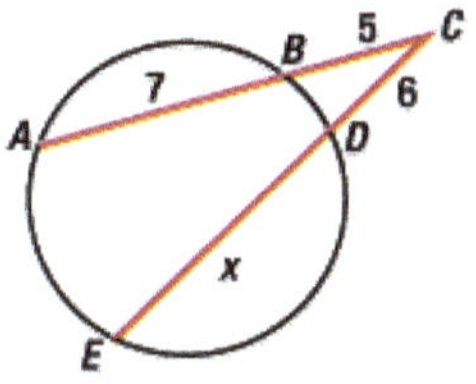

Secant lengths: $BC = 5$ and $AC = 12$

$AC \cdot BC = EC \cdot DC$	*Secant-Secant Product Thm.*
$12 \cdot 5 = (x + 6)6$	*Substitute the given values.*
$60 = 6x + 36$	*Simplify*
$24 = 6x$	*Subtract 36 from both sides.*
$x = 4$	*Divide both sides by 6.*

Secant lengths: $DC = 6$ and $EC = 4 + 6 = 10$

The Secant-Secant Product Theorem is used in this example to determine the lengths of secant segments. The lengths of three of the segments formed by two secants intersecting a circle are given.

Substitute the given values into the equation for the Secant-Secant Product Theorem and solve for x. The solution yields $x = 4$. Therefore, BC = 5, AC = 12, DC = 6, and EC = 10.

Example 4 Applying the Secant-Tangent Product Theorem

Secant-Tangent Product Theorem

If a secant and a tangent intersect in the exterior of a circle, then the product of the lengths of the secant segment and its external segment equals the length of the tangent segment squared.
(Secant-Tangent Product Thm.)

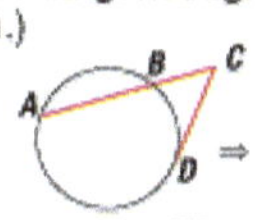

$\Rightarrow AC \cdot BC = DC^2$

Secant $\overline{AC}$ and tangent $\overline{DC}$ intersect at C.

If a secant and a tangent intersect in the exterior of a circle, then the product of the lengths of the secant segment and its external segment equals the length of the tangent segment squared.

(whole · outside = tangent^2)

Find the value of x.

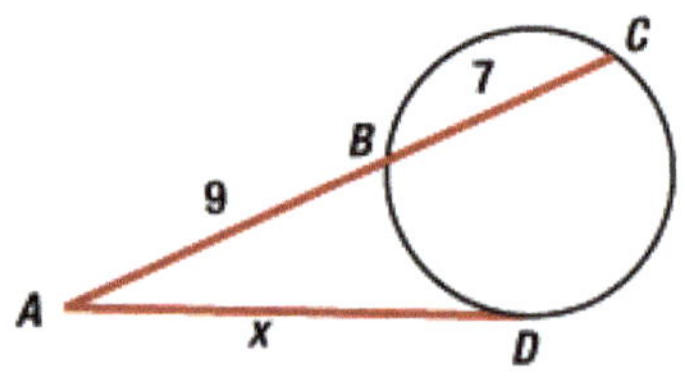

$AD^2 = AC \cdot AB$	*Secant-Tangent Product Thm.*
$x^2 = (9+7)(9)$	*Substitute the given values.*
$x^2 = 16 \cdot 9$	*Simplify.*
$x = \sqrt{16 \cdot 9}$	*Take the positive square root of both sides.*
$= 4 \cdot 3 = 12$	*Simplify.*

The Secant-Tangent Product Theorem is used in this example to determine the length of a tangent segment. The lengths of the segments of the secant are given.

Substitute the given values into the equation from the Secant-Tangent Product Theorem and solve for x, the length of the tangent. The solution yields $x = 12$.

12

12 Transformational Geometry

12.1 Congruence Transformations

Reflections

Key Objectives

- Identify and draw reflections.

Key Terms

- An **isometry** is a transformation that does not change the shape or size of a figure.

Example 1 Identifying Reflections

Tell whether each transformation appears to be a reflection. Explain.

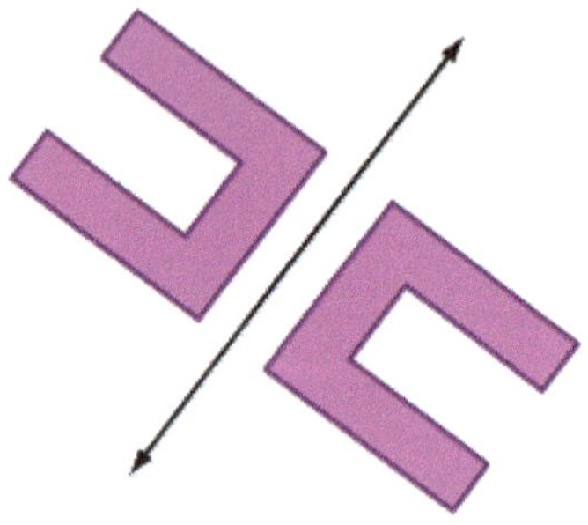

Yes; the image appears to be flipped across a line.

Whether a transformation appears to be a reflection is determined in this example.

A reflection is a transformation across the line of reflection in which each point in the preimage is the same distance from the line of reflection as the corresponding point in the image.

The two figures do appear to be mirror images of each other across the line of reflection. This transformation is a reflection.

Tell whether each transformation appears to be a reflection. Explain.

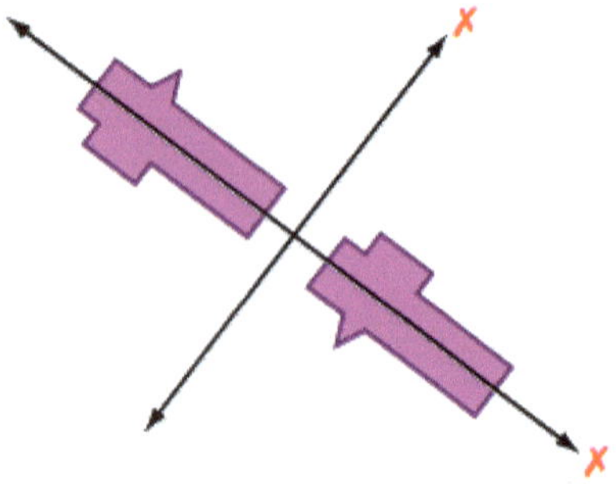

No; the figure does not appear to be flipped.

Whether a transformation appears to be a reflection is determined in this example.

A reflection is a transformation across the line of reflection in which each point in the preimage is the same distance from the line of reflection as the corresponding point in the image.

The two figures do not appear to be mirror images of each other across the potential line of reflection that runs from the lower left to upper right. Points on one image do not have corresponding points equidistant from the line of reflection in the other image. The same is true of a second potential line of reflection, which runs from the upper left to the lower right. This transformation is a not reflection.

Example 2 Drawing Reflections

Copy the quadrilateral and the line of reflection. Draw the reflection of the quadrilateral across the line.

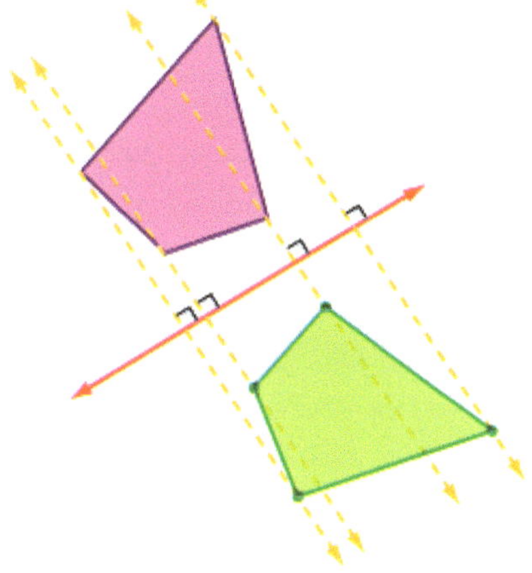

The reflection of a figure is drawn in this example. The preimage and the line of reflection are given.

To draw the reflected image, draw lines that are perpendicular to the line of reflection from the vertices of the preimage figure. Measure the distance from the vertex of the preimage to the line of reflection and then measure the same distance on the other side of the line of reflection. Mark the point for the vertex of the image. When all the vertices are transformed across the line of reflection, connect the vertices with line segments to draw the image figure.

Example 3 Landscape Architecture Application

A trail designer is planning two trails that connect picnic areas *A* and *B* to a point on the river. He wants the total length of the trails to be as short as possible. Where should the trail meet the river?

Reflect B across the river. Let B' be the reflection of B.

Draw $\overline{AB'}$ and locate X at the intersection of $\overline{AB'}$ and the river.

For any point X on the river, $\overline{XB'} \cong \overline{XB}$, so $AX + XB = AX + XB'$. $AX + XB'$ is least when A, X, and B' are collinear.

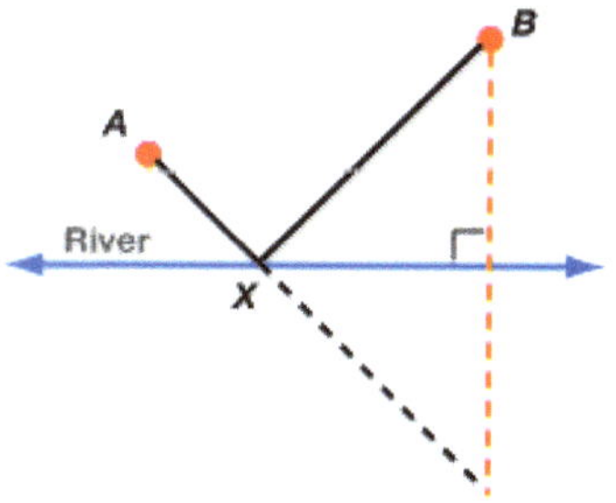

A reflection is used to construct lines representing the shortest path from one point to a line and then a second point in this application example.

To find this line, recognize that the shortest distance between two points is a straight line. However, the path between the points A and B must touch the river, so the shortest path will be the one that joins the reflection of one point with the other point through the line for the river (which is a straight line).

To construct the line, find the reflection of B by drawing a line perpendicular to the river. Extend the line past the river the same distance between B and the river. Draw a line between B' and A. Draw a line that extends from the intersection of AB', point X, with the river line back to B. Since $AX + XB = AX + XB'$, and the shortest distance between A and B' is a straight line, the solution shown is also the shortest distance between A and B that includes contact with the river.

Example 4 Drawing Reflections in the Coordinate Plane

Reflect the figure with the given vertices across the given line.

A(1, 3), *B*(1, 5), *C*(2, 2); *x*-axis

The reflection of (*x*, *y*) is (*x*, −*y*).

A(1, 3) → *A*'(1, −3)

B(1, 5) → *B*'(1, −5)

C(2, 2) → *C*'(2, −2)

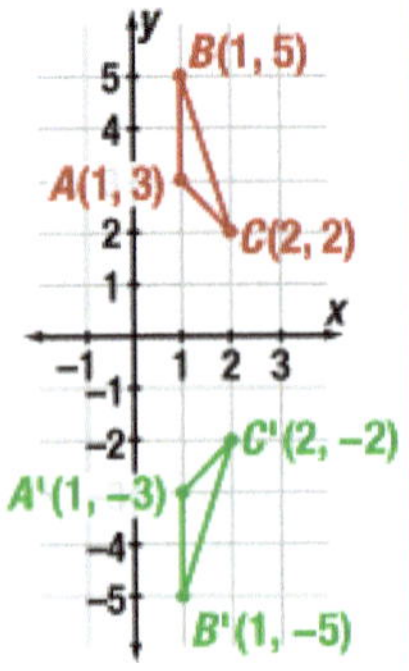

The reflection of a figure is drawn in the coordinate plane in this example. The coordinates of the vertices of the preimage figure and the line of reflection are given.

To draw the reflection across the *x*-axis, count the distance to the *x*-axis in the *y* direction of each vertex of the preimage. Count the same distance below the *x*-axis to locate the image vertex. Draw lines joining the three image vertex points to construct the reflected triangle.

It is also possible to apply the rule for the reflection, which is described as the transformation of the point (x, y) to $(x, -y)$. So, for example, $A(1, 3)$ is transformed in the reflection to $A'(1, -3)$.

Reflect the figure with the given vertices across the given line.

D(1, 3), *E*(1,5), *F*(4, 5), *G*(4, 4); *y* = *x*

The reflection of (*x*, *y*) is (*y*, *x*).

D(1, 3) → *D*'(3, 1)

E(1, 5) → *E*'(5, 1)

F(4, 5) → *F*'(5, 4)

G(4, 4) → *G*'(4, 4)

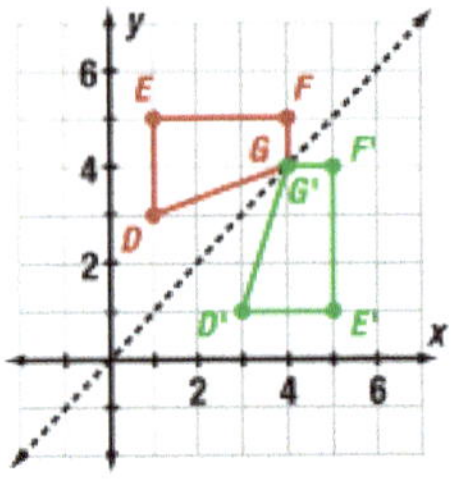

The reflection of a figure is drawn in the coordinate plane in this example. The coordinates of the vertices of the preimage figure and the line of reflection are given.

To find the image as it is reflected from the preimage across the $x = y$ line, apply the transformation rule, $(x, y) \rightarrow (y, x)$, to each of the preimage points. For example, $D(1, 3) \rightarrow D'(3, 1)$. Find each of the vertices of the image in this way and draw lines between them to construct the reflected figure.

Translations

Key Objectives

- Identify and draw translations.

Example 1 Identifying Translations

Tell whether each transformation appears to be a translation. Explain.

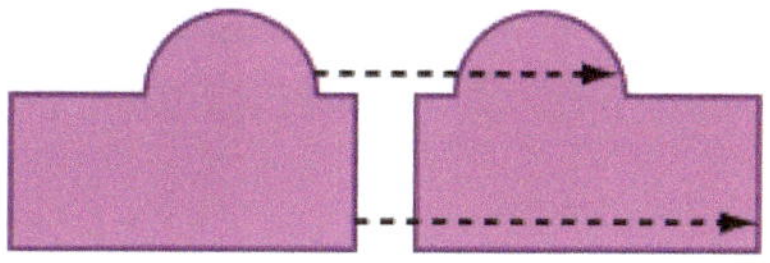

No; not all the points have moved the same distance.

Whether a transformation is a translation is determined in this example.

A translation is a transformation along a vector such that each segment joining a point and its image has the same length as the vector and is parallel to the vector.

In this example, the two vectors shown are parallel but not the same length. Therefore, this transformation is not a translation.

Tell whether each transformation appears to be a translation. Explain.

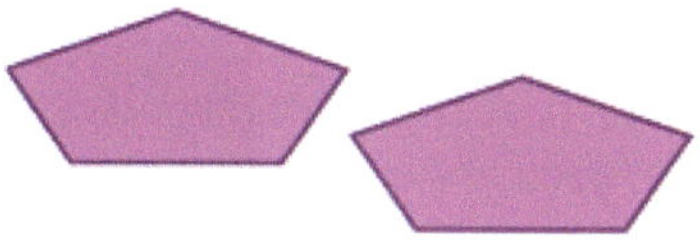

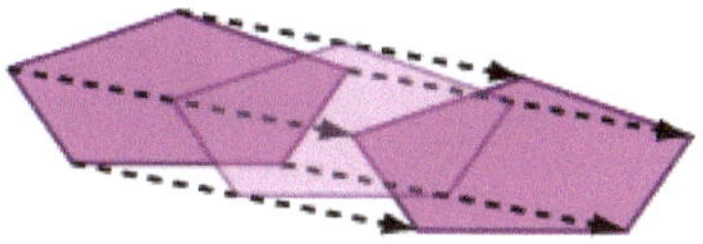

Yes; all the points have moved the same distance in the same direction.

Whether a transformation is a translation is determined in this example.

A translation is a transformation along a vector such that each segment joining a point and its image has the same length as the vector and is parallel to the vector.

In this example, all the vectors shown are parallel and the same length. Therefore, this transformation is a translation.

Example 2 Drawing Translations

Copy the triangle and the translation vector. Draw the translation of the triangle along $\vec{v}$.

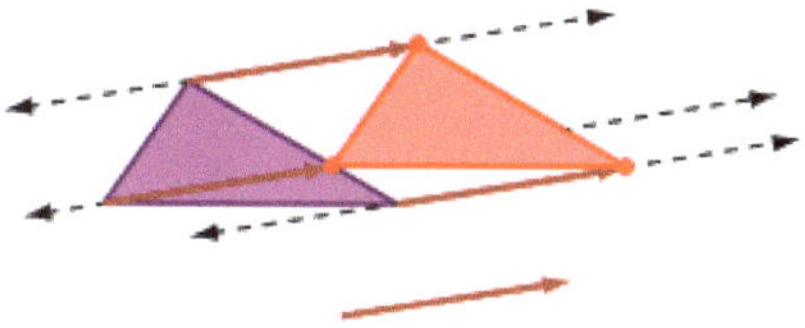

Draw a line parallel to the vector through each vertex of the triangle.

Measure the length of the vector. Then, from each vertex mark off this distance in the same direction as the vector, on each of the parallel lines.

Connect the images of the vertices.

A translation of a figure is drawn in this example. The preimage figure and the translation vector are given.

To draw the image, the translation of the preimage, draw a line parallel to the vector through each of the vertices of the preimage.

Take the measure of the vector and mark off this distance on each of the lines from the preimage vertex in the direction of the vector. Draw line segments between the marks, which are the vertices of the image, to form the image figure.

Example 3 Drawing Translations in the Coordinate Plane

Translate the triangle with vertices $A(-4, 1)$, $B(-1, 0)$, $C(-2, 4)$ along the vector $\langle 1, -5 \rangle$.

The image of (x, y) is $(x + 1, y - 5)$.

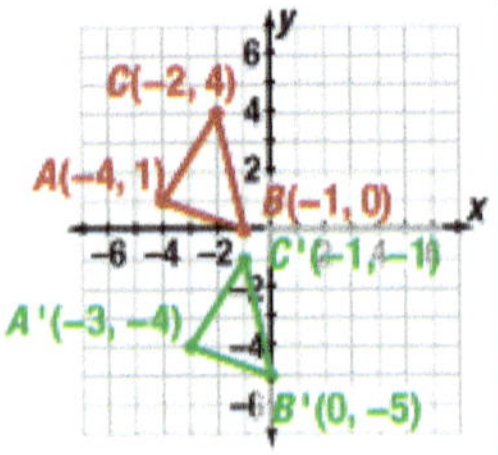

$A(-4, 1) \rightarrow A'(-3, -4)$

$B(-1, 0) \rightarrow B'(0, -5)$

$C(-2, 4) \rightarrow C'(-1, -1)$

A figure in the coordinate plane is translated in this example. The coordinates of the vertices of the preimage and the translation vector are given.

The translation vector, $\langle 1, -5 \rangle$, can be used to derive a rule for the translation of the coordinates: $(x, y) \rightarrow (x + 1, y - 5)$. Apply the rule to translate each of the three preimage vertices to the image vertices. For example, $A(-4, 1) \rightarrow A'(-4 + 1, 1 - 5) = A'(-3, -4)$. Draw line segments between the image vertices to form the translated figure.

Example 4 Entertainment Application

In a marching drill, it takes 7 steps to march 4 yards. A drummer starts 7 steps to the right and 7 steps down from the center of the field. She marches 14 steps to the left to her second position. Then she marches 21 steps up the field to her final position. What is the drummer's final position? What single translation vector moves her from the starting position to her final position?

The drummer's starting coordinates are $(7, -7)$.

Her second position is $(7 - 14, -7) = (-7, -7)$.

Her final position is $(-7, -7 + 21) = (-7, 14)$.

The vector that moves her directly from her starting position to her final position is $\langle -14, 21 \rangle$.

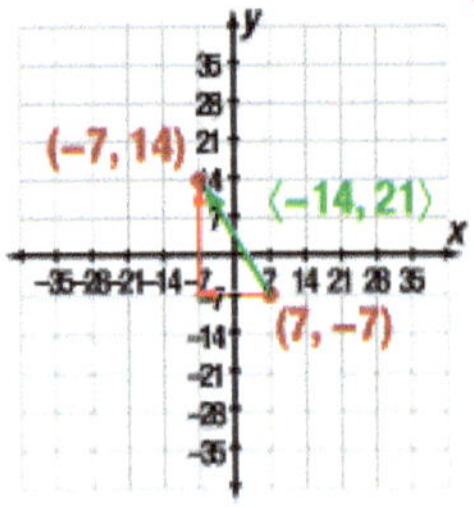

The translation vector for the sum of a sequence of position changes in the coordinate plane is determined in this application example. The starting point coordinates and the changes in x or y of each movement are given.

Draw the given information in the coordinate plane to better visualize it. The starting point coordinates are given as $(7, -7)$. The drummer's movements take her to the end point with coordinates $(-7, 14)$. To find the vector that gives the translation $(7, -7) \rightarrow (-7, 14)$, subtract the x and y coordinates of the starting point from the x and y coordinates of the end point. The vector is: $\langle -7 - 7, 14 - (-7) \rangle = \langle -14, 21 \rangle$.

Rotations

Key Objectives

- Identify and draw rotations.

Example 1 Identifying Rotations

Tell whether each transformation appears to be a rotation. Explain.

Yes; the figure appears to be turned around a point.

Whether a transformation is a rotation is determined in this example.

A rotation is a transformation about a point *P*, called the center of rotation, such that each point and its image are the same distance from *P*, and such that all angles with vertex *P* formed by a point and its image are congruent.

To determine if a transformation is a rotation, look for a center of rotation. There appears to be a center of rotation and the preimage and image are congruent figures. The transformation appears to be a rotation.

Tell whether each transformation appears to be a rotation. Explain.

No; the figure appears to be flipped and rotated, not just rotated.

Whether a transformation is a rotation is determined in this example.

To determine if a transformation is a rotation, look for a center of rotation. There does not appear to be a center of rotation and the preimage and image are not congruent figures. The transformation appears not to be a rotation.

Example 2 Drawing Rotations

Copy the figure and the angle of rotation. Draw the (counter-clockwise) rotation of the triangle about point *P* by m∠*A*.

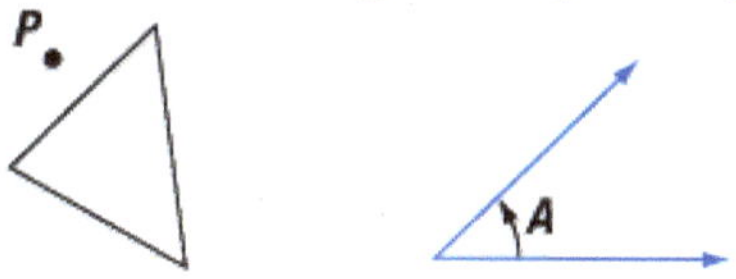

Draw a segment from the vertex to point *P*.

Construct an angle congruent to ∠*A* onto the segment. Measure the distance from the vertex to point *P* and mark off this distance on the corresponding ray to locate the image of the vertex.

Repeat for the other two vertices.

Connect the images of the vertices.

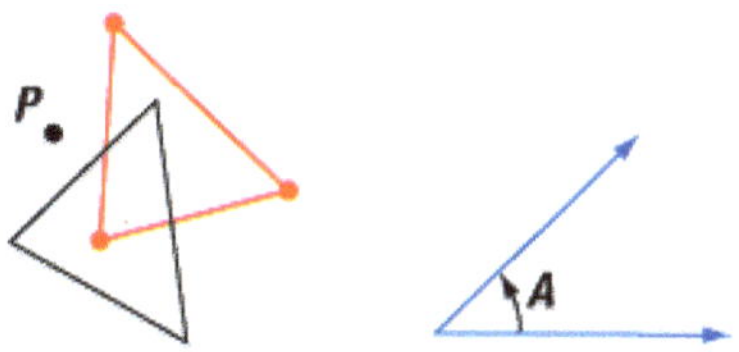

A method for drawing the counter-clockwise rotation of a preimage is demonstrated in this example. The figure, the angle of rotation, and the center of rotation are given.

To draw a vertex of the image, draw a line segment from a preimage vertex to the center of rotation, *P*. Construct an angle congruent to the angle of rotation using the line segment as one side of the angle and with the angle vertex at *P*. Measure the distance from *P* to the preimage vertex and mark off the same distance from *P* along the other side, or ray, of the angle. The mark is the corresponding image vertex.

Follow the same steps with each of the other vertices. Draw line segments between the image vertices to draw the rotated figure.

Example 3 Drawing Rotations in the Coordinate Plane

Rotate $\triangle ABC$ with vertices $A(-1, 1)$, $B(-4, 2)$, and $C(-2, 5)$ by 90° about the origin.

The rotation of (x, y) is $(-y, x)$.

$A(-1, 1) \rightarrow A'(-1, -1)$

$B(-4, 2) \rightarrow B'(-2, -4)$

$C(-2, 5) \rightarrow C'(-5, -2)$

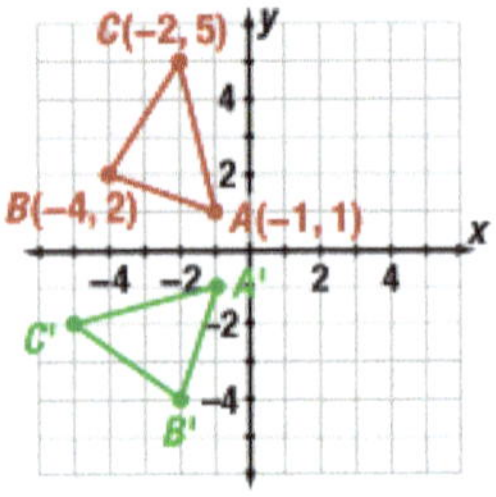

Remember

Unless otherwise stated, all rotations are counterclockwise.

A method for drawing the counter-clockwise rotation of a figure in the coordinate plane is demonstrated in this example. The coordinates of the vertices and the angle of rotation are given. The center of rotation is given as the origin.

The rule for this particular rotation is: $(x, y) \rightarrow (-y, x)$. To visualize why this is the case, the rotation can also be seen as the 90° clockwise rotation of the x- and y-axes about the origin while the figure remains in place. This rotation changes all x distances into y distances of the opposite sign, and all y distances into x distances of the same sign.

Apply the rule to each of the vertices of the triangle. For example, $A(-1, 1) \rightarrow A'(-1, -1)$. Draw line segments between the image vertices to draw the image figure.

Example 4 Engineering Application

The water grist mill has a radius of 6 ft and takes 5 secs to make a complete rotation. A bucket starts at position (6, 0). What are the coordinates of the bucket's location after 2 seconds?

2 seconds is $\frac{2}{5}$ of a complete rotation.

$\frac{2}{5}(360^\circ) = 144^\circ$

Draw a right triangle to represent the bucket's location after a rotation of 144° about the origin.

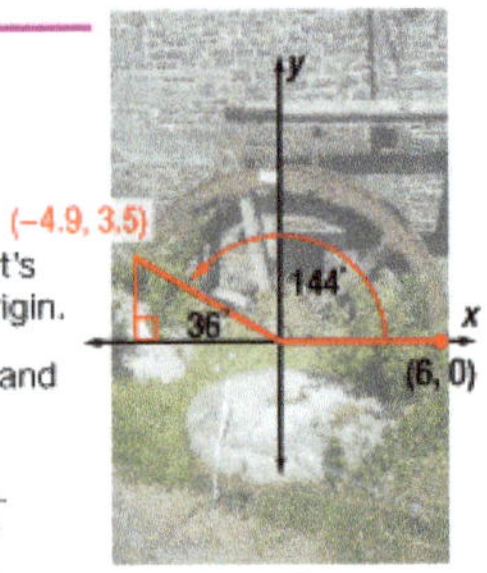

Use the cosine and sine ratios to find the x and y-coordinates.

$\cos 36^\circ = \frac{-x}{6}$ $\cos = \frac{adj.}{hyp.}$

$x = -6\cos 36^\circ \approx -4.9$ Solve for x.

$\sin 36^\circ = \frac{y}{6}$ $\sin = \frac{opp.}{hyp.}$

$y = 6\sin 36^\circ \approx 3.5$ Solve for y.

The bucket's location after 2 seconds is approximately $(-4.9, 3.5)$.

The coordinates of a point rotating at a particular rate is determined after a time interval in this application example. The beginning coordinates of the point (a bucket on a mill wheel) and its rate of rotation are given. The center of rotation is given as the origin.

A complete rotation about the origin takes 5 seconds, so after 2 seconds the point will have rotated through $(2/5)360^\circ = 144^\circ$. To find the coordinates of the bucket at this point, construct a right triangle with leg lengths equal to the x and y coordinates and hypotenuse equal to the radius, 6 ft. The angle at the origin of the triangle is $180^\circ - 144^\circ = 36^\circ$. Calculate the x and y coordinates using the sine and cosine of the 36° angle. The coordinates of the bucket's location after 2 seconds of rotation are approximately $(-4.9, 3.5)$.

13

13 Homeschool Algebra Review

Simplifying Expressions

Key Objectives

- Use the Commutative, Associative, and Distributive Properties to simplify expressions.
- Combine like terms.

Key Terms

- The **terms** of an expression are the parts that are added together.
- **Like terms** are terms that contain the same variables raised to the same powers.
- A **coefficient** is a number multiplied by a variable.

Theorems, Postulates, Corollaries, and Properties

- **Commutative Property** Two numbers can be added in any order and two numbers can be multiplied in any order.
  $a + b = b + a$
  $ab = ba$
- **Associative Property** In the sum or product of three (or more) numbers, any pair of numbers can be grouped together.
  $a + b + c = (a + b) + c = a + (b + c)$
  $abc = (ab)c = a(bc)$
- **Distributive Property** The product of a number and a sum can be computed by finding the sum first and then multiplying the sum by the number or by multiplying each term in the sum by the number first and then adding those two products.
  $a(b + c) = ab + ac$ and $a(b - c) = ab - ac$

Example 1 Using the Commutative and Associative Properties

The Commutative and Associative Properties can be used to rearrange or regroup the numbers in an expression so that it can be more easily simplified. Remember, the Commutative and Associative Properties apply to expressions containing either addition or multiplication only. They do not apply to expressions that contain subtraction or division, nor do they apply to expressions that contain addition *and* multiplication.

Simplify the expression.

$2 \cdot 20 \cdot 5$

$2 \cdot 20 \cdot 5$

$2 \cdot 5 \cdot 20$ *Use the Commutative Property.*

$(2 \cdot 5) \cdot 20$ *Use the Associative Property to make groups of compatible numbers.*

$10 \cdot 20$

200

The expression in this example contains only one operation, multiplication. Therefore, the Commutative and Associative Properties can be applied to the expression.

Notice that there are three factors in this multiplication expression, 2, 20, and 5. (*Factors* are the numbers being multiplied.) From these three factors, the easiest pair of numbers to multiply is 2 and 5. However, 2 and 5 are separated by 20 in the expression. So, first use the Commutative Property to rearrange the factors so that 5 is after 2. Then use the Associative Property to group 2 and 5 to show that those two numbers will be multiplied first. However, since the order of operations states that numbers are multiplied from left to right, using the Associative Property is unnecessary.

Example 2 Using the Distributive Property with Mental Math

The Distributive Property can be used when a number is multiplied by a sum or when a number is multiplied by a difference. Basically, the Distributive Property states that it doesn't matter if the sum is found first and then that sum is multiplied by the number, or if each term in the sum is multiplied by the number and then the sum of those two products is found. Either way, the result is the same.

Write the product using the Distributive Property. Then simplify. 7(18) 7(18) 7(20 − 2) — *Rewrite 18 as 20 − 2.* 7 · 20 − 7 · 2 — *Use the Distributive Property.* 140 − 14 — *Multiply.* 126 — *Subtract.*	The Distributive Property can be used to make mental multiplication of two numbers easier. In this example, 7 is multiplied by 18. Since 18 is not the easiest number to multiply by, write 18 as either a sum or difference of two numbers that are easier to multiply. Here, 18 is written as the difference of 20 and 2. Any pair of numbers whose difference is 18 could have been used (such as 21 and 3 or 30 and 12), but 20 and 2 are easy to multiply, so those two numbers are used. Once 18 is written as a difference, use the Distributive Property, then multiply and subtract.
Write the product using the Distributive Property. Then simplify. 14(105) 14(105) 14(100 + 5) — *Rewrite 105 as 100 + 5.* 14(100) + 14(5) — *Use the Distributive Property.* 1400 + 70 — *Multiply.* 1470 — *Add.*	In this example, 14 is multiplied by 105. Since 105 is not easy to multiply by, write 105 as a sum or difference of two easier numbers. Here, 105 is written as the sum of 100 and 5, which are two numbers that are easy to multiply by. Once 105 has been written as 100 + 5, use the Distributive Property, then multiply and add.

Example 3 Combining Like Terms

Terms are the numbers in an expression that are either added or subtracted. For example, the expression 1 + 8 contains two terms, 1 and 8. Consider the expression 1 + 2(4). This expression contains three numbers, 1, 2, and 4, but only two terms, 1 and 2(4), since these are the two numbers that are being added. A term can be a constant or it may contain one or more variables.

Like terms are terms that contain exactly the same variable(s) to exactly the same power(s). For example, in the expression $2x + 3 + 5x$, the terms $2x$ and $5x$ are like terms. All terms that are constants are also like terms.

Like terms can be combined (added or subtracted) by adding or subtracting their coefficients. A coefficient is the number multiplied by a variable (or product of variables). All variables have a coefficient. If no coefficient is written, then the coefficient is understood to be 1.

Simplify the expression by combining like terms. $8y + 42y$ $8y + 42y$ — *8y and 42y are like terms.* $50y$ — *Add the coefficients.*	The expression given in this example contains two terms, $8y$ and $42y$. Since these two terms contain exactly the same variable to exactly the same power, y, they are like terms and can be combined. The operation between the terms is addition, so the terms can be combined by adding their coefficients. The coefficient of $8y$ is 8 and the coefficient of $42y$ is 42. So, add 8 and 42 to combine the terms. The variable does not change when like terms are combined.

Simplify the expression by combining like terms.

$5.4w^2 - w^2$

$5.4w^2 - w^2$	*$5.4w^2$ and $-1w^2$ are like terms.*
$4.4w^2$	*Subtract the coefficients, $5.4 - 1 = 4.4$.*

Remember

If a variable is written without a coefficient, the coefficient is understood to be 1.

The expression given in this example also contains two terms, $5.4w^2$ and $-w^2$. Since these two terms contain exactly the same variable to exactly the same power, w^2, they are like terms and can be combined by adding the coefficients.

The coefficient of the first term is 5.4. The second term is preceded by only a subtraction sign, so the coefficient is –1. (Note that if a variable is preceded by only an addition sign, then the coefficient is 1).

Add the coefficients to combine the terms.
$5.4 + -1 = 5.4 - 1 = 4.4$
Remember, the variable stays the same in the simplified expression.

Simplify the expression by combining like terms.

$5x^2 + 5x$

$5x^2 + 5x$	*$5x^2$ and $5x$ are not like terms.*

Do not combine the terms.

The expression given in this example again contains two terms, $5x^2$ and $5x$. However, these two terms contain exactly the same variable, but the power is not the same. So, they are not like terms and cannot be combined.

Notice that the two terms have the same coefficient, 5, but the coefficient does not effect whether terms are like terms.

Example 4 Simplifying Algebraic Expressions

The expression given below contains more than one operation. It contains multiplication, subtraction, and addition. Therefore, the order of operations must be used to simplify it.

Remember, the Commutative and Associative Properties cannot be applied to expressions with subtraction. So, before applying these properties, write expressions so that subtraction is addition of a negative number.

Simplify the expression. Justify each step with an operation or property.

$3(x - 2) + 10x$

$3(x - 2) + 10x$	
$3(x) - 3(2) + 10x$	*Distributive Property*
$3x - 6 + 10x$	*Multiply.*
$3x + 10x - 6$	*Commutative Property*
$(3x + 10x) - 6$	*Associative Property*
$13x - 6$	*Combine like terms.*

simplified

According to the order of operations, multiplication should be completed after operations within parentheses. This expression contains subtraction within parentheses, but the terms to be subtracted, x and 2, are not like terms. So, begin with multiplication. Use the Distributive Property to multiply 3 by $(x - 2)$, resulting in $3x - 6 + 10x$.

The next operations to complete are addition and subtraction from left to right. However, $3x$ and 6 cannot be combined because they are not like terms. The terms $3x$ and $10x$ are like. So, use the Commutative Property to write the expression with $10x$ before the 6, and then combine the like terms. The expression contains subtraction, so before applying the Commutative Property, mentally consider the expression with subtraction written as addition of a negative number.
$3x - 6 + 10x$
$3x + -6 + 10x$
$3x + 10x + -6$
$3x + 10x - 6$
Now combine the coefficients of $3x$ and $10x$ to complete the simplification.

Writing and Solving Two-Step Equations

Key concepts

- Learn to solve two-step equations.

Key terms

- **Two-step equations** are equations that are solved using two operations.

Example 1

Example 1 demonstrates translating a sentence into a two-step equation.

To translate a sentence into an equation, break the sentence into words or phrases where each word or phrase describes a number, variable, operation, or equality. Then, write the number, variable, operation, or equality that is indicated by each word or phrase. Simplify, if possible.

The words and phrases commonly used for operations include "product" for multiplication, "quotient" for division, "sum", "total", or "increased by" for addition, and "difference" or "decreased by" for subtraction. The word "is" represents an equals sign and the phrase "a number" represents a variable.

Translating Sentences into Two-Step Equations

Translate each sentence into an equation.

The product of 2 and a number, increased by 5, is 1.

The product of 2 and a number, increased by 5, is 1.

$2 \cdot x + 5 = 1$

$2x + 5 = 1$

The sum of a number and −2, divided by 3, is 4.

The sum of a number and −2, divided by 3, is 4.

$\frac{x + (-2)}{3} = 4$

$\frac{x + (-2)}{3} = 4$

Consider the first exercise. Notice that the sentence contains the words "product" and "increased by". So, the equation must contain multiplication and addition.

Consider the second exercise. Notice that the sentence must contain the words "sum" and "divided by". So, the equation contains addition and division.

Example 2

Example 2 demonstrates solving two-step equations that contain multiplication and either addition or subtraction.

To solve a two-step equation, undo the two operations that are performed on the variable in the reverse order of the order of operations. For example, consider the equation $3x + 5 = 17$. The two operations performed on the variable are multiplication by 3 and addition of 5. If the expression $3x + 5$ was evaluated for some value of the variable, the first operation completed would be multiplication and then addition. Use the opposite of these two operations and apply them in the reverse of the order of operations to solve the equation $3x + 5 = 17$. The opposite operation of multiplication is division and the opposite operation of addition is subtraction. So, subtract by 5 first and then divide by 3 to solve.

Solving Two-Step Equations

Solve.

$3n + 1 = 19$

$$\begin{aligned} 3n + 1 &= 19 \\ -1 \quad & \quad -1 \\ 3n &= 18 \end{aligned}$$

Subtract 1 from both sides.

$$\frac{3n}{3} = \frac{18}{3}$$

Divide both sides by 3.

$$n = 6$$

In this equation, identify the operations that are performed on the variable and the order that those operations would be completed if the expression was evaluated for some number. The operations are multiplication by 3 and addition of 1. And, the first operation completed would be multiplication and then addition.

So, to solve this equation, undo these operations in the reverse order. Begin solving by subtracting 1 from each side to undo the addition of 1. Next, divide both sides by 3 to undo the multiplication by 3. The equation is now solved.

$21 = -2p - 5$ $21 = -2p - 5$ $+5 \quad\quad +5$ *Add 5 to both sides.* $26 = -2p$ $\frac{26}{-2} = \frac{-2p}{-2}$ *Divide both sides by −2.* $-13 = p$ **Check Your Work** $21 = -2p - 5$ $21 \stackrel{?}{=} -2(-13) - 5$ *Substitute −13 for p.* $21 \stackrel{?}{=} 26 - 5$ $21 = 21$ ✓ *−13 is a solution.*	Here, the variable is on the right side of the equation and the coefficient is a negative number. To solve, reverse and undo the operations that would be used to evaluate the expression. Therefore, first add 5 to each side to undo the subtraction by 5. Then, divide each side by –2 to undo the multiplication by –2. A solution of a two-step equation can be checked just as the solution of a one-step equation is checked. Be sure to follow the order of operations.

Example 3

Example 3 demonstrates solving a two-step equation that contains division.

Solving Two-Step Equations Using Multiplication **Solve.** $2 + \frac{m}{3} = 11$ $2 + \frac{m}{3} = 11$ $-2 \quad\quad -2$ *Subtract 2 from both sides.* $\frac{m}{3} = 9$ $3\left(\frac{m}{3}\right) = (9)3$ *Multiply both sides by 3.* $m = 27$ **Check Your Work** $2 + \frac{m}{3} = 11$ $2 + \frac{27}{3} \stackrel{?}{=} 11$ *Substitute 27 for m.* $2 + 9 \stackrel{?}{=} 11$ $11 = 11$ ✓ *27 is a solution.*	To solve, reverse and undo the operations that would be used to evaluate the expression. To evaluate the expression, the substituted number would be divided by 3 and then 2 would be added to it. Thus, begin by subtracting 2 from each side to undo the addition of 2. Next, multiply each side by 3 to undo the division by 3.
$2 = \frac{u}{5} - 5$ $2 = \frac{u}{5} - 5$ $+5 \quad\quad +5$ *Add 5 to both sides.* $7 = \frac{u}{5}$ $5(7) = \left(\frac{u}{5}\right)5$ *Multiply both sides by 5.* $35 = u$	This equation is similar to the equation given in the last exercise. It is a two-step equation where the variable is divided by a number. However, the expression with the variable is now on the right side of the equation and the second operation is subtraction. To solve, reverse and undo the operations that would be used to evaluate the expression. Thus, add 5 to both sides to undo the subtraction of 5. Then, multiply both sides by 5 to undo the division by 5.

Example 4

Example 4 demonstrates writing and solving a two-step equation.

Consumer Application

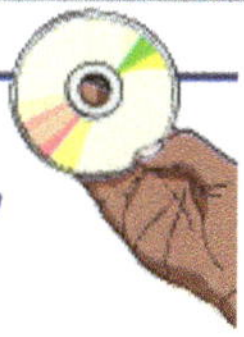

The total charge for a yearly internet DVD rental membership is $231. Each year, a registration fee of $15 is paid up front, and the rest is paid monthly. How much do members pay each month?

Let m be the amount members pay each month.

total yearly charge	is	registration fee	plus	monthly cost
231	=	15	+	$12m$

$231 = 15 + 12m$

$$231 = 15 + 12m$$
$$-15 \quad -15$$
$$216 = 12m$$

Subtract 15 from both sides.

$$\frac{216}{12} = \frac{12m}{12}$$

Divide both sides by 12.

$$18 = m$$

Members pay $18 per month for a one-year membership.

Consider this real-life situation. The charge for the one-year membership is composed of two parts: the registration fee of $15 and the monthly fee, which is unknown. Thus, the total yearly charge, $231, is 12 times the monthly fee (since there are 12 months in a year) plus the registration fee.

Write an equation to express this information and then solve the equation to find the monthly charge.

Example 5

Example 5 demonstrates two methods for solving two-step equations where the sum of a variable and a constant is divided by a number.

Solving Multi-Step Equations

Solve the equation.

Method 1: Work backward to isolate the variable.

$\frac{n+5}{2} = 4$

$$\frac{n}{2} + \frac{5}{2} = 4$$

Rewrite the fraction as the sum of two fractions.

$$\frac{n}{2} + \frac{5}{2} - \frac{5}{2} = 4 - \frac{5}{2}$$

Subtract $\frac{5}{2}$ from both sides.

$$\frac{n}{2} = \frac{4}{1} - \frac{5}{2}$$

$$\frac{n}{2} = \frac{2}{2} \cdot \frac{4}{1} - \frac{5}{2}$$

Multiply by the fraction $\frac{2}{2}$ to create a common denominator.

$$\frac{n}{2} = \frac{8}{2} - \frac{5}{2}$$

Simplify.

$$\frac{n}{2} = \frac{3}{2}$$

Subtract.

$$2\left(\frac{n}{2}\right) = 2\left(\frac{3}{2}\right)$$

Multiply both sides by 2.

$$n = 3$$

Simplify.

The first method begins by separating the quotient into like fractions; the resulting two-step equation is solved using familiar steps.

The calculations appear to be complicated at first glance, but the steps are completed in exactly the same way as in previous exercises. However, since fractions are being subtracted, a common denominator must be found.

Method 2: Multiply both sides of the equation by the denominator.

$$\frac{n+5}{2} = 4$$

$$2\left(\frac{n+5}{2}\right) = 2(4)$$ *Multiply both sides by the denominator, 2.*

$$\begin{aligned} n + 5 &= 8 \\ -5 &\quad -5 \end{aligned}$$ *Subtract 5 from both sides.*

$$n = 3$$ *Same answer!*

Instead of separating the quotient into two fractions, multiply both sides by 2. The result of multiplying the left side by 2 is $n + 5$. The result of multiplying the right side by 2 is 8. Since the variable is not isolated, another step must be completed to solve the equation. Subtract 5 from both sides and then the equation is solved.

So, if the entire expression on one side of an equation is divided by a number, begin by multiplying both sides of the equation by that divisor (the denominator). This method is shorter and more efficient than Method 1.

Solving Two-Step Inequalities

Key concepts

- Learn to solve simple two-step inequalities.

The same process that was used when solving two-step equations can be used when solving two-step inequalities. However, whenever both sides of the inequality are multiplied or divided by a *negative* number, the inequality symbol must be *reversed*.

Example 1

Example 1 demonstrates solving a two-step inequality.

Solving Two-Step Inequalities

Solve. Then graph each solution set on a number line.

$\frac{x}{2} + 7 < 27$

$$\frac{x}{2} + 7 < 27$$
$$-7 \quad -7$$ *Subtract 7 from both sides.*
$$\frac{x}{2} < 20$$
$$2\left(\frac{x}{2}\right) < (20)2$$ *Multiply both sides by 2.*
$$x < 40$$

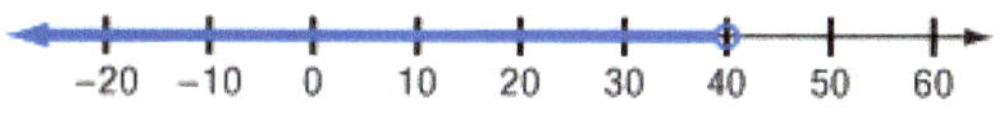

Consider this inequality. Notice that the variable in the inequality is divided by a positive number. Therefore, the inequality symbol will not need to be reversed when solving.

The first solving step is to subtract 7 from both sides. Then, multiply both sides by 2, but do *not* reverse the inequality symbol since the multiplication was not by a negative number.

$32 \leq \frac{y}{-5} + 30$

$$32 \leq \frac{y}{-5} + 30$$
$$-30 \quad -30$$ *Subtract 30 from both sides.*
$$2 \leq \frac{y}{-5}$$
$$(-5)(2) \geq \left(\frac{y}{-5}\right)(-5)$$ *Multiply both sides by −5 and reverse the inequality symbol.*
$$-10 \geq y$$
or
$$y \leq -10$$

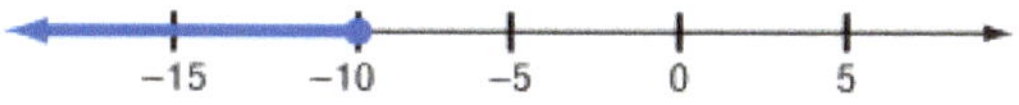

Consider this inequality. The variable is divided by a negative number. Therefore, the inequality will be multiplied by a negative number as the second step in the solving process. So, the inequality symbol will have to be reversed.

$5x - 1 \geq 49$

$$5x - 1 \geq 49$$
$$+1 \quad +1$$ *Add 1 to both sides.*
$$5x \geq 50$$
$$\frac{5x}{5} \geq \frac{50}{5}$$ *Divide both sides by 5.*
$$x \geq 10$$

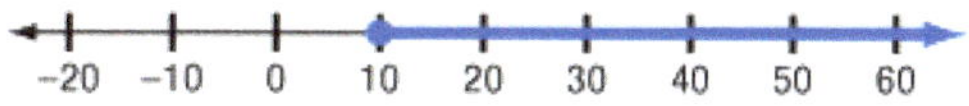

In this inequality, notice that the variable is multiplied by a positive number. Therefore, the inequality symbol will not need to be reversed when solving.

To solve this inequality, add 1 to both sides to undo the subtraction of 1. Then, divide both sides by 5 to undo the multiplication by 5.

$-3y + 10 > -50$

$$-3y + 10 > -50$$
$$-10 \quad -10$$ *Subtract 10 from both sides.*
$$-3y > -60$$
$$\frac{-3y}{-3} < \frac{-60}{-3}$$ *Divide both sides by −3 and reverse the inequality symbol.*
$$y < 20$$

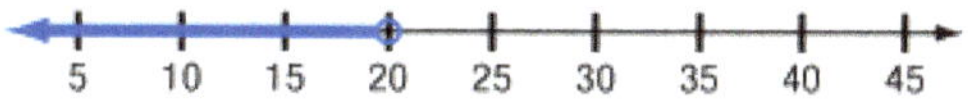

Consider this inequality. The variable is multiplied by a negative number. Therefore, the inequality symbol will be reversed at some point in the solving process.

To solve this inequality, begin by subtracting 10 from both sides. Then, divide both sides by −3. At this point, reverse the inequality symbol. Complete the division and the inequality will be solved.

Example 2

Example 2 demonstrates writing an inequality to describe a situation given in a word problem and then solving that inequality.

School Application

The 70 members of the Newman Middle School Band are trying to raise at least $2,000 to buy new percussion instruments. If they have already raised $250, how much should each student still raise, on average, to meet the goal?

Let x represent the average amount each student should still raise.

$70x + 250 \geq 2000$ *Write the inequality.*

$\underline{\quad -250 \quad -250}$ *Subtract 250 from both sides.*

$70x \geq 1750$

$\frac{70x}{70} \geq \frac{1750}{70}$ *Divide both sides by 70.*

$x \geq 25$

On average, each band member should raise at least $25.

Consider the situation presented in this word problem. The amount raised, $250, plus the average amount that each of the 70 members of the band should still raise, must be at least $2,000 in order to meet their goal. Write an inequality from this information. Let x be the average amount that each member should raise. So, $70x$ is the expression for the amount that will be raised by the 70 members. Since the amount raised by the 70 members plus the amount they have already raised, $250, must be at least $2,000, the inequality is written as $70x + 250 \geq 2000$. Solve this inequality as usual.

Writing and Solving Multi-Step Equations

Key concepts

- Learn to solve multi-step equations.

Key terms

- The **distributive property** states that for any numbers, a, b, and c, $a(b + c) = ab + ac$ and $a(b - c) = ab - ac$.

A multi-step equation will contain more than two operations. Typically, a multi-step equation should be simplified before applying the solving steps. Expressions on either side of the equation can be simplified if they contain parentheses or like terms (terms with exactly the same variable to exactly the same power).

Example 1

Example 1 demonstrates solving a multi-step equation by first combining like terms.

Before performing any solving steps, simplify the expressions on each side of an equation whenever possible. Remember, if an expression contains like terms, these like terms can be combined by adding or subtracting the coefficients.

Combining Like Terms to Solve Equations

Solve.

$8n + 2 - n = 16$

$7n + 2 = 16$ — *Combine like terms.*

$\quad -2 \quad -2$ — *Subtract 2 from both sides.*

$7n = 14$

$\frac{7n}{7} = \frac{14}{7}$ — *Divide both sides by 7.*

$n = 2$

Consider the equation presented here. The expression on the left side of the equation has three terms, $8n$, 2, and n. Two of those are like terms, $8n$ and n. Therefore, the equation can be simplified by combining the like terms. Notice that the term n is preceded by a subtraction symbol. So, subtract the coefficient of n, 1, from the coefficient of $8n$, 8, to combine these terms. Since $8n - n = 7n$, the equation simplifies to $7n + 2 = 16$.

Once the equation is simplified, it is a two-step equation and can be solved as such.

Example 2

Example 2 demonstrates using the distributive property to remove parentheses from an equation before solving.

Remember, the distributive property states that if a number is multiplied by a sum or difference, that number can be multiplied by each value in the sum or difference. If one side of an equation contains a sum or difference that is multiplied by a number, use the distributive property to simplify that side of the equation.

Using the Distributive Property to Solve Equations

Solve.

$5(m - 3) + 1 = 21$

$5(m) - 5(3) + 1 = 21$ — *Distribute 5 on the left side.*

$5m - 15 + 1 = 21$ — *Simplify.*

$5m - 14 = 21$ — *Combine like terms.*

$\quad +14 \quad +14$ — *Add 14 to both sides.*

$5m = 35$

$\frac{5m}{5} = \frac{35}{5}$ — *Divide both sides by 5.*

$m = 7$

The **distributive property** states that $a(b + c) = ab + ac$.

Notice that this equation contains parentheses and that the terms inside the parentheses, m and 3, cannot be combined because they are not like terms. There may appear to be like terms on that side of the equation since 5, 3, and 1 are all constants. However, only the number 1 is a term, so these are not like terms.

Since the left side of the equation contains a difference, $m - 3$, multiplied by a number, 5, use the distributive property to simplify the equation. The result is $5m - 15 + 1$. Notice that the left side of the equation now contains like terms, 15 and 1, and that 15 is preceded by a subtraction symbol. So, combine the like terms to simplify the equation. Then, the equation is a two-step equation to be solved.

Example 3

In Example 3, two equations are written to represent the situation described in the word problem.

Before writing equations to represent situations, define the unknowns. Consider the situation presented below. Three people have action figures, Jef, Kim, and Isaac. Kim has 15 action figures. The number of action figures that Jef and Isaac have are unknown. So, let J be the number of action figures that Jef has, and let I be the number of action figures that Isaac has. Since there are two unknowns, two equations will be solved.

Problem-Solving Application

Jef owns three times as many action figures as Isaac owns. Subtracting 3 from the number of action figures Jef owns and then dividing by 2 gives the number Kim owns. Kim owns 15 action figures. How many does Isaac own?

Let J be the number of action figures Jef owns, and I be the number of action figures Isaac owns.

$\frac{J-3}{2} = 15$	*Write an equation relating action figures owned by Jef to a known value.*
$2\left(\frac{J-3}{2}\right) = (15)2$	*Multiply both sides by 2.*
$J - 3 = 30$	
$+3 \quad +3$	*Add 3 to both sides.*
$J = 33$	*Jef owns 33 action figures.*
$J = 3I$	*Relate action figures owned by Jef to those owned by Isaac.*
$33 = 3I$	*Substitute 33 for J in the initial equation.*
$\frac{33}{3} = \frac{3I}{3}$	*Divide both sides by 3.*
$11 = I$	

Isaac owns 11 action figures.

An equation can be written from the information given in the first sentence. Since "Jef owns three times as many action figures as Isaac", it follows that Jef's number must be equal to 3 times Isaac's number, or $J = 3I$. Notice that this equation contains two variables. Therefore, a single solution cannot be found.

Use the second and third sentences to write another equation. Since "subtracting 3 from the number of action figures Jef owns and then dividing by 2 gives the number Kim owns" and "Kim owns 15", it follows that subtracting 3 from J and then dividing that difference by 2 is equal to 15, or $(J - 3)/2 = 15$. Notice that this equation contains only one variable. Therefore, a single solution can be found.

Use the usual solving steps to solve the equation for J. The solution is 33, so Jef owns 33 action figures. Now, use the relationship between J and I, $J = 3I$, and the fact that J is 33 to solve for I. Substitute 33 into $J = 3I$ for J and then solve the equation. The solution is 11, so Isaac owns 11 action figures.

Solving Equations with Variables on Both Sides

Key Objectives

- Learn to solve equations with variables on both sides of the equals sign.

Key Terms

- The **LCD (lowest common denominator)** is the least common multiple of two or more given denominators.

An equation with variables on both sides refers to an equation where the same variable appears on each side of the equation. To solve an equation where the variable occurs on both sides, the terms that contain a variable must be combined on one side of the equation. Accomplish this by adding or subtracting one of the terms to both sides of the equation after the equation has been simplified.

Example 1 Example 1 demonstrates solving simplified equations with variables on both sides of the equation.

Solving Equations with Variables on Both Sides

Solve.

$5a + 7 = 6a$

$$\begin{array}{rcl} 5a + 7 & = & 6a \\ -5a & & -5a \\ \hline 7 & = & a \end{array}$$

Subtract 5a from both sides.

$2v - 9 = 3 + 5v$

$$\begin{array}{rcl} 2v - 9 & = & 3 + 5v \\ -2v & & -2v \\ \hline -9 & = & 3 + 3v \\ -3 & & -3 \\ \hline \frac{-12}{3} & = & \frac{3v}{3} \\ -4 & = & v \end{array}$$

Subtract 2v from both sides.

Subtract 3 from both sides.

Divide both sides by 3.

Consider the first equation. The variable, a, appears on both sides of the equation. To solve, begin by subtracting (since both terms are positive) one of the terms containing the variable, either $5a$ or $6a$, from both sides. Since $6a$ is the only term on the right side of the equation, the most efficient step is to subtract $5a$ from each side. After subtracting, the equation becomes $7 = a$, so solving is complete and the solution is 7.

Notice that the second equation has two terms on each side and that these terms cannot be combined since they are not like terms. Notice that each term with a variable is positive. So, begin by subtracting either of the variable terms, either $2v$ or $5v$, from both sides. Here, $2v$ is subtracted. The result of subtracting $2v$ on the left side is $2v - 9 - 2v$, or -9. The result of subtracting $2v$ on the right side is $3 + 5v - 2v$, or $3 + 3v$. Now, the equation is a two-step equation, so solve it as such.

So far, every equation has had a solution. However, this is not always the case. If an equation has the variable on both sides, it is possible for the equation to have no solution. Consider the equation $n + 1 = n$. This equation states that some number, n, plus 1 is equal to that same number, n. However, no such number exists. So, the equation is said to have "no solution".

$4 + x = x - 5$

$$\begin{array}{rcl} 4 + x & = & x - 5 \\ -x & & -x \\ \hline 4 & & -5 \end{array}$$

Subtract x from both sides.

$$4 \neq -5$$

There is no solution. There is no number that can be substituted for x to make the equation true.

TIP

If the variables in an equation are eliminated and the resulting statement is false, the equation has no solution.

Here, the term x is subtracted from both sides. The result of subtracting x on the left side is $4 + x - x$, or 4. The result of subtracting x on the right side is $x - 5 - x$, or -5. Notice that the remaining expressions are two constants and that these constants are not equal. So, the result is a false statement and the equation has no solution.

If the variables cancel out and the result is a true statement (i.e., two equal constants, such as $4 = 4$), then all real numbers are solutions to the equation.

Example 2 Example 2 demonstrates solving an unsimplified equation with variables on both sides.

Whenever possible, it is best to simplify before applying the solving steps to any equation. Remember, an equation can be simplified by combining the like terms that appear on the same side of the equation and by using the distributive property.

Solving Multi-Step Equations with Variables on Both Sides

Solve.

$5c + 1 - 6c = -7 + 2c + 2$

$-c + 1 = -5 + 2c$ *Combine like terms.*

$+c \qquad\quad + c$ *Add c to both sides.*

$1 = -5 + 3c$

$+5 \quad +5$ *Add 5 to both sides.*

$6 = 3c$

$\frac{6}{3} = \frac{3c}{3}$ *Divide both sides by 3.*

$2 = c$

Here, notice that the equation has three terms on each side and that some of these terms are like terms. Therefore, the equation can be simplified. Once the equation is simplified, the solving steps are used.

If an equation contains a fraction, the fraction can be replaced by a whole number by multiplying *each term* by the denominator of that fraction. Furthermore, if an equation contains more than one fraction, the fractions can be replaced by a whole numbers by multiplying *each term* by the LCD. Actually, multiplying each term by any of the denominators' common multiples will clear the fractions, but the LCD is the easiest because it is the smallest common multiple.

$\frac{3w}{4} - \frac{8w}{5} + \frac{1}{8} = -w + \frac{1}{2}$

$40\left(\frac{3w}{4} - \frac{8w}{5} + \frac{1}{8}\right) = 40\left(-w + \frac{1}{2}\right)$ *Multiply by the LCD, 40.*

$30w - 64w + 5 = -40w + 20$ *Distributive Property*

$-34w + 5 = -40w + 20$ *Combine like terms.*

$+40w \qquad + 40w$ *Add 40w to each side.*

$6w + 5 = 20$

$-5 \qquad -5$ *Subtract 5 from each side.*

$6w = 15$

$\frac{6w}{6} = \frac{15}{6}$ *Divide both sides by 6.*

$w = \frac{15}{6} = \frac{5}{2}$

This equation appears to be extremely complicated because it has many fractions. So, clear the fractions by multiplying each term in the equation by the LCD. The denominators are 4, 5, 8, and 2. The LCD of 4, 5, 8, and 2 is 40. So, multiply each term in the equation by 40. It is worth noting that all of the terms in the equation must be multiplied by the LCD, not just the terms with fractions. If any of the terms are left out of the multiplication process, the equation will no longer be an equivalent equation.

To multiply $3w/4$ by 40, simply divide 40 by the denominator, 4, and then multiply that quotient, 10, by the numerator, $3w$. So, the new term is $30w$. Be careful to keep the subtraction symbols in front of the appropriate terms.

After the fractions are cleared from the equation, simplify by combining like terms and then solve as usual.

Example 3 Example 3 demonstrates writing and solving an equation with variables on both sides.

Business Application

Flex Gym charges a membership fee of \$150.00 plus \$40.50 per month to join the gym. A rival gym, Able Gym, charges a membership fee of \$120.00 plus \$46.75 per month. Find the number of months for which you would pay the same total fee to both gyms.

Let x be the number of months.

$150 + 40.50x = 120 + 46.75x$

$-40.50x \qquad -40.50x$ *Subtract 40.50x from both sides.*

$150 = 120 + 6.25x$

$150 = 120 + 6.25x$

$-120 \quad -120$ *Subtract 120 from both sides.*

$30 = 6.25x$

$\frac{30}{6.25} = \frac{6.25x}{6.25}$ *Divide both sides by 6.25.*

$4.8 = x$

The two gyms cost the same when used for 4.8 months.

Consider this example. To find the number of months for which the total fee is equivalent at both gyms, first write expressions to represent the total fee at each gym. Then, set these expressions equal to each other and solve.

Since the number of months is unknown, let the variable x represent the number of months. Since the number of months will be the same at both gyms, the same variable can be used with each expression.

At Flex Gym, the total fee is \$150 plus \$40.50 per month, or $150 + 40.50x$. At Able Gym, the total fee is \$120 plus \$46.75 per month, or $120 + 46.75x$. Write an equation where the first expression is the left side and the second expression is the right side, then solve this equation as usual.

Systems of Equations

Key Objectives

- Learn to solve systems of equations.

Key Terms

- A **system of equations** is a set of two or more equations that contain two or more variables.
- A **solution of a system of equations** is a set of values that are solutions of all of the equations.

Example 1 Solving Systems of Equations

Solve the system of equations.

$y = 2x - 5$
$y = x + 7$

The expressions $2x - 5$ and $x + 7$ both equal y.
So, by the Transitive Property, they equal each other.

$y = \boxed{2x - 5}$ $y = \boxed{x + 7}$

$2x - 5 = x + 7$

Solve the equation to find x.

$$\begin{aligned} 2x - 5 &= x + 7 \\ -x \quad & \quad -x \quad \text{Subtract } x \text{ from both sides.} \\ x - 5 &= 7 \\ +5 \quad & \quad +5 \quad \text{Add 5 to both sides.} \\ x &= 12 \end{aligned}$$

To find y, substitute 12 for x in one of the original equations.

$$\begin{aligned} y &= x + 7 \\ y &= 12 + 7 \\ y &= 19 \end{aligned}$$

The solution is (12, 19).

The solution for a system of equations is determined in this example. The equations are given.

Set the two expressions that are equal to y equal to each other and solve for x. Solving the resulting equation for x gives $x = 12$.

To solve for y, substitute the solution for x back into one of the equations. The solution (from either equation) yields $y = 19$.

The solution to the system of equations is (12, 19).

Solve the system of equations.

$y = 5x - 1$
$y = 3 + 5x$

$$\begin{aligned} 5x - 1 &= 3 + 5x \quad \text{Transitive Property} \\ -5x \quad & \quad -5x \quad \text{Subtract } 5x \text{ from both sides.} \\ -1 &\neq 3 \end{aligned}$$

The system of equations has no solution.

The solution for a system of equations is determined in this example. The equations are given.

Set the two expressions that are equal to y equal to each other and solve for x. Solving the resulting equation for x gives a false result of $-1 = 3$. This means that there is no solution to this system of equations; there are no values of x and y that make both equations true.

Example 2 Solving Systems of Equations by Solving for a Variable

Solve the system of equations.

$x + y = 7$
$3x - y = 1$

Solve both equations for y.

$$\begin{aligned} x + y &= 7 \\ -x \quad & \quad -x \\ y &= -x + 7 \end{aligned} \qquad \begin{aligned} 3x - y &= 1 \\ y &= 3x - 1 \end{aligned}$$

$$\begin{aligned} -x + 7 &= 3x - 1 \\ +x \quad & \quad +x \quad \text{Add } x \text{ to both sides.} \\ 7 &= 4x - 1 \\ 7 &= 4x - 1 \\ +1 \quad & \quad +1 \quad \text{Add 1 to both sides.} \\ 8 &= 4x \\ \frac{8}{4} &= \frac{4x}{4} \quad \text{Divide both sides by 4.} \\ 2 &= x \end{aligned}$$

$$\begin{aligned} y &= -x + 7 \\ y &= -(2) + 7 \quad \text{Substitute 2 for } x. \\ y &= 5 \end{aligned}$$

The solution is (2, 5).

The solution for a system of equations is determined in this example. The equations are given.

Begin by isolating one of the variables in both the equations. It is usually easiest to do this by isolating the variable with a coefficient of 1. In this case, that is y. Then set the two expressions that are equal to y equal to each other and solve for x. Solving the resulting equation for x gives $x = 2$.

To solve for y, substitute the solution for x back into one of the equations. The solution yields $y = 5$.

The solution to the system of equations is (2, 5).

Solve the system of equations.

$$5x + 15y = 10$$
$$x + 3y = 2$$

$$\frac{5x}{5} + \frac{15y}{5} = \frac{10}{5}$$ *Divide both sides by 5.*

$$x + 3y = 2 \qquad x + 3y = 2$$ *The two equations are the same.*

Because the two equations are the same, the system of equations has an infinite number of solutions.

The solution for a system of equations is determined in this example. The equations are given.

Begin by isolating one of the variables in both the equations. It is usually easiest to do this by isolating the variable with a coefficient of 1. In this case, that is x. Simplifying one of the equations shows that both equations are the same. Since the equations are the same, any values of x and y that make one equation true, and there are an infinite number, also make the other equation true. Another way to look at it is that this is not a system of equations; it is only one equation.

The Coordinate Plane

Key concepts

- Learn to plot and identify ordered pairs on a coordinate plane.

Key terms

- A **coordinate plane** is a plane containing a horizontal number line, the x-axis, and a vertical number line, the y-axis.
- The axes (plural of axis) divide the coordinate plane into four regions called **quadrants**.
- All points (locations) on a coordinate plane correspond with a specific value on the x-axis and a specific value on the y-axis, known as the **x-coordinate** and the **y-coordinate**.
- The x-coordinate and y-coordinate that correspond with a specific point on a coordinate plane are called a coordinate pair, or **ordered pair**. An ordered pair is always written as (x, y).

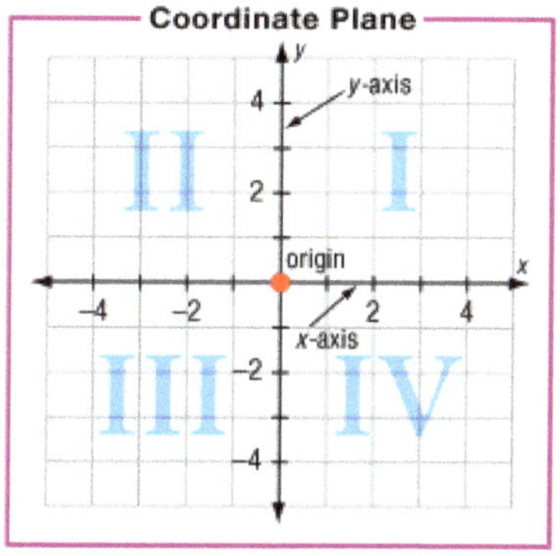

An example of a coordinate plane is given above. A coordinate plane contains an x-axis and a y-axis. The x-axis is the horizontal axis and the y-axis is the vertical axis. The axes are just number lines that intersect at zero. The axes divide the plane into four quadrants, defined above. Each location on the coordinate plane, otherwise known as a point, can be described with a coordinate pair that contains an x- and a y-value. One point is marked in the coordinate plane given above. This point is located at the center of the plane, or where the x-axis intersects the y-axis. This point is called the origin.

Example 1

Example 1 illustrates identifying the general location of a given point on a coordinate plane.

Any point on a plane must be in one of seven areas. The seven areas on a plane are the four quadrants, the two axes, and the origin. The origin is the point at which the axes intersect.

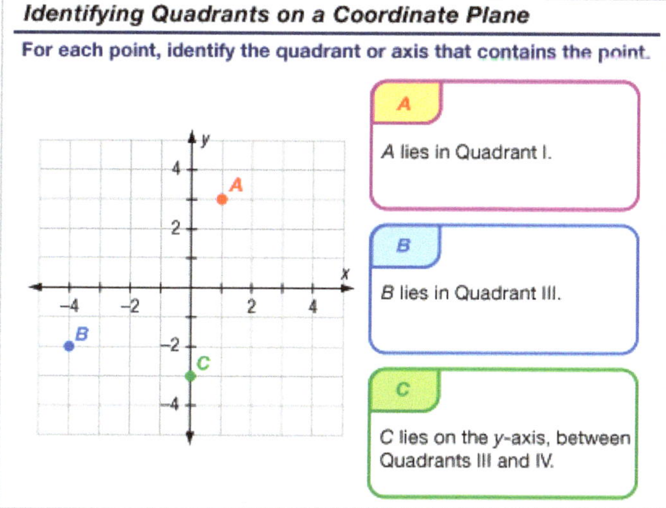

Consider the points on the plane. Point A is not on an axis, therefore it must be in a quadrant. Use the image above to determine the name of the quadrant.

Point B is also not on an axis, therefore it must be in a quadrant as well.

Point C is on an axis, specifically it is on the y-axis.

A point cannot be in quadrant and on an axis.

Example 2

Example 2 illustrates plotting a point on a coordinate plane given a coordinate pair.

Plotting a point means to find the specific location of that point on a coordinate plane. Consider the coordinate pair that defines a point. The first value in the coordinate pair is the x-value. The second value in the coordinate pair is the y-value. To plot a point, start at the origin, then move on the x-axis to the x-value given in the coordinate pair. Next, move up or down the number of units defined by the y-value in the coordinate pair. Move up if the y-value is positive or move down if the y-value is negative.

Plotting Points on a Coordinate Plane

Plot each point on a coordinate plane.

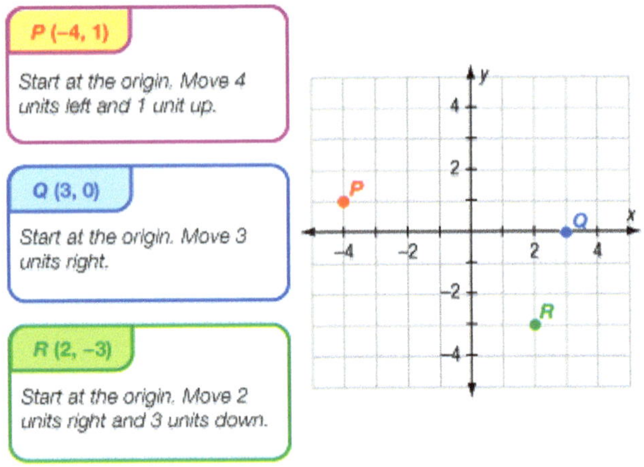

Consider the first ordered pair, (–4, 1). This point is identified with the letter P. The x-coordinate is –4 and the y-coordinate is 1. To plot (–4, 1), begin at the origin and move left 4 units on the x-axis since the x-coordinate is –4. Next, move up 1 unit since the y-coordinate is 1. This point corresponds with –4 on the x-axis and 1 on the y-axis. Point P is located in quadrant II.

The second ordered pair is (3, 0). To plot this point, begin at the origin and move 3 units right on the x-axis since the x-coordinate is 3. Notice that the y-coordinate is 0. So, no movement up or down is required. Thus, point Q is located on the x-axis.

Furthermore, any point where the y-coordinate is 0 will be located on the x-axis, and any point where the x-coordinate is 0 will be located on the y-axis.

Example 3

Example 3 illustrates writing an ordered pair to define a given point on a coordinate plane.

To write the coordinate pair of a specific point, first identify the value on the x-axis that corresponds with that point. This is the x-value in the coordinate pair. Next, identify the value on the y-axis that corresponds with that point. This is the y-value in the coordinate pair.

Identifying Points on a Coordinate Plane

Give the coordinates of each point.

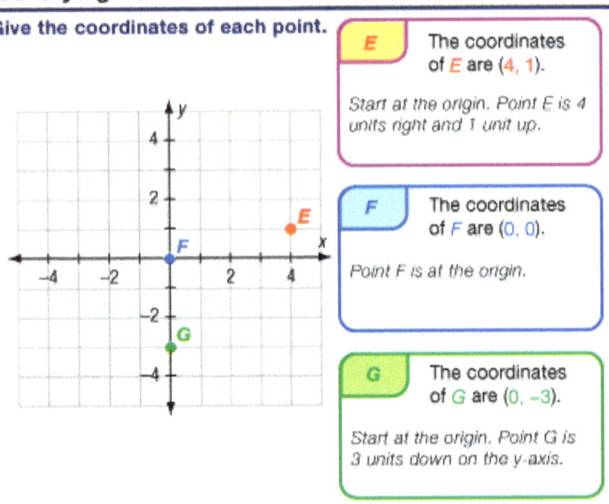

Consider point E given on the coordinate plane. This point is located in quadrant I. So, the x-coordinate and the y-coordinate must both be positive. To find the coordinates of the point (or the ordered pair that corresponds with this point), identify values on the x- and y-axis that correspond with the point. The corresponding value on the x-axis is 4. So, the x-coordinate is 4. The corresponding value on the y-axis is 1. So, the y-coordinate is 1. Now, write the coordinates as an ordered pair: (4, 1).

Notice that the point F is at the center of the plane. So, the x-coordinate is 0 and the y-coordinate is 0. Thus, the corresponding ordered pair is (0, 0).

Notice that the point G is located on the y-axis. So, the x-coordinate must be 0. The corresponding y-value is –3. Thus, the corresponding ordered pair is (0, –3).

Equations, Tables, and Graphs

Key Objectives

- Learn to generate different representations of the same data.

Example 1 Using Equations to Generate Different Representations of Data

The distance a car travels is represented by the equation $d = 60t$, where d is the distance and t is the number of hours. Make a table and sketch a graph of the equation.

t	$60t$	d
0	60(0)	0
1	60(1)	60
2	60(2)	120
3	60(3)	180
4	60(4)	240

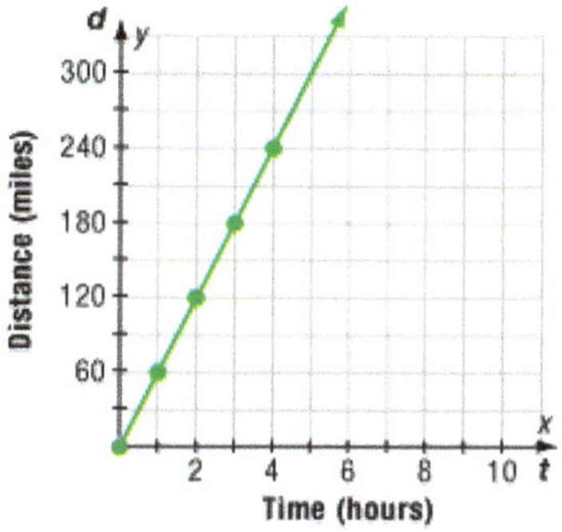

An equation is presented as data in a table and as a graph in this example. The equation is given.

To make a table, put the variables (and the function that relates one variable to the other) as column headings. Choose values of the input variable (x, or in this case, t) and calculate the corresponding value of the output variable (y, or in this case, d).

To make a graph, label a coordinate plane with the appropriate units on the axes. The input variable typically goes on the x-axis. Plot the paired values from the table as the coordinates of points on the graph. Join the points with a smooth line to visualize the solutions to the equation that lie between the calculated points.

Example 2 Using Tables to Generate Different Representations of Data

Use the table to make a graph and to write an equation.

x	y
0	0
1	3
2	6
3	9

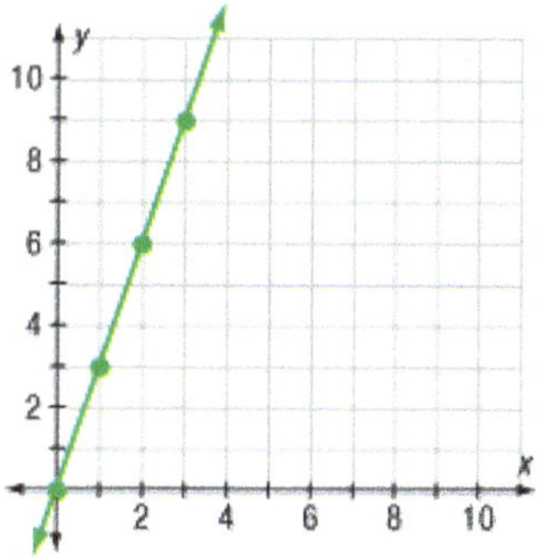

Look for a pattern in the values:

$3 = 3 \cdot 1$
$6 = 3 \cdot 2$
$9 = 3 \cdot 3$

Each value of y is three times the value of x.

The equation is $y = 3x$.

The data in a table are presented as a graph and an equation in this example. The data table is given.

To make a graph, plot the paired values from the table as the coordinates of points on the graph. Join the points with a smooth line to visualize the values that lie between the table data points.

To develop an equation from the data in the table, look for a pattern in the relationship of the x and y values. In each case, the value of y is three times the value of x. The equation that describes this relationship is $y = 3x$.

Example 3 Using Graphs to Generate Different Representations of Data

Use the graph to make a table and to write an equation.

x	y
−1	1
0	2
1	3
2	4
3	5
4	6

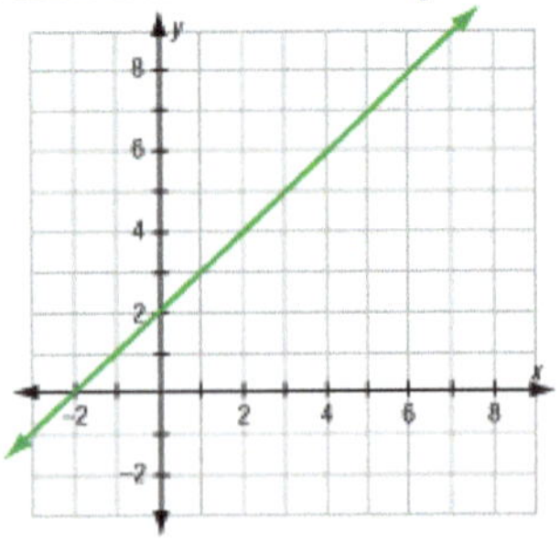

Look for a pattern in the values:

$1 = -1 + 2$
$2 = 0 + 2$
$3 = 1 + 2$
$4 = 2 + 2$
$5 = 3 + 2$

Each value of y is two more than the value of x.

The equation is $y = x + 2$.

A graph is presented as data in a table and an equation in this example. The graph is given.

To make a table, put the variables as column headings. Choose values of the input variable, x, to put in the x column of the table. Read the corresponding value of the output variable, y, from the graph and enter it into the y column of the table.

To develop an equation from the data in the table, look for a pattern in the relationship of the x and y values. In each case, the value of y is equal to the value of x plus 2. The equation that describes this relationship is $y = x + 2$.

Graphing Using Intercepts

Key concepts

- Learn to use intercepts to graph linear equations.

Key terms

- The **x-intercept** of a line is the value of x where the line intersects the x-axis (where $y = 0$).
- The **y-intercept** of a line is the value of y where the line intersects the y-axis (where $x = 0$).

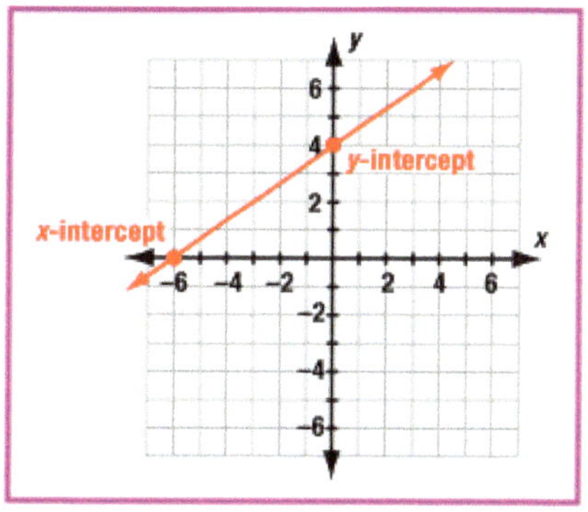

A line is graphed in the coordinate plane given above. Notice that this line intersects the x-axis and it also intersects the y-axis. Since the graph is a line, there must be a linear equation that corresponds with this graph.

Consider the point at which the line intersects the x-axis, $(-6, 0)$. The x-coordinate of this point is known as the x-intercept. So, the x-intercept of the line above is -6. A line's x-intercept is always located on the x-axis. The y-coordinate of any point located on the x-axis is always 0. So, the x-intercept is the x-value that corresponds with a y-value of 0. It follows that to find a line's x-intercept from a given equation, substitute 0 into the equation for y and solve for x. This x-value is the line's x-intercept.

Consider the point at which the line intersects the y-axis, $(0, 4)$. The y-coordinate of this point is known as the y-intercept. So, the y-intercept of the line above is 4. A line's y-intercept is always located on the y-axis. The x-coordinate of any point located on the y-axis is always 0, so the y-intercept is the y-value that corresponds with a x-value of 0. It follows that to find a line's y-intercept from a given equation, substitute 0 into the equation for x and solve for y. This y-value is the line's y-intercept.

Example 1

In Example 1, a line is graphed from a given linear equation by finding and graphing the intercepts.

Steps for Using Intercepts to Graph a Line

1. Substitute 0 into the given equation for y and solve for x to find the x-intercept.

2. Substitute 0 into the given equation for x and solve for y to find the y-intercept.

3. Plot the x-intercept and the y-intercept on a coordinate plane and draw the line that passes through these two points.

Finding x-intercepts and y-intercepts to Graph Linear Equations

Find the x-intercept and y-intercept of the line $2x + 3y = 6$. Use the intercepts to graph the equation.

Find the x-intercept ($y = 0$).

$$2x + 3y = 6$$
$$2x + 3(0) = 6$$
$$2x = 6$$
$$\frac{2x}{2} = \frac{6}{2}$$
$$x = 3$$

The x-intercept is 3.

Find the y-intercept ($x = 0$).

$$2x + 3y = 6$$
$$2(0) + 3y = 6$$
$$3y = 6$$
$$\frac{3y}{3} = \frac{6}{3}$$
$$y = 2$$

The y-intercept is 2.

The graph of $2x + 3y = 6$ is the line that crosses the x-axis at the point (3, 0) and the y-axis at the point (0, 2).

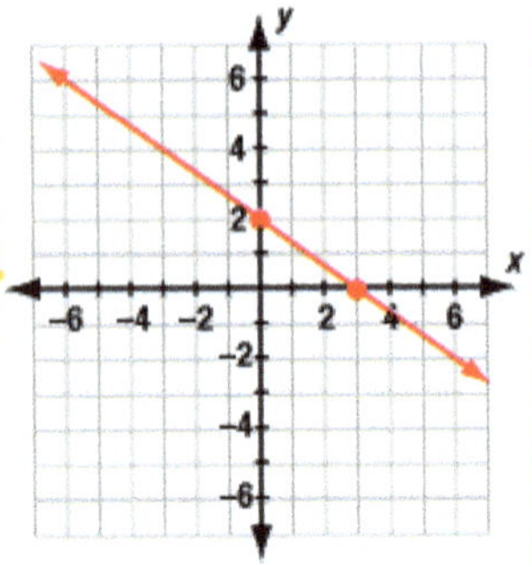

Notice that the two variables are both on the left side of the equation. If the variables are on opposite sides of an equation, the line is graphed using intercepts by following the same process.

To find the x-intercept, 0 is substituted into $2x + 3y = 6$ for y: $2x + 3(0) = 6$. Next, the equation is solved for x by simplifying and then dividing both sides by 2. The result is 3. Thus, the x-intercept is 3.

To find the y-intercept, 0 is substituted into $2x + 3y = 6$ for x: $2(0) + 3y = 6$. Next, the equation is solved for y by simplifying and then dividing both sides by 3. The result is 2. Thus, the y-intercept is 2.

Rates, Ratios, and Proportions

Key Objectives

- Write and use ratios, rates, and unit rates.
- Write and solve proportions.

Key Terms

- A **ratio** is a comparison of two quantities by division.
- A statement that two ratios are equivalent, such as $\frac{1}{12} = \frac{2}{24}$ is called a **proportion**.
- A **rate** is a ratio of two quantities with different units.
- A **unit rate** is a rate with a second quantity of 1 unit.
- A rate in which the two quantities are equal but use different units is called a **conversion factor**.
- In the proportion $\frac{a}{b} = \frac{c}{d}$, the products a · d and b · c are called **cross products**.
- A **scale** is a ratio between two sets of measurements.
- A **scale drawing** or **scale model** uses a scale to represent an object as smaller or larger than the actual object.

Example 1 Using Ratios

The ratio of faculty members to students at a college is 1:12. There are 1440 students. How many faculty members are there?

faculty → 1
students → 12

$$\frac{1}{12} = \frac{F}{1440}$$ *Write a proportion. Let F be the total faculty.*

$$1440\left(\frac{1}{12}\right) = \left(\frac{F}{\cancel{1440}}\right)\cancel{1440}$$ *Solve for F. Multiply both sides by 1440.*

$$\frac{1440}{12} = F$$

$$120 = F$$

There are about 120 total faculty members at the school.

A ratio is used to determine the number of faculty at a college. The faculty-student ratio and the number of students are given.

To find the number of faculty, set up a proportion with the number of faculty as an unknown. Make sure that the unknown is in the appropriate place corresponding to the ratio value for faculty. In this case, the ratio number for faculty is 1, so the unknown should be the numerator of the other ratio in the proportion.

Solve the equation for F to find $F = 120$. There are 120 faculty at the school.

Example 2 Finding Unit Rates

The price for 8.5 pounds of compost starter is $15. Find the unit rate.

price → p
pounds → 1

$$\frac{p}{1} = \frac{15}{8.5}$$ *Write a proportion. Let p be the price per unit.*

$$p \approx \$1.76$$ *Divide.*

The price per pound (unit price) for compost starter is $1.76.

A unit ratio is found for the cost per weight of compost starter in this example. The cost for a given number of pounds of compost starter is given.

To find the unit rate, set up a proportion with the unknown price, p, over 1 dollar as equal to the actual number of pounds over their actual cost. Since a per-dollar unit rate is being sought, make sure that the dollars are on the bottoms of the ratios.

Divide both ratios to find the unit rate, $p \approx \$1.76$ per pound.

Example 3 Converting Rates

As you go deeper underground, the earth's temperature increases. In some places, it may increase by 20°C per kilometer. What is this rate in degrees per meter?

$$\left(\frac{20^\circ\text{C}}{1\ \text{km}}\right) \cdot \frac{1\ \text{km}}{1000\ \text{m}} = \frac{20^\circ\text{C}}{1000\ \text{m}}$$ *Convert kilometers to meters.*

$$= \frac{1^\circ\text{C}}{50\ \text{m}}$$ *Simplify.*

$$= \frac{0.02^\circ\text{C}}{1\ \text{m}}$$ *Simplify.*

A unit rate is converted to different units using a conversion factor in this example. In this example one of the units in the unit rate is converted. The unit rate is given.

To make the conversion, set up the unit rate as a ratio including the units. Write the unit rate with the conversion factor. Make sure that the units for the conversion factor are in the correct place to make the cancellation shown. Multiply the unit rate and the conversion factor and cancel the units.

Simplify the resulting rate so that it is again a unit rate by dividing the top and bottom by the same factors (or divide the numerator by the denominator). The converted rate is 0.02 °C per m.

The record running speed of a giraffe is 34.7 miles per hour. What is the speed in feet per minute?

$$\frac{34.7\ \text{mi}}{1\ \text{h}} \cdot \frac{1\ \text{h}}{60\ \text{min}} = \frac{34.7\ \text{mi}}{60\ \text{min}}$$ *Convert hours to minutes.*

$$= \frac{34.7\ \text{mi}}{60\ \text{min}} \cdot \frac{5280\ \text{ft}}{1\ \text{mi}} = \frac{183{,}216\ \text{ft}}{60\ \text{min}}$$ *Convert miles to feet.*

Miles to feet

$$= \frac{3{,}053.6\ \text{ft}}{1\ \text{min}}$$ *Simplify.*

conversion factor
1 mile = 5,280 feet

A unit rate is converted to different units using a conversion factor in this example. In this example both of the units in the unit rate are converted. The unit rate is given.

To make the first conversion (miles per hour to miles per minute), set up the unit rate as a ratio including the units. Write the unit rate with the conversion factor. Make sure that the units for the conversion factor are in the correct place to make the cancellation shown. Multiply the unit rate and the conversion factor and cancel the units. Repeat this with the second conversion (miles per minute to feet per minute).

Simplify the resulting rate so that it is again a unit rate by dividing the top and bottom by the same factors (or divide the numerator by the denominator). The converted rate is 3,053.6 feet per minute. This rate is equal to 34.7 miles per hour.

Example 4 Solving Proportions

Solve the proportion.

$$\frac{2}{7} = \frac{5}{w}$$

$$2(w) = 7(5)$$ *Use cross products.*

$$\frac{2w}{2} = \frac{35}{2}$$ *Divide both sides by 2.*

$$w = \frac{35}{2}$$

A proportion is solved to find an unknown in this example.

To solve the proportion, take the cross products by multiplying the numerators with the denominators and setting the two products equal. Isolate the unknown and simplify. In this example, $w = 35/2 = 17.5$.

Solve the proportion.

$$\frac{4}{x+2} = \frac{1}{3}$$

$1(x + 2) = 3(4)$ *Use cross products.*

$x + 2 = 12$ *Subtract 2 from both sides.*

$\quad -2 \quad -2$

$x = 10$

A proportion is solved to find an unknown in this example.

To solve the proportion, take the cross products by multiplying the numerators with the denominators and setting the two products equal. Isolate the unknown and simplify. In this example, $x = 10$.

Example 5 Scale Drawings and Scale Models

The actual distance between Fresno and Bakersfield is 110 mi. What is the distance between these two locations on the map?

$$\frac{1 \text{ in.}}{25 \text{ mi}} = \frac{l \text{ in.}}{110 \text{ mi}}$$ *Write a proportion. Let l be the distance on the map.*

$$\frac{25l}{25} = \frac{110}{25}$$ *Use the cross products. Divide both sides by 25.*

$l = 4.4$ in.

A map distance is calculated using a real distance and a scale. The scale and the real distance are given.

Set up a proportion using the scale and the known distance between the two cities. Make sure the values are in their correct places in the proportion so that inches are compared to inches and miles to miles. Cross multiply and set the products equal to each other. Solve for the unknown map distance. The distance between the two cities on the map is 4.4 inches.

Applications of Proportions

Key concepts

- Learn to use proportions to solve application problems.

Key terms

- The **cross products** are the results of multiplying along the diagonals of a proportion.

Proportions may be used to solve a variety of real-world applications. When writing proportions, be sure to be consistent with the units. For example, if the first ratio is written as inches to feet, the second ratio must also be written as inches to feet, not feet to inches. Remember, a ratio is a comparison of two values and a proportion is two equal ratios.

Example 1

Example 1 demonstrates using a proportion to find a unit rate.

Consumer Application

The cost for 3 pounds of apples is \$5.95. Find the cost per pound for these apples.

Step 1 Write a proportion.

\$5.95 is to 3 pounds as \$c is to 1 pound.

$$\frac{\$5.95}{3\text{ lb}} = \frac{\$c}{1\text{ lb}}$$ *Let c represent the cost per pound for apples.*

Step 2 Find the unknown value in the proportion.

$$\frac{\$5.95}{3\text{ lb}} \times \frac{\$c}{1\text{ lb}}$$ *This is a proportion, so the cross products must be equal.*

$3 \cdot c = 5.95$ *Find the cross products.*

$\frac{3 \cdot c}{3} = \frac{5.95}{3}$ *c is multiplied by 3, so divide each side by 3 to undo the multiplication.*

$c = 1.9833\ldots$ *Simplify each side.*

$c \approx 1.98$ *Round to the nearest cent.*

The apples cost about \$1.98 per pound.

Consider the example illustrated here. To find the cost per pound, write a proportion from the given information. Remember, the "cost per pound" is the cost for one pound of apples, so the number of pounds is 1.

Since the first ratio is written as dollars to pounds, the second ratio must also be written as dollars to pounds.

After the proportion is written, set the cross products equal to each other and solve the resulting equation to find the cost per pound.

Example 2

In Example 2, a proportion is written and then solved for an unknown value.

A scale is used when making a map so that the distance between two locations can be determined. For example, if a map uses the scale 2 inches = 150 miles, it means that if two locations are 2 inches apart on the map, then the actual distance between these two locations is 150 miles. This scale may be written as a ratio.

Problem-Solving Application

The scale on Ashley's map of Arkansas is 3 inches to 100 miles. She determines that the distance on the map between Hot Springs and West Memphis is $5\frac{1}{2}$ inches. What is the actual distance between the two cities? Round the answer to the nearest mile.

Step 1 Write a proportion.

100 miles is to 3 inches as D miles is to $5\frac{1}{2}$ inches.

$\frac{100 \text{ miles}}{3"} = \frac{D \text{ miles}}{5.5"}$ *Let D represent the actual distance between Hot Springs and West Memphis.*

Step 2 Find the unknown value in the proportion.

$\frac{100 \text{ miles}}{3"} \times \frac{D \text{ miles}}{5.5"}$ *This is a proportion, so the cross products must be equal.*

$3 \cdot D = 100 \cdot 5.5$ *Find the cross products.*

$3 \cdot D = 550$ *Simplify.*

$\frac{3 \cdot D}{3} = \frac{550}{3}$ *D is multiplied by 3, so divide each side by 3 to undo the multiplication.*

$D = 183.33\ldots$ *Simplify each side.*

$D \approx 183$ *Round to the nearest mile.*

The distance between Hot Springs and West Memphis is about 183 miles.

Consider the map described in this exercise. The map's scale is "3 inches to 100 miles". As a ratio, this is 3 inches/100 miles or 100 miles/3 inches. Here, the ratio is written as miles to inches.

In addition to the scale, the distance on the map between the two cities is given. To find the unknown distance, write a second ratio that includes the unknown. Since the first ratio is written as miles to inches, the second ratio must also be written as miles to inches. So, the second ratio is D miles/5.5 inches, where D is the unknown distance in miles between the two cities. Write a proportion from these two ratios.

Once the proportion is written, it can be solved for the unknown by using the cross products.

Example 3

In Example 3, a proportion involving time is written.

Consumer Application

Claire can wash 5 dogs in 1 hour and 30 minutes. At this rate, how many dogs could she wash in 7 hours?

Step 1 Write a proportion.

5 dogs is to 1 hour and 30 min as D dogs is to 7 hours.

$\frac{5 \text{ dogs}}{1.5 \text{ hr}} = \frac{D \text{ dogs}}{7 \text{ hr}}$ *Let D represent the number of dogs that Claire can wash in 7 hours.*

Step 2 Find the unknown value in the proportion.

$\frac{5 \text{ dogs}}{1.5 \text{ hr}} \times \frac{D \text{ dogs}}{7 \text{ hr}}$ *This is a proportion, so the cross products must be equal.*

$(1.5)D = 5 \cdot 7$ *Find the cross products.*

$(1.5)D = 35$ *Simplify.*

$\frac{(1.5)D}{1.5} = \frac{35}{1.5}$ *D is multiplied by 1.5, so divide each side by 1.5 to undo the multiplication.*

$D = 23.33\ldots$ *Simplify each side.*

$D \approx 23$ *Round to the nearest dog.*

Claire can wash about 23 dogs in 7 hours.

Consider this example. Since the first ratio is written as dogs to hours, the second ratio must also be dogs to hours, not hours to dogs.

To express 1 hour and 30 minutes as a numerical value, remember that 30 minutes can be expressed as 1/2 an hour, or as 0.5 hours in decimal form. Thus, 1 hour and 30 minutes is 1.5 in decimal form.

Round the answer to the nearest whole number, or nearest dog.

Finding and Using Percents

Key concepts

- Learn to find percents and to find a number when a percent is known.

Example 1

In Example 1, two numbers are given, a part and a whole, and the percent that the part is of the whole is found using two methods.

An equation can often be used to find a percent. To write the equation, remember that "of" indicates multiplication and "is" indicates equality. Translate the given information into an equation using a variable to represent the unknown.

Finding the Percent One Number Is of Another

What percent of 162 is 90?

Method 1: Set up an equation to find the percent.

$p \cdot 162 = 90$	*Set up an equation.*
$\frac{162p}{162} = \frac{90}{162}$	*Divide both sides by 162.*
$p = 0.\overline{5}$, or approximately 0.555	*Simplify.*
$p \approx 56\%$	

90 is approximately 56% of 162.

To find the percent that 90 is of 162, translate the question "What part of 162 is 90?" into an equation. In the equation, "What part" is represented by p, "of" is represented by a multiplication symbol, and "is" is represented by an equals sign. Now, solve the equation by dividing both sides by 162.

A second method for finding a percent is to solve the proportion *number*/100 = *part*/*whole*.

NOTE: In the proportion *number*/100 = *part*/*whole*, the *number* represents the percent. It should not be confused with a decimal that must be converted to a percent.

To find the percent using this proportion, substitute the given information into the proportion and solve for the unknown value.

Earth has a surface area of approximately 197 million square miles. About 58 million square miles of that surface area is land. Find the percent of Earth's surface area that is land.

Total surface area	197 million mi^2
Land surface area	58 million mi^2

Method 2: Set up a proportion to find the percent.

$\frac{\text{number}}{100} = \frac{\text{part}}{\text{whole}}$	*Set up a proportion.*
$\frac{n}{100} = \frac{58}{197}$	*Substitute.*
$n \cdot 197 = 100 \cdot 58$	*Find the cross products.*
$197n = 5800$	*Simplify.*
$\frac{197n}{197} = \frac{5800}{197}$	*Divide both sides by 197.*
$n \approx 29.44 \approx 29\%$	*Simplify.*

Approximately 29% of the Earth's surface area is land.

In the exercise demonstrated here, the *whole* is the entire surface area of the Earth, 187 million square miles, and the *part* is the Earth's surface area that is covered by land, 58 million square miles. Substitute these values into the proportion and represent the unknown percent with n, and solve. Notice that the millions can be canceled from the *part* and the *whole* before finding the cross products.

After dividing, the value of n is found to be approximately 29.44, which rounds to 29. This value is *not* converted to a percent because it already *is* a percent.

Example 2

Example 2 demonstrates finding a part when a percent and the whole are known by using two methods: an equation and a proportion.

The statement "*part* is *percent* of *whole*" or "*part* = *percent* · *whole*" can be used to write an equation. If two of the three values (*part*, *percent*, and *whole*) are known, the third one can be found by solving the equation.

Finding the Percent of a Number **Andy can run about 11 kilometers per hour. Her 10-year-old son Jack can run about $66\frac{2}{3}\%$ of the speed of his mother. To the nearest tenth, how fast can Jack run?** Method 1: Set up an equation. Find the speed that is $66\frac{2}{3}\%$ of 11 km/h. $s = 66\frac{2}{3}\% \cdot 11$ *Set up an equation.* $s = \frac{2}{3} \cdot 11$ *$66\frac{2}{3}\%$ is equivalent to $\frac{2}{3}$.* $s = \frac{22}{3}$ *Simplify.* $s \approx 7.3$ *Round to the nearest tenth.* Jack can run about 7.3 kilometers per hour.	In this exercise, the *whole* is the mother's running speed and the *part* is the son's running speed. The son's speed is unknown, so use a variable to represent that value. Substitute the known percent and the mother's running speed into the equation "*part* = *percent* · *whole*" and multiply to find the unknown value.
The Chrysler Building in NYC is about 1,046 feet tall. The height of the Empire State Building is about 120% of the height of the Chrysler Building. To the nearest foot, find the height of the Empire State Building. Method 2: Set up a proportion. Set up as: 120 is to 100 as what height is to 1,046 ft? $\frac{120}{100} = \frac{E}{1046}$ *Set up a proportion.* $100 \cdot E = 120 \cdot 1046$ *Find the cross products.* $100E = 125520$ *Simplify.* $\frac{100E}{100} = \frac{125520}{100}$ *Divide both sides by 100.* $E = 1255.2$ *Simplify.* $E \approx 1255$ *Round to the nearest whole number.* The Empire State Building is about 1,255 feet tall.	Here, the height of the Chrysler Building, 1,046 feet, is the *whole* and the percent, 120%, is the *number*. Substitute these values into the proportion and solve to find the *part*, or the height of the Empire State Building. Since the percent is greater than 100%, the *part* must be greater than the *whole*, or the height of the Empire State Building must be greater than the height of the Chrysler Building.

Example 3

In Example 3, an equation to find a number when the percent is known is written and then solved.

An equation can often be used to find a percent. To write the equation, remember that "of" indicates multiplication and "is" indicates equality. Translate the given information into an equation using a variable to represent the unknown.

Finding a Number When the Percent is Known **52 is 8% of what number?** $52 = 8\% \cdot N$ *Set up an equation.* $52 = 0.08N$ *8% = 0.08* $\frac{52}{0.08} = \frac{0.08N}{0.08}$ *Divide both sides by 0.08.* $650 = N$ *Simplify.* 52 is 8% of 650.	Consider this example. First, translate the question "52 is 8% of what number" into an equation. In the equation, "is" is represented by an equals sign, "what number" is represented by *N*, and "of" is represented by a multiplication symbol. Now write the percent as a decimal and then solve the equation by dividing both sides by 0.08.

Example 4

In Example 4, an equation is written and solved for a real-world application.

Architecture Application

The Taipei building in Taipei, Taiwan is 1,670 feet high. This is 206.7% of the height of the MetLife building in NYC. Find the height of the MetLife building to the nearest foot.

$1670 = 206.7\% \cdot M$	*Set up an equation.*
$1670 = 2.067M$	*206.7% = 2.067*
$\frac{1670}{2.067} = \frac{\cancel{2.067}M}{\cancel{2.067}}$	*Divide both sides by 2.067.*
$808 \approx M$	*Simplify.*

The height of the MetLife building is approximately 808 feet.

In this example, the information can be summarized to produce the question "1,670 is 206.7% of what number?" Translate this question into an equation, write the percent as a decimal, and solve.

Notice that the height of the Taipei building is over 200% of the height of the MetLife building. Therefore, the height of the Taipei building must be over twice the height of the MetLife building.

Example 5

In Example 5, an equation is written and solved for a real-world application.

Recreation Application

The Blue Hills adventure race at Lake Clinton covers a 22-mile course. This is only about 17% of the length of the Wild Onion urban adventure race in Chicago. Find the length of the Wild Onion race to the nearest mile.

$22 = 17\% \cdot w$	*Set up an equation.*
$22 = 0.17w$	*17% = 0.17*
$\frac{22}{0.17} = \frac{\cancel{0.17}w}{\cancel{0.17}}$	*Divide both sides by 0.17.*
$129 \approx w$	*Simplify.*

The length of the Wild Onion race is approximately 129 miles.

Here again, summarize the information given in the application exercise into the question "22 is 17% of what number?" Translate this question into an equation, write the percent as a decimal, and solve.

Adding and Subtracting Polynomials

Key Objectives

- Add polynomials horizontally.
- Write polynomial expressions for perimeter.
- Find the opposite of polynomials.
- Subtract polynomials horizontally.
- Write polynomial expressions for profit.
- Evaluate polynomial expressions.

Formulas

- **Perimeter of a Rectangle**
  $P = 2L + 2w$, where L and w are the length and width of the rectangle, respectively.
- **Profit Formula**
  $P = R - C$, where R is revenue and C is cost.

Operations can be performed on polynomials, just as they can be performed on numbers.

Example 1
Adding Polynomials Horizontally

To add polynomials, combine like terms. Recall that like terms are two or more terms with the same variable raised to the same power.

In Example 1, Prof. Burger adds polynomials by applying the Associative Property (to remove the parentheses) and then combining like terms.

Add.

$(8x^2 - 2x + 3) + (9x - 5)$

$= 8x^2 - 2x + 3 + 9x - 5$ *Associative Property*

$= 8x^2 + 7x - 2$ *Combine like terms.*

$(ab^2 + 3a) + (2ab^2 + 3a - 2) + (2a + 4)$

$= ab^2 + 3a + 2ab^2 + 3a - 2 + 2a + 4$ *Associative Property*

$= 3ab^2 + 8a + 2$ *Combine like terms.*

Example 2
Art Application

In Example 2, Prof. Burger writes a polynomial expression to represent the perimeter of a rectangle. The Distributive Property is used to simplify the expression. Recall that by the Distributive Property, $a(b + c) = ab + ac$.

Libby is putting a mat of width m and a frame of width f around an 8-inch by 10-inch picture. Find an expression for the perimeter of the entire frame.

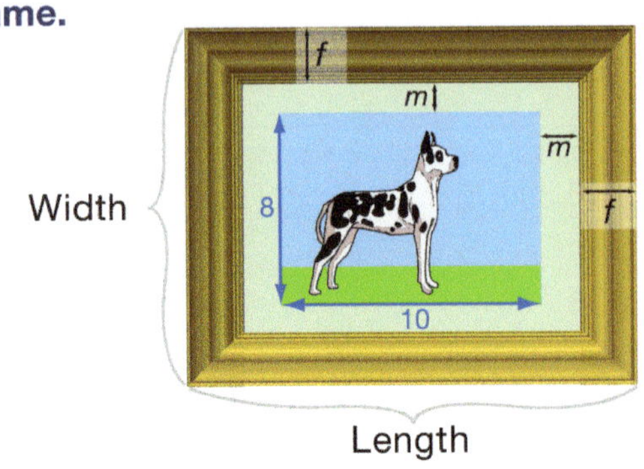

Find the outer dimensions of the frame.

Length $= f + m + 10 + m + f$ Width $= f + m + 8 + m + f$

$P = 2(f + m + 10 + m + f) + 2(f + m + 8 + m + f)$

$= 2(2f + 2m + 10) + 2(2f + 2m + 8)$

$= 4f + 4m + 20 + 4f + 4m + 16$ *Distributive Property*

$= 8f + 8m + 36$ *Combine like terms.*

The perimeter of the frame is $8f + 8m + 36$ inches.

Example 3
Finding the Opposite of a Polynomial

To find the opposite of a polynomial, multiply the polynomial by -1, or write the opposite of each term in the polynomial.

Find the opposite of each polynomial.

$9x^2y^4z$

$-(9x^2y^4z) = -9x^2y^4z$

$2ab^2 - 3ab + 2$

$-(2ab^2 - 3ab + 2) = -2ab^2 + 3ab - 2$ *Distributive Property*

Example 4
Subtracting Polynomials Horizontally

Recall that subtracting a number is the same as adding its opposite (e.g., $3 - 5 = 3 + -5$). To subtract a polynomial, add the opposite of that polynomial and then combine like terms, as demonstrated by Prof. Burger in Example 4.

Subtract.

$(n^3 - n + 4n^2) - (6n - 3n^2 + 8)$

$= (n^3 - n + 4n^2) + (-6n + 3n^2 - 8)$	*Add the opposite.*
$= n^3 - n + 4n^2 - 6n + 3n^2 - 8$	*Associative Property*
$= n^3 + 7n^2 - 7n - 8$	*Combine like terms.*

$(-3cd^2 + cd + 6) - (-9cd^2 + 2 - 7cd)$

$= (-3cd^2 + cd + 6) + (9cd^2 - 2 + 7cd)$	*Add the opposite.*
$= -3cd^2 + cd + 6 + 9cd^2 - 2 + 7cd$	*Associative Property*
$= 6cd^2 + 8cd + 4$	*Combine like terms.*

Example 5
Business Application

In Example 5, Prof. Burger writes a polynomial expression for the profit generated by selling x model kits, using the fact that profit is equal to revenue minus cost. He then evaluates that expression for $x = 300{,}000$ to find the profit earned for selling 300,000 model kits.

Suppose the cost in dollars of producing x model kits is given by the polynomial $500{,}000 + 2x$ and the revenue generated from sales is given by the polynomial $30x - 0.00005x^2$. Find a polynomial expression for the profit from making and selling x model kits, and evaluate the expression for $x = 300{,}000$.

$30x - 0.00005x^2 - (500{,}000 + 2x)$	*revenue – cost*
$= 30x - 0.00005x^2 + (-500{,}000 - 2x)$	*Add the opposite.*
$= 30x - 0.00005x^2 - 500{,}000 - 2x$	*Associative Property*
$= 28x - 0.00005x^2 - 500{,}000$	*Combine like terms.*

The profit is given by the polynomial $28x - 0.00005x^2 - 500{,}000$.

For $x = 300{,}000$,

$28(300{,}000) - 0.00005(300{,}000)^2 - 500{,}000 = 3{,}400{,}000$

The profit is \$3,400,000 or \$3.4 million.

Multiplying and Dividing Monomials

Key concepts

- Learn to multiply and divide monomials and to take powers of monomials.

Monomials can be added or subtracted if they are like terms. Monomials do not need to be like terms in order to be multiplied or divided. Any two monomials can be multiplied or divided using the properties of exponents.

Properties of Exponents

Product of Powers Property: $a^m \cdot a^n = a^{m+n}$

Power of a Power Property: $(a^m)^n = a^{m \cdot n}$

Power of a Product Property: $(ab)^n = a^n \cdot b^n$

Quotient of Powers Property: $a^m/a^n = a^{m-n}$

Negative Exponent Property: $a^{-n} = 1/a^n$

Example 1

In Example 1, monomials are multiplied by using the product of powers property.

Remember, the product of powers property states that when powers with like bases are multiplied, the product is the base to the sum of the powers. In other word, to multiply powers with like bases, keep the base and add the exponents.

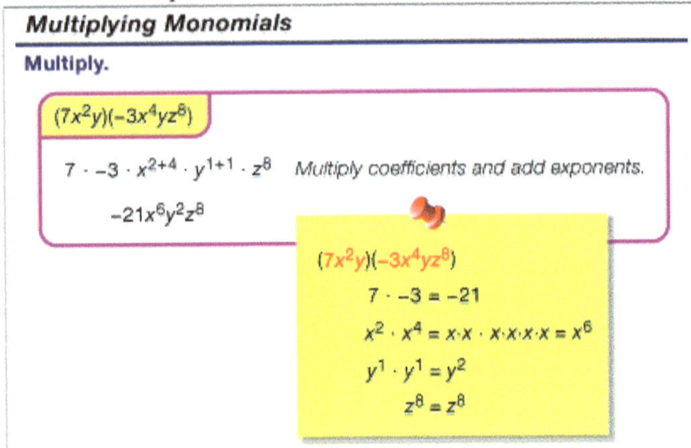

To multiply these monomials, first use the commutative property of multiplication to rearrange the factors so that the coefficients are grouped together and the powers with like bases are grouped together. Remember, y can be written as y^1.

Then, multiply the coefficients and add the exponents of the like powers.

Example 2

In Example 2, monomials are divided by using the quotient of powers property.

Remember, the quotient of powers property states that when powers with like bases are divided, the quotient is the base to the difference of the powers. In other word, to divide powers with like bases, keep the base and subtract the exponents.

Dividing Monomials

Divide. Assume that no denominator equals zero.

$\frac{15a^6}{5a^2}$

$\frac{15}{5} \cdot \frac{a^6}{a^2}$	*Write as two fractions.*
$3a^{6-2}$	*Divide coefficients. Subtract exponents that have the same base.*
$3a^4$	*Simplify the exponent.*

$\frac{20w^5z}{16w^5z^4}$

$\frac{20}{16} \cdot \frac{w^5}{w^5} \cdot \frac{z}{z^4}$	*Write as three fractions.*
$\frac{5}{4} w^{5-5} z^{1-4}$	*Divide coefficients. Subtract exponents that have the same base.*
$\frac{5}{4} w^0 z^{-3}$	*Simplify the exponent.*
$\frac{5}{4}(1)\left(\frac{1}{z^3}\right)$	*Think:* $w^0 = 1$ *and* $z^{-3} = \frac{1}{z^3}$
$\frac{5}{4z^3}$	*Multiply.*

Quotient of Powers

$$\frac{a^n}{a^m} = a^{n-m}$$

In the first exercise, write the coefficients as a single fraction, 15/5, and write the powers as a single fraction, a^6/a^2. Then, divide the coefficients and subtract the exponents of the powers: $15/5 = 3$ and $a^6/a^2 = a^{6-2} = a^4$.

To simplify the second expression, begin by writing three fractions: the first fraction contains the coefficients, the second fraction contains the w powers, and the third fraction contains the z powers. Then, divide the coefficients and subtract the exponents.

Notice that the result of subtracting the exponents of the w powers is 0. Remember, any nonzero number to the 0 power is 1. Furthermore, notice that the result of subtracting the exponents of the z powers is -3. Use the negative exponent property to write this power with a positive exponent. Last, multiply the three simplified fractions.

In the second expression, notice that the numerator and the denominator each contain w^5. So, all factors of w can be removed. Furthermore, the denominator contains 3 more factors of z than the numerator, so only 3 factors of z will remain in the denominator after removing common factors.

Example 3

In Example 3, a monomial is raised to a power.

Use the power of a product property and the power of a power property to raise a monomial that contains more than one factor to a power. Apply the power to each factor in the monomial and then multiply the exponents.

Raising a Monomial to a Power

Simplify.

$(5y)^3$

$5^3 \cdot y^3$	*Raise each factor to the power.*
$125y^3$	*Simplify the power.*

$(-3m^2p)^4$

$(-3)^4 \cdot (m^2)^4 \cdot (p)^4$	*Raise each factor to the power.*
$(-3)^4 \cdot (m^{2 \cdot 4}) \cdot (p^{1 \cdot 4})$	*Multiply exponents.*
$81m^8p^4$	*Simplify the power.*

In the first expression, the monomial contains two factors, 5 and y. Use the power of a product property to apply the exponent, 3, to each of these factors, and then simplify 5^3.

The monomial in the second expression contains three factors, -3, m^2, and p. Apply the exponent, 4, to each of these factors and then use the power of a power property to multiply the exponents.

Multiplying Monomials by Polynomials

Key Objectives

- Learn to multiply monomials by polynomials.

Use the distributive property to multiply a monomial by a polynomial. Remember to distribute the monomial to each of the terms in the polynomial.

Example 1 In Example 1, a monomial is multiplied by a binomial and a monomial is multiplied by a trinomial.

Multiplying a Polynomial by a Monomial

Multiply.

$\frac{1}{2}h(b_1 + b_2)$

$\frac{1}{2}h(b_1 + b_2)$ *Multiply each term in the parentheses by $\frac{1}{2}h$.*

$= \frac{1}{2}hb_1 + \frac{1}{2}hb_2$

$4rs^2(r^2s^4 + 2rs^3 - 3rst)$

$4rs^2(r^2s^4 + 2rs^3 - 3rst)$ *Multiply each term in the parentheses by $4rs^2$.*

$= 4rs^2(r^2s^4) + 4rs^2(2rs^3) - 4rs^2(3rst)$

$= 4r^3s^6 + 8r^2s^5 - 12r^2s^3t$

In the first expression, a monomial is multiplied by the binomial $b_1 + b_2$. Use the distributive property to multiply the monomial by each of the two terms in the binomial.

Since the monomial is multiplied by a trinomial in the second exercise, this monomial will be used as a factor three times when the distributive property is used.

Example 2 In Example 2, the formula for the volume of a rectangular prism is used.

Remember, to find the volume of a rectangular prism, multiply the three dimensions.

Problem-Solving Application

Carlos is making a stained-glass box with a square base. He wants the height of the box to be 2 inches less than the side length of the base. If he wants the volume of the box to be 32 cubic inches, what should the side length of the base be?

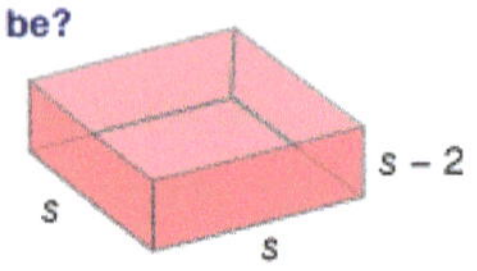

$$\text{Volume} = s \cdot s \cdot (s - 2)$$
$$= s^2(s - 2)$$
$$= s^3 - 2s^2 \quad \textit{Distributive Property}$$
$$32 = s^3 - 2s^2$$

The answer will be a value of s that makes the volume of the box equal to 32.

Solve for s.

s	1 ✗	2 ✗	3 ✗	4 ✓
$s^3 - 2s^2$	$(1)^3 - 2(1)^2 = -1$	$(2)^3 - 2(2)^2 = 0$	$(3)^3 - 2(3)^2 = 9$	$(4)^3 - 2(4)^2 = 32$

The side length of the base should be 4 inches.

Expressions must be written for the dimensions of this box. Since the base is square, let s be the side of the square. Therefore, the length and width of the box can be represented by s. The height of the box is "2 inches less than the side length of the base". So, subtract 2 from s to represent the height. Therefore, the height of the box can be represented by $s - 2$. Multiply these three expressions, s, s, and $s - 2$ to find the expression for the volume of the box.

To find the side length when the volume is 32 cubic inches, set the expression for the volume equal to 32. So, the side length is the number such that its cube minus twice its square is 32, or $s^3 - 2s^2 = 32$. Test several possible values for s by substituting each value into the expression. Notice that when $s = 4$, the value of the expression is 32. So, the side length must be 4 in order for the volume to be 32 cubic inches.

Multiplying Binomials

Key Objectives

- Multiply two binomials.
- Write polynomial expressions to represents lengths and areas.
- Find special products of binomials.

Key Terms

- **FOIL** is an acronym for the terms used when multiplying two binomials; the First, Inner, Outer, and Last terms.

To multiply a binomial by a binomial, use the Distributive Property twice.

$(x+y)(x+z)$
$= x(x+z)+y(x+z)$
$= x(x)+x(z)+y(x)+y(z)$
$= x^2+xz+xy+yx$

Alternatively, use the FOIL method to multiply a binomial by a binomial.

The Distributive Property can be used to multiply two binomials. The product can be simplified using the FOIL method.

FOIL: the First terms, the Outer terms, the Inner terms, the Last terms

First, Last, Outer, Inner

$(x + y)(x + z) = x^2 + xz + xy + yz$

Example 1
Multiplying Two Binomials

In Example 1, Prof. Burger demonstrates the two methods for multiplying a binomial by a binomial.

Multiply.

$(p + 3)(4 - q)$

Method 1: Distribute twice.

$(p + 3)(4 - q)$
$= (p + 3)4 + (p + 3)(-q)$ — *Distribute $(p + 3)$.*
$= 4p + 12 - pq - 3q$

Method 2: Use FOIL.

$(p + 3)(4 - q)$ — *FOIL*

$= 4p - pq + 12 - 3q$

$(a + b)(c + d)$

$(a + b)(c + d)$ — *FOIL*

$= ac + ad + bc + bd$

$(x + 3)(x + 7)$

$(x + 3)(x + 7)$ FOIL

$= x^2 + 7x + 3x + 21$

$= x^2 + 10x + 21$ Combine like terms.

$(2m + n)(m - 3n)$

$(2m + n)(m - 3n)$ FOIL

$= 2m^2 - 6mn + mn - 3n^2$

$= 2m^2 - 5mn - 3n^2$ Combine like terms.

Example 2
Problem Solving Application

In Example 2, Prof. Burger uses multiplying binomials to answer a real-world question.

Find the area of a cement walkway of width x meters around a 50 m by 25 m pool.

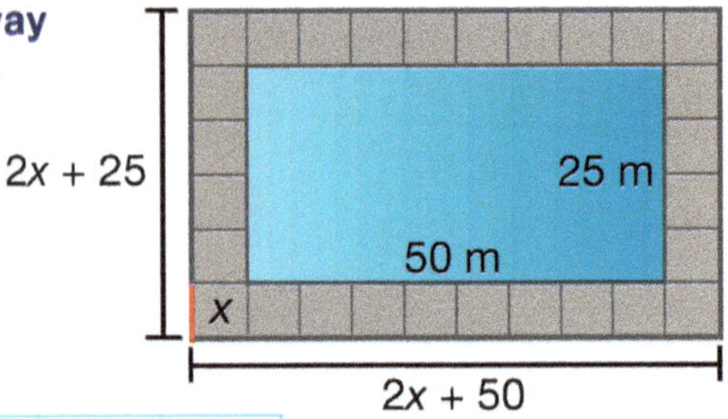

Area of walkway = Total area − Area of pool

$= (2x + 50)(2x + 25)$ − Area of pool

$= 4x^2 + 50x + 100x + 1250$ − Area of pool

$= 4x^2 + 150x + 1250$ − Area of pool

$= 4x^2 + 150x + 1250 - 1250$

$= 4x^2 + 150x$

The area of the cement walkway is $(4x^2 + 150x)$ m^2.

Example 3
Special Products of Binomials

In Example 3, Prof. Burger finds the square of a binomial and the product of two binomials where they are the sum and difference of the same term. To square a binomial, expand the power into the product of two binomials and then use the FOIL method to multiply.

Multiply.

$(x-4)^2$

$(x-4)(x-4)$ *FOIL*

$= x^2 - 4x - 4x + 16$

$= x^2 - 8x + 16$ *Combine like terms.*

$(a+b)^2$

$(a+b)(a+b)$ *FOIL*

$= a^2 + ab + ab + b^2$

$= a^2 + 2ab + b^2$ *Combine like terms.*

$(n+4)(n-4)$

$(n+4)(n-4)$ *FOIL*

$= n^2 - 4n + 4n - 16$

$= n^2 - 16$ *Combine like terms: $-4n + 4n = 0$.*

Factoring with the GCF

Key concepts

- Learn to write the prime factorizations of monomials, to find the GCF of monomials, and to factor polynomials by using the GCF.

Recall the distributive property: $a(b + c) = ab + ac$. When a monomial is multiplied by a polynomial, the distributive property is used to write $a(b + c)$ as $ab + ac$. Factoring a polynomial by using its GCF means to write a polynomial as the product of its factors, such as writing $ab + ac$ as $a(b + c)$, where a is the GCF.

Example 1

In Example 1, the GCF of a pair of monomials is determined.

Remember, the GCF (greatest common factor) of two numbers is the greatest factor that is common to the two numbers. Similarly, the GCF of two monomials is the greatest factor that is common to the two monomials. To find the GCF of two monomials, write the prime factorization of each monomial and then identify the common factors. The product of these common factors is the GCF.

Finding the GCF of Monomials **Find the GCF of each pair of monomials.** **$8x$ and $4x^2$** $8x = 2 \cdot 2 \cdot 2 \cdot x$ — *Write the prime factorization of each coefficient and write powers as products.* $4x^2 = 2 \cdot 2 \cdot x \cdot x$ — *Align the common factors.* $2 \cdot 2 \cdot x = 4x$ — *Find the product of the common factors.* The GCF of $8x$ and $4x^2$ is $4x$.	To find the GCF of $8x$ and $4x^2$, first write each prime factorization. The factors common to each prime factorization are 2, 2, and x. The product of these common factors, $4x$, is the GCF.
$3a^2$ and $5b^4$ $3a^2 = 3 \cdot a \cdot a$ — *Write the prime factorization of each coefficient and write powers as products.* $5b^4 = 5 \cdot b \cdot b \cdot b \cdot b$ — *Align the common factors.* The GCF of $3a^2$ and $5b^4$ is 1. — *There are no common factors other than 1.*	Notice that each of the monomials in the second pair contain a power, but that the two powers do not have a common base. So, there are no common variable factors. Furthermore, there are no factors common to the coefficients either. So, the GCF of these two monomials is 1.

Example 2

In Example 2, polynomials are factored by using the GCF.

To find the GCF of a polynomial, find the GCF of the monomials in that polynomial. To factor the GCF from a polynomial, first write each term in the polynomial as a product where one of the two factors is the GCF. Then, use the distributive property to remove the GCF from each of the products.

Factoring by Using the GCF

Factor each polynomial. Check your answer.

$5x^3 - 7x$

$5x^3 = 5 \cdot x \cdot x \cdot x$		*Find the GCF.*
$7x = 7 \cdot x$		
x		*The GCF of $5x^3$ and $7x$ is x.*
$5x^2(x) - 7(x)$		*Write the terms as products using the GCF as a factor.*
$x(5x^2 - 7)$		*Use the distributive property to factor out the GCF.*

Check Your Work

$x(5x^2 - 7) = 5x^3 - 7x$ ✓ *Multiply to check the answer. The product is the original polynomial.*

This polynomial contains two terms, $5x^3$ and $7x$. The only factor common to these two terms is x, so x is the GCF. Now, write each term as a product where one of the factors is x. So, write $5x^3$ as $5x^2(x)$, and write $7x$ as $7(x)$. Last, use the distributive property to factor x out of each of the two products. When x is factored from $5x^2(x)$, the result is $5x^2$. When x is factored from $7(x)$, the result is 7.

Notice that the factored form is the product of a monomial and a binomial. So, the factored form can be checked by multiplying the factors. The product must equal the original polynomial.

$8a^5 - 4a^2 + 20a$

$8a^5 = 2 \cdot 2 \cdot 2 \cdot a \cdot a \cdot a \cdot a \cdot a$	*Find the GCF.*
$4a^2 = 2 \cdot 2 \cdot a \cdot a$	
$20a = 2 \cdot 2 \cdot 5 \cdot a$	
$2 \cdot 2 \cdot a = 4a$	*The GCF of $8a^5 - 4a^2 + 20a$ is $4a$.*
$2a^4(4a) - a(4a) + 5(4a)$	*Write the terms as products using the GCF as a factor.*
$4a(2a^4 - a + 5)$	*Use the distributive property to factor out the GCF.*

Check Your Work

$4a(2a^4 - a + 5) = 8a^5 - 4a^2 + 20a$ ✓ *Multiply to check the answer. The product is the original polynomial.*

There are three terms in this expression; find the GCF of these three terms. Since the only factors common to the three terms are 2, 2, and a, the GCF is $4a$. Next, write each term as a product where one of the factors is $4a$. Then, use the distributive property to remove $4a$ from each product.

Notice that the expression within parentheses contains three terms, as did the original expression.

$-15w^2 - 10w^3$

$-1[15w^2 + 10w^3]$	*Both coefficients are negative. Factor out −1.*
$15w^2 = 3 \cdot 5 \cdot w \cdot w$	*Find the GCF.*
$10w^3 = 2 \cdot 5 \cdot w \cdot w \cdot w$	
$5 \cdot w \cdot w$	*The GCF of $15w^2 + 10w^3$ is $5w^2$.*
$-1[3(5w^2) + 2w(5w^2)]$	
$-1[5w^2(3 + 2w)]$	
$-5w^2(3 + 2w)$	*Use the distributive property to factor out the GCF.*

Check Your Work

$-5w^2(3 + 2w) = -15w^2 - 10w^3$ ✓ *Multiply to check the answer. The product is the original polynomial.*

Notice that both of the terms in this expression are negative. Therefore, –1 may be factored out first to simplify the process. However, be sure to include the –1 in the final factored form.

$10n^2 + 3$

$10n^2 = 2 \cdot 5 \cdot n \cdot n$	*Find the GCF.*
$3 = 3$	
$10n^2 + 3$	*There are no common factors other than 1.*

The polynomial cannot be factored further.

Notice that there are two terms in this polynomial and that there are no factors common to these terms. Therefore, the polynomial cannot be factored by using the GCF.

Example 3

In Example 3, a given binomial is factored by using the GCF.

Science Application

The area of the slides for Hank's microscope can be represented by ($5x^2 + 2x$) square centimeters. Factor this polynomial to find expressions for the dimensions of the slides.

$5x^2 + 2x$

$A = 5x^2 + 2x$	*The GCF of $5x^2$ and $2x$ is x.*
$= 5x(x) + 2(x)$	*Write the terms as products using the GCF as a factor.*
$= x(5x + 2)$	*Use the distributive property to factor out the GCF.*

Possible expressions for the dimensions of the microscope slide are x centimeters and $(5x + 2)$ centimeters.

The expression $5x^2 + 2x$ represents the area of a rectangle. Therefore, this expression is the product of the dimensions of the rectangle. The factored form of this expression will give the possible expressions for the dimensions of the rectangle.

Example 4

In Example 4, a common binomial is factored from a polynomial.

The GCF of a polynomial is not always a monomial; the GCF may be a binomial (or a trinomial, or a polynomial). If the GCF of a polynomial is a binomial, that binomial must be a factor of each term in the polynomial.

Factoring Out a Common Binomial Factor

Factor each expression.

$6(x - 5) - x(x - 5)$

$6(x - 5) - x(x - 5)$	*The terms have a common binomial factor of $(x - 5)$.*
$(x - 5)(6 - x)$	*Factor out $(x - 5)$.*

The first polynomial is the difference of two terms: $6(x - 5)$ and $x(x - 5)$. Notice that each of these terms is written as a product where one of the factors is $x - 5$. Since $x - 5$ is the only factor common to both terms, the GCF is $x - 5$. Remove this factor from each term to factor the binomial. The factored form is the product of $x - 5$ and the polynomial formed by the remaining factors.

$(x^2 + 3) - x(x^2 + 3)$

$(x^2 + 3) - x(x^2 + 3)$	*The terms have a common binomial factor of $(x^2 + 3)$.*
$1(x^2 + 3) - x(x^2 + 3)$	*$(x^2 + 3) = 1(x^2 + 3)$*
$(x^2 + 3)(1 - x)$	*Factor out $(x^2 + 3)$.*

In the second polynomial, notice that the first term is not written as a product. Since 1 is the multiplicative identity, write the first term as the product of 1 and itself. Then, factor as usual.

$3(2 + x) - 5x(x + 2)$

$3(2 + x) - 5x(x + 2)$	*$(x + 2) = (2 + x)$, so the terms have a common binomial factor of $(2 + x)$.*
$3(2 + x) - 5x(2 + x)$	
$(2 + x)(3 - 5x)$	*Factor out $(2 + x)$.*

The third polynomial may appear to contain no common factors. However, notice that $x + 2 = 2 + x$. So, write $x + 2$ as $2 + x$, and then factor as usual.

$2x(x - 1) + 7(x - 3)$

$2x(x - 1) + 7(x - 3)$	*There are no common factors.*

The expression cannot be factored.

The fourth polynomial contains two terms, but these two terms have no common factors since $x - 1$ and $x - 3$ are not equivalent and $2x$ and 7 have no common factors. Therefore, the polynomial cannot be factored.

Factoring $x^2 + bx + c$

Key Objectives

- Factor quadratic trinomials of the form $x^2 + bx + c$.

Example 1 Factoring Trinomials by Guess and Check

Some, but not all, trinomials can be factored into two binomials. One method for factoring a trinomial is called the "guess and check" method. In the "guess and check" method, the first step is to write the skeleton of each of the two binomials (the parentheses and the addition symbol with a blank on each side). Then, write factors of the trinomial's first term as the first term of each binomial. If the trinomial's first term is x^2, the first term in each binomial must be x. Next, write factors of the trinomial's last term as the last term of each binomial. If the trinomial's last term is a constant, the last term in each binomial will also be a constant. That completes the "guess" part of the process. Now, multiply the binomials to "check." If the product of the binomials is not equivalent to the original trinomial, then choose different factors for the terms and check again. Repeat this process until the correct factors are determined, or until all combinations of factors have been checked. Remember, some trinomials cannot be factored.

Factor.

$x^2 + 9x + 20$

$(\square + \square)(\square + \square)$ *Write two sets of parentheses.*

$(x + \ \ )(x + \ \ )$ *The first term is x^2, so the variable terms have a coefficient of 1.*

The constant term in the trinomial is 20.

$(x + 1)(x + 20) = x^2 + 21x + 20$ ✗ *Try factors of 20 for the constant terms in the binomials.*

$(x + 2)(x + 10) = x^2 + 12x + 20$ ✗

$(x + 4)(x + 5) = x^2 + 9x + 20$ ✓

The factors of $x^2 + 9x + 20$ are $(x + 4)$ and $(x + 5)$.

$x^2 + 9x + 20 = (x + 4)(x + 5)$

This trinomial's first term is x^2. Therefore, the first term in each of its binomial factors must be x, since $x \cdot x = x^2$ and there are no other factors of x^2. The trinomial's last term is 20. Therefore, the last term in each of its binomial factors must be a factor of 20. So, the possible last terms are 1 and 20, 2 and 10, and 4 and 5. There is no need to test the reverse order of these pairs, for example 20 and 1, since the first terms in each of the binomials, x, are the same. (If the first terms are not the same, the reverse order should be tested.) Write the factors 1 and 20 as the last term in each binomial. Then, use FOIL to check the factoring. Since $x^2 + 21x + 20$ is not equivalent to the original trinomial, try another pair of factors. The factors 4 and 5 complete the correct binomials. Notice that the sum of these factors is 9 and that 9 is also the coefficient of the trinomial's middle term.

Example 2 Factoring $x^2 + bx + c$ When c is Positive

A second method for factoring a trinomial is to list the factors of c and then to find the sum of each pair of factors. The pair whose sum is b will be used as the constant terms in the binomial factors. Since the first term is always x^2 in a trinomial of the form $x^2 + bx + c$, the first term of each binomial factor will be x.

Factor the trinomial. Check your answer.

$x^2 + 7x + 10$

$(x + \square)(x + \square)$ *$b = 7$ and $c = 10$; look for factors of 10 whose sum is 7.*

Factors of 10	Sum	
1 and 10	11	✗
2 and 5	7	✓ *The factors are 2 and 5.*

$(x + 2)(x + 5)$

Check Your Work

$(x + 2)(x + 5) = x^2 + 5x + 2x + 10$ *Use the FOIL method.*

$= x^2 + 7x + 10$ ✓

In this trinomial, the coefficient of x is 7 and the constant term is 10. So, $b = 7$ and $c = 10$. Find the factors of 10 whose sum is 7. The number 10 has two pairs of factors: 1 and 10 and 2 and 5. Since $2 + 5 = 7$, use 2 and 5 as the constant term in each binomial. Of course, the first term in each binomial is x since the trinomial's first term is x^2.

Factor the trinomial. Check your answer.

$x^2 - 11x + 18$

$(x + \square)(x + \square)$ — *$b = -11$ and $c = 18$; look for factors of 18 whose sum is -11.*

Factors of 18	Sum	
−1 and −18	−19	✗
−3 and −6	−9	✗
−2 and −9	−11	✓

The factors are −2 and −9.

$(x - 2)(x - 9)$

Check Your Work

$(x - 2)(x - 9) = x^2 - 9x - 2x + 18$ *Use the FOIL method.*

$= x^2 - 11x + 18$ ✓

Notice that subtraction precedes the x-term in this trinomial. Therefore, the value of b is negative. When b is negative and c is positive, the factors of c used in the binomial factors must both be negative (since the sum of two negative numbers is negative and the product of two negative numbers is positive).

Example 3 Factoring $x^2 + bx + c$ When c is Negative

The operation that precedes a term determines the sign of the term. So, if addition precedes a term, the term is considered to be positive and if subtraction precedes a term, the term is considered to be negative.

Consider a trinomial of the form $x^2 + bx + c$. If both operations in the trinomial are addition, the factors of c used for the binomial factors will both be positive. If the first operation in the trinomial is subtraction and the second operation is addition, the factors of c used for the binomial factors will both be negative. (Examples of trinomials in those two forms were factored in the last example.)

However, if the second operation in a trinomial of the form $x^2 + bx + c$ is subtraction, the factors of c used for the binomial factors will have opposite signs. In this case, if the first operation is subtraction, the factor with the greater absolute value will be negative, and if the first operation is addition, the factor with the greater absolute value will be positive.

Factor the trinomial. Check your answer.

$x^2 + 6x - 16$

$(x + \square)(x + \square)$ — *$b = 6$ and $c = -16$; look for factors of -16 whose sum is 6. The factor with the greater absolute value is positive.*

Factors of −16	Sum	
−1 and 16	15	✗
−4 and 4	0	✗
−2 and 8	6	✓

The factors are −2 and 8.

$(x - 2)(x + 8)$

Check Your Work

$(x - 2)(x + 8) = x^2 + 8x - 2x - 16$ *Use the FOIL method.*

$= x^2 + 6x - 16$ ✓

In this trinomial, the first operation is addition and the second operation is subtraction. So, $b = 6$ and $c = -16$. Therefore, find factors of -16 whose sum is 6. Because the second operation in the trinomial is subtraction (in other words, c is negative), the factors will have opposite signs. Since the first operation is addition (in other words, since b is positive), the factor of -16 with the greater absolute value must be positive.

Factor the trinomial. Check your answer.

$x^2 - x - 20$

$(x + \square)(x + \square)$ — *b = −1 and c = −20; look for factors of −20 whose sum is −1. The factor with the greater absolute value is negative.*

Factors of −20	Sum	
−20 and 1	−19	✗
2 and −10	−8	✗
4 and −5	−1	✓

The factors are −5 and 4.

$(x + 4)(x - 5)$

Check Your Work

$(x + 4)(x - 5) = x^2 - 5x + 4x - 20$ *Use the FOIL method.*

$= x^2 - x - 20$ ✓

In this trinomial, both operations are subtraction. So, $b = -1$ and $c = -20$. Therefore, find factors of -20 whose sum is -1. Because the second operation is subtraction, the factors must have opposite signs. Since the first operation is subtraction (in other words, since b is negative), the factor of -20 with the greater absolute value must be negative.

Factoring $ax^2 + bx + c$

Key Objectives

- Factor quadratic trinomials of the form $ax^2 + bx + c$.

Example 1 Factoring $ax^2 + bx + c$ by Guess and Check

Factor $6x^2 + 19x + 10$.

$(\square + \square)(\square + \square)$ *Write two sets of parentheses.*

$(\square x + \square)(\square x + \square)$ *The first term is $6x^2$, so at least one variable term has a coefficient other than 1.*

The coefficient of the x^2 term is 6. The constant term in the trinomial is 10.

Try factors of 6 for the coefficients and factors of 10 for the constant terms.

$(1x + 10)(6x + 1) = 6x^2 + 61x + 10$ ✗

$(1x + 5)(6x + 2) = 6x^2 + 32x + 10$ ✗

$(1x + 2)(6x + 5) = 6x^2 + 17x + 10$ ✗

$(2x + 10)(3x + 1) = 6x^2 + 32x + 10$ ✗

$(2x + 5)(3x + 2) = 6x^2 + 19x + 10$ ✓

The factors of $6x^2 + 19x + 10$ are $(2x + 5)$ and $(3x + 2)$.

$6x^2 + 19x + 10 = (2x + 5)(3x + 2)$

The first term of this trinomial is $6x^2$, so the product of the first terms in the factored form must not only give x^2, but they must also have coefficients that are factors of 6. In addition, those coefficients will contribute to the value of b. The bx term is the product of the outside terms plus the product of the inside terms.

In the guess and check method, set up two sets of parentheses with addition signs between terms. For the first and third terms put factors of 6, for example, 1 and 6, as coefficients of x. For the second and third terms put factors of 10, such as 10 and 1. Check this solution by FOIL. This factoring does not produce the original trinomial, so try another combination. Proceed in an orderly sequence, modifying one set of factors in all possible combinations, then modifying the second set. The correct factorial is $(2x + 5)(3x + 2)$.

Example 2 Factoring $ax^2 + bx + c$ When c is Positive

Product = a Product = c

$ax^2 + bx + c = (\square x + \square)(\square x + \square)$

Sum of outer and inner products = b

The principles behind factoring a trinomial of the form $ax^2 + bx + c$ can be summarized.

1. The coefficients of x in each binomial must be factors of a.

2. The constant terms in each binomial must be factors of c.

3. The sum of the outer and inner products must equal b.

Factor the trinomial. Check your answer.

$3x^2 + 29x + 18$

$(\square x + \square)(\square x + \square)$ *$a = 3$ and $c = 18$; Outer + Inner = 29.*

Factors of 3	Factors of 18	Outer + Inner	
1 · 3	1 · 18	1 · 18 + 3 · 1 = 21	✗
1 · 3	18 · 1	1 · 1 + 3 · 18 = 55	✗
1 · 3	2 · 9	1 · 9 + 3 · 2 = 15	✗
1 · 3	9 · 2	1 · 2 + 3 · 9 = 29	✓

$(x + 9)(3x + 2)$

Check Your Work

$(x + 9)(3x + 2) = 3x^2 + 2x + 27x + 18$ *Use the FOIL method.*

$= 3x^2 + 29x + 18$

In this trinomial the coefficient of x^2 is 3, so the coefficients in the factorization must be factors of 3. The value of c is 18, so the constant values in the factorization must be factors of 18.

Set up a table to list the factors and the sum of the products of the inner and outer terms.

For the first and third terms put the only factors of 3, 1 and 3, as coefficients of x. For the second and third terms put factors of 18, such as 1 and 18. Check this solution by FOIL. This factoring does not produce the original trinomial, so try another combination. Notice that using the same factors (1 and 18) in a different order does produce a different result. Be sure to use each combination. The correct factorial is $(x + 9)(3x + 2)$.

Factor the trinomial. Check your answer.

$7x^2 + 17x + 6$

$(\square x + \square)(\square x + \square)$ *$a = 7$ and $c = 6$; Outer + Inner = 17.*

Factors of 7	Factors of 6	Outer + Inner	
$1 \cdot 7$	$1 \cdot 6$	$1 \cdot 6 + 7 \cdot 1 = 13$	✗
$1 \cdot 7$	$6 \cdot 1$	$1 \cdot 1 + 7 \cdot 6 = 43$	✗
$1 \cdot 7$	$2 \cdot 3$	$1 \cdot 3 + 7 \cdot 2 = 17$	✓

$(x + 2)(7x + 3)$

Check Your Work

$(x + 2)(7x + 3) = 7x^2 + 3x + 14x + 6$ *Use the FOIL method.*

$= 7x^2 + 17x + 6$

In this trinomial the coefficient of x^2 is 7, so the coefficients in the binomials must be factors of 7. The value of c is 6, so the constant values in the binomials must be factors of 6.

Set up a table to list the factors and the sum of the products of the inner and outer terms.

For the first and third terms put the only factors of 7, 1 and 7, as coefficients of x. For the second and third terms put factors of 6, such as 1 and 6. Check this solution by FOIL. This factoring does not produce the original trinomial, so try another combination. The correct factorial is $(x + 2)(7x + 3)$.

Example 3 Factoring $ax^2 + bx + c$ When c is Negative

Factor the trinomial. Check your answer.

$2x^2 + 5x - 3$

$(\square x + \square)(\square x + \square)$ *$a = 2$ and $c = -3$; Outer + Inner = 5.*

Factors of 2	Factors of −3	Outer + Inner	
$1 \cdot 2$	$1(-3)$	$1(-3) + 2 \cdot 1 = -1$	✗
$1 \cdot 2$	$(-1)3$	$1 \cdot 3 + 2(-1) = 1$	✗
$1 \cdot 2$	$3(-1)$	$1(-1) + 2 \cdot 3 = 5$	✓

$(x + 3)(2x - 1)$

Check Your Work

$(x + 3)(2x - 1) = 2x^2 - x + 6x - 3$ *Use the FOIL method.*

$= 2x^2 + 5x - 3$

In this trinomial the coefficient of x^2 is 2, so the coefficients in the factorization must be factors of 2. The value of c is −3, so the constant values in the factorization must be factors of −3. One of these factors must be positive and one must be negative to make c negative.

Set up a table to list the factors and the sum of the products of the inner and outer terms.

For the first and third terms put the factors of 2, 1 and 2, as coefficients of x. For the second and third terms put factors of −3, such as 1 and −3. Check this solution by FOIL. This factoring does not produce the original trinomial, so try another combination. In the next trial only the location of the sign is changed, which produces a different result. The correct factorial is $(x + 3)(2x - 1)$.

Factor the trinomial. Check your answer.

$3x^2 - 10x - 8$

$(\square x + \square)(\square x + \square)$ *$a = 3$ and $c = -8$; Outer + Inner = −10.*

Factors of 3	Factors of −8	Outer + Inner	
$1 \cdot 3$	$1(-8)$	$1(-8) + 3 \cdot 1 = -5$	✗
$1 \cdot 3$	$(-1)8$	$1 \cdot 8 + 3(-1) = 5$	✗
$1 \cdot 3$	$8(-1)$	$1(-1) + 3 \cdot 8 = 23$	✗
$1 \cdot 3$	$(-8)1$	$1 \cdot 1 + 3(-8) = -23$	✗
$1 \cdot 3$	$2(-4)$	$1(-4) + 3 \cdot 2 = 2$	✗
$1 \cdot 3$	$(-2)4$	$1 \cdot 4 + 3(-2) = -2$	✗
$1 \cdot 3$	$4(-2)$	$1(-2) + 3 \cdot 4 = 10$	✗
$1 \cdot 3$	$(-4)2$	$1 \cdot 2 + 3(-4) = -10$	✓

$(x - 4)(3x + 2)$

In this trinomial the coefficient of x^2 is 3, so the coefficients in the binomials must be factors of 3. The value of c is −8, so the constant values in the binomials must be factors of −8. One of the factors must be positive and one must be negative. In addition, the sum of the products of the inner and outer terms must be −10.

Set up a table to list the factors and the sum of the products of the inner and outer terms.

For the first and third terms put the only factors of 3, 1 and 3, as coefficients of x. For the second and third terms put factors of −8, such as 1 and −8. Check this solution by FOIL. This factoring does not produce the original trinomial, so try another combination. The correct factorial is $(x - 4)(3x + 2)$.

Example 4 Factoring $ax^2 + bx + c$ When a is Negative

Factor $-5x^2 - 11x - 2$.

$-1(5x^2 + 11x + 2)$ *Factor out −1.*

$-1(\square x + \square)(\square x + \square)$ *a = 5 and c = 2; Outer + Inner = 11.*

Factors of 5	Factors of 2	Outer + Inner	
5 · 1	2 · 1	5 · 1 + 1 · 2 = 7	✗
5 · 1	1 · 2	5 · 2 + 1 · 1 = 11	✓

$-1(5x + 1)(x + 2)$

$(-5x - 1)(x + 2)$ *Distribute negative sign into the first factor.*

Check Your Work

$(-5x - 1)(x + 2) = -5x^2 - 10x - x - 2$ *Use the FOIL method.*

$= -5x^2 - 11x - 2$

In this trinomial the coefficient of x^2 is −5. To simplify factoring this trinomial, take the negative signs out of the trinomial by removing a factor of −1 and factoring the new trinomial. The coefficients in the binomials must be 1 and 5, the factors of 5. The value of c is 2, so the constant values in the binomials must be 1 and 2 and because c is positive they must have the same sign. Because b is also positive, the constant terms must both be positive.

A table can be used to test each of the combinations, but it is also possible to work through the combinations mentally, especially when there are very few, as in this case. Since b is 11, it makes sense to allow the 2 to be multiplied with 5 to put the sum of the inner and outer terms closer to 11. The factorization that does this is the correct one. The correct factorial is $-(5x + 1)(x + 2)$. To put the negative back into the factorization, distribute it into one of the binomials only. The final factorization is $(-5x - 1)(x + 2)$.

Radical Expressions

Key Objectives

- Simplify radical expressions.

Key Terms

- An expression that contains a radical sign ($\sqrt{}$) is a **radical expression**.
- The expression under a radical sign is the **radicand**.

Theorems, Postulates, Corollaries, and Properties

- **Product Property of Square Roots** For any nonnegative real numbers a and b, the square root of ab is equal to the square root of a times the square root of b.
- **Quotient Property of Square Roots** For any real numbers a and b ($a \geq 0$ and $b > 0$), the square root of $\frac{a}{b}$ is equal to the square root of a divided by the square root of b.

An expression containing a radical is in its simplest form when the radicand has no perfect square factors other than 1, the radicand has no fractions, and there are no square roots in any denominator.

Example 1 Simplifying Square Root Expressions

Simplify each expression.

$\sqrt{\frac{3}{27}}$

$$\sqrt{\frac{3}{27}} = \sqrt{\frac{1}{9}} = \frac{1}{3}$$

$\sqrt{8^2 + 6^2}$

$$\sqrt{8^2 + 6^2} = \sqrt{64 + 36} = \sqrt{100} = 10$$

$\sqrt{x^2 + 10x + 25}$

$$\sqrt{x^2 + 10x + 25} = \sqrt{(x + 5)(x + 5)} = \sqrt{(x + 5)^2} = |x + 5|$$

To simplify the first radical expression, begin by simplifying the fraction under the square root sign. Then determine if the term is a perfect square. In this case it is: the square root of 1/9 is 1/3.

To simplify the second radical expression, first simplify what is under the square root sign. Do not be tempted to simply remove the squares. Evaluating gives the square root of 100, which is simply 10.

To simplify the third radical expression, factor the trinomial under the square root sign. Because this expression has a variable it will not be possible to evaluate it, but it may be possible to simplify it. The trinomial factors into two binomials that are identical. The radical can be simplified to $x + 5$, but the answer must be qualified because it is not possible to have the result of a square root be negative, and some values of x could make the expression $x + 5$ negative. Thus, the answer is $|x + 5|$, the absolute value of $x + 5$.

Example 2 Using the Product Property of Square Roots

Product Property of Square Roots

$\sqrt{ab} = \sqrt{a} \cdot \sqrt{b}$, where $a \geq 0$ and $b \geq 0$

For any nonnegative real numbers a and b, the square root of ab is equal to the square root of a times the square root of b.

Simplify. All variables represent nonnegative numbers.

$\sqrt{50}$

$\sqrt{50} = \sqrt{25 \cdot 2}$ — *Factor the radicand using perfect squares.*
$= \sqrt{25}\sqrt{2}$ — *Product Property of Square Roots*
$= 5\sqrt{2}$ — *Simplify.*

$\sqrt{x^3y^8}$

$\sqrt{x^3y^8} = \sqrt{x^3}\sqrt{y^8}$ — *Product Property of Square Roots*
$= \sqrt{x^2}\sqrt{x}\sqrt{y^8}$ — *Product Property of Square Roots*
$= xy^4\sqrt{x}$ — *Since x is nonnegative, $\sqrt{x^2} = x$.*

Some radicands are not perfect squares but can still be simplified using the Product Property of Square Roots.

In the first example, 50 can be factored into 25, a perfect square, and 2. The square root of 25 times 2 is equal to the square root of 25 times the square root of 2, according to the Product Property of Square Roots. Taking the square root of 25, the radical can be written as $5\sqrt{2}$.

In the second example, begin by separating the expressions in x and y using the Product Property of Square Roots. Each of these expressions can be further decomposed to yield $xy^4\sqrt{x}$. Remember that x and y must both be greater than or equal to zero to avoid taking the square root of a negative number.

Example 3 Using the Quotient Property of Square Roots

Quotient Property of Square Roots

$\sqrt{\frac{a}{b}} = \frac{\sqrt{a}}{\sqrt{b}}$, where $a \geq 0$ and $b \geq 0$

For any real numbers a and b ($a \geq 0$ and $b > 0$), the square root of $\frac{a}{b}$ is equal to the square root of a divided by the square root of b.

Simplify. All variables represent nonnegative numbers.

$\sqrt{\frac{3}{16}}$

$\sqrt{\frac{3}{16}} = \frac{\sqrt{3}}{\sqrt{16}}$ — *Quotient Property of Square Roots*
$= \frac{\sqrt{3}}{4}$ — *Simplify.*

$\sqrt{\frac{x^{11}}{49x^3}}$

$\sqrt{\frac{x^{11}}{49x^3}} = \sqrt{\frac{x^8}{49}}$ — *Simplify.*
$= \frac{\sqrt{x^8}}{\sqrt{49}}$ — *Quotient Property of Square Roots*
$= \frac{x^4}{7}$ — *Simplify.*

Some radicands can be simplified using the Quotient Property of Square Roots.

In the first example, the square root of 3/16 is equal to the square root of 3 over the square root of 16, according to the Quotient Property of Square Roots. Taking the square root of 16, the radical can be written as $\frac{\sqrt{3}}{4}$.

In the second example, begin by simplifying within the square root sign. Then apply the Quotient Property of Square Roots. Each of these expressions is a perfect square, so the radical can be simplified to $x^4/7$. Remember that x must be greater than or equal to zero to avoid taking the square root of a negative number.

Example 4 Using the Product and Quotient Properties Together

Simplify. All variables represent nonnegative numbers.

$\sqrt{\frac{75}{81}}$

$\sqrt{\frac{75}{81}} = \frac{\sqrt{75}}{\sqrt{81}}$ — *Quotient Property*
$= \frac{\sqrt{25(3)}}{\sqrt{81}}$ — *Write 75 as 25(3).*
$= \frac{\sqrt{25}\sqrt{3}}{\sqrt{81}}$ — *Product Property*
$= \frac{5\sqrt{3}}{9}$ — *Simplify.*

In this example, begin by applying the Quotient Property of Square Roots. The square root of 81 is just 9. Apply the Product Property of Square Roots to simplify the square root of 75 to $5\sqrt{3}$.

Simplify. All variables represent nonnegative numbers.

$\sqrt{\frac{9x^3}{16}}$

$$\sqrt{\frac{9x^3}{16}} = \frac{\sqrt{9x^3}}{\sqrt{16}} \quad \text{Quotient Property}$$

$$= \frac{\sqrt{9}\sqrt{x^3}}{\sqrt{16}} \quad \text{Product Property}$$

$$= \frac{\sqrt{9}\sqrt{x^2}\sqrt{x}}{\sqrt{16}}$$

$$= \frac{3x\sqrt{x}}{4} \quad \text{Simplify.}$$

In this example, begin by applying the Quotient Property of Square Roots. The square root of 16 is 4. The numerator can be simplified using the Product Property of Square Roots to the square root of 9 times the square root of x^3. The square root of x^3 can be further decomposed to the square root of x^2 times the square root of x. Evaluating the perfect squares yields $\frac{3x\sqrt{x}}{4}$.

Example 5 Sports Application

A little league baseball diamond is a square with sides of 60 feet. How far is a throw from third base to first base? Give the answer as a radical expression in simplest form. Then estimate the length to the nearest tenth of a foot.

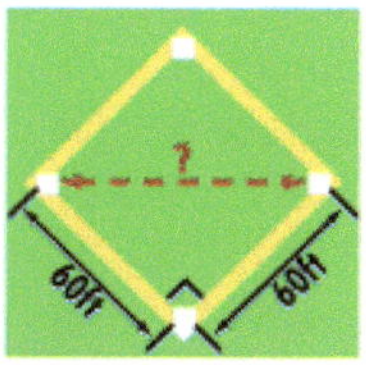

The distance from third base to first base is the hypotenuse of a right triangle. Use the Pythagorean Theorem: $c^2 = a^2 + b^2$.

$$c = \sqrt{a^2 + b^2} \quad \text{Solve for } c.$$

$$= \sqrt{60^2 + 60^2} \quad \text{Substitute 60 for } a \text{ and } b.$$

$$= \sqrt{2 \cdot 60^2} \quad \text{Simplify.}$$

$$= \sqrt{2} \cdot \sqrt{60^2} \quad \text{Use the Product Property of Square Roots.}$$

$$= 60 \cdot \sqrt{2} \quad \text{Simplify.}$$

$$\approx 84.9 \quad \text{Use a calculator and round to the nearest tenth.}$$

The distance is $60\sqrt{2}$, or about 84.9 feet.

A distance on a baseball field is determined by simplifying a radical in this application example.

The unknown distance forms the hypotenuse of a right triangle. The lengths of the legs of the triangle are known, and the length of the hypotenuse can be determined using the Pythagorean Theorem. Substitute 60 for a and b in $c^2 = a^2 + b^2$ and take the square root of both sides to solve for c. The Product Property of Square Roots can be used to rewrite the square root of 2 times 60^2 as the square root of 2 times the square root of 60^2. This expression simplifies to 60 times the square root of 2. This is the number of feet from first to third base. Using a calculator, the distance is found to be about 84.9 feet.

Solving Quadratic Equations by Factoring

Key Objectives

- Solve quadratic equations by factoring.

Theorems, Postulates, Corollaries, and Properties

- **Zero Product Property** If the product of two quantities equals zero, at least one of the quantities equals zero.

Example 1 Using the Zero Product Property

Use the Zero Product Property to solve the equation. Check your answer.

Zero Product Property: If the product of two quantities equals zero, at least one of the quantities equals zero.

$(x + 5)(x - 2) = 0$

$x + 5 = 0$ or $x - 2 = 0$ *Use the Zero Product Property.*

$x = -5$ or $x = 2$ *Solve each equation.*

The solutions are -5 and 2.

Check Your Work

Substitute each solution for x into the original equation.

$(x + 5)(x - 2)$	0
$(-5 + 5)(-5 - 2)$	0
$(0)(-7)$	0
0	0 ✓

$(x + 5)(x - 2)$	0
$(2 + 5)(2 - 2)$	0
$(7)(0)$	0
0	0 ✓

One way to solve quadratic equations is to use the Zero Product Property, which states that if the product of two quantities is zero then one or both of the quantities must be zero.

To apply the Zero Product Property to a factored quadratic equation such as the one shown, find the two conditions that satisfy the Zero Product Property. One of those conditions is that $x + 5 = 0$, which is true when $x = -5$. The second condition is that $x - 2 = 0$, or $x = 2$. These are the two solutions to the quadratic equation.

To check that the solutions are correct, substitute each for x in the original equation to determine if they satisfy the equation. In both cases the solutions work.

Example 2 Solving Quadratic Equations by Factoring

Solve the quadratic equation by factoring.

$-3x^2 = 6x + 3$

$$\begin{aligned} -3x^2 &= 6x + 3 \\ +3x^2 & \quad +3x^2 \\ 0 &= 3x^2 + 6x + 3 \end{aligned}$$

The equation must be written in standard form. So add $3x^2$ to both sides.

$3x^2 + 6x + 3 = 0$

$\dfrac{3(x^2 + 2x + 1)}{3} = \dfrac{0}{3}$ *Factor out 3 to simplify the trinomial. Divide both sides by 3.*

$x^2 + 2x + 1 = 0$

$(x + 1)(x + 1) = 0$ *Factor the trinomial.*

$x + 1 = 0$ or $x + 1 = 0$ *Use the Zero Product Property.*

$x = -1$ or $x = -1$ *Solve each equation.*

Both factors result in the same solution, so there is one solution, -1.

To solve this quadratic equation, begin by factoring it. To set the equation up for factoring, rearrange so all of the terms are on one side and zero is on the other side of the equation.

First, take out the factor of 3 common to all the terms in the quadratic. Divide both sides by 3 and the factor of 3 disappears. Second, factor the equation by setting up two binomials and determining that the constant term in each should multiply to equal 1, the c term in the quadratic, and add to equal 2, the b term in the quadratic. The quadratic equation factors into $(x + 1)(x + 1) = 0$.

To find the solutions to the equation find the two conditions the satisfy the Zero Product Property. These are $x = -1$ and $x = -1$. Both solutions are the same. To check, substitute -1 as x in the original equation.

The single solution can be visualized using a graph. The quadratic is an equation for an upside down parabola that touches the x-axis (the solution) once at its vertex (rather than crossing it twice).

Example 3 Sports Application

The diver's height above the water during a dive can be modeled by $h = -16t^2 + 8t + 8$, where h is height in feet and t is time in seconds. Find the time it takes the diver to reach the water.

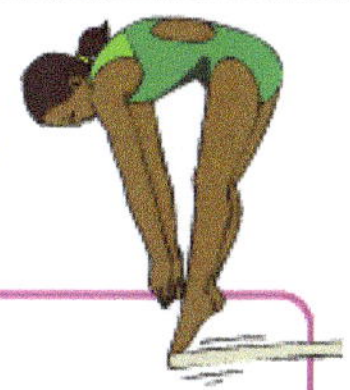

$h = -16t^2 + 8t + 8$

$-16t^2 + 8t + 8 = 0$ — *The diver reaches the water when $h = 0$.*

$\frac{-2(8t^2 - 4t - 4)}{-2} = \frac{0}{-2}$ — *Factor out −2 to simplify the trinomial. Divide both sides by −2.*

$8t^2 - 4t - 4 = 0$

$\frac{4(2t^2 - t - 1)}{4} = \frac{0}{4}$ — *Factor out 4 to simplify the trinomial. Divide both sides by 4.*

$2t^2 - t - 1 = 0$

$(2t + 1)(t - 1) = 0$ — *Factor the trinomial.*

$2t + 1 = 0$ or $t - 1 = 0$ — *Use the Zero Product Property. Solve each equation.*

$\frac{2t}{2} = \frac{-1}{2}$

~~$t = -\frac{1}{2}$~~ or $t = 1$ — *Since time cannot be negative, $-\frac{1}{2}$ does not make sense in this situation.*

Are they both solutions? **NO!**

It takes the diver 1 second to reach the water.

A quadratic equation is solved to find the amount of time until a diver hits the water in this application example.

To solve the given quadratic, begin by factoring it. As a first step, recognize that there is a common factor of −8 that can be factored out (this is done in two steps on the left). Then, dividing both sides by −8 removes the common factor. Factor the resulting equation by setting up two binomials. The factored equation is $(2t + 1)(t - 1) = 0$.

Use the Zero Product Property to find the solutions to the equation. The solutions are $t = -1/2$ and $t = 1$. Only one of these solutions makes sense in the context of the original question, since time cannot take on a negative value in this context. The answer to the question is that it will take 1 second for the diver to reach the water.

Solving Quadratic Equations by Using Square Roots

Key Objectives

- Solve quadratic equations by using square roots.

Key Terms

- A **quadratic equation** is an equation that can be written in the standard form $ax^2 + bx + c = 0$, where a, b, and c are real numbers and $a \neq 0$.

Example 1 Using Square Roots to Solve $x^2 = c$

Solve using square roots.

$x^2 = 25$

$x^2 = 25$

$x = \pm\sqrt{25}$

$x = \pm 5$

Solve for x by taking the square root of both sides. Use ± to show both square roots.

The solutions are −5 and 5.

Check Your Work

Substitute the solutions into the original equation.

$x^2 = 25$	
$(-5)^2$	25
25	25 ✓

$x^2 = 25$	
$(5)^2$	25
25	25 ✓

Equations in the form $x^2 = c$ can be solved by taking the square root of each side. Remember, both the positive and the negative root must be included as solutions when the steps for solving an equation include taking the square root of both sides.

The equation given in this example is in the form $x^2 = c$. So, solve the equation by taking the square root of each side.

On the left side, taking the square root of x^2 results in x.

On the right side, taking the square root of 25 results in 5 and −5. Therefore, the solutions are 5 and −5.

Solve using square roots.

$x^2 = -9$

$x^2 = -9$

$x \neq \pm\sqrt{-9}$

There is no real number whose square is negative.

There is no real solution.

Again, this equation is in the form $x^2 = c$. So, solve the equation by taking the square root of each side.

On the left side, taking the square root of x^2 results in x.

On the right side, the square root of −9 is taken. There is no real number whose square is negative. Therefore, there is no real solution to this equation.

Example 2 Using Square Roots to Solve Quadratic Equations

If a quadratic equation does not include a bx-term (i.e., $b = 0$ when the equation is written in standard form), then that equation can be written in the form $x^2 = c$ by solving for x^2. So, any quadratic equation that does not include a bx-term can be solved using square roots.

Solve using square roots.

$9x^2 - 49 = 0$

$$9x^2 - 49 = 0$$
$$\underline{\quad + 49 \quad + 49}$$
$$9x^2 = 49$$
$$\frac{9x^2}{9} = \frac{49}{9}$$
$$x^2 = \frac{49}{9}$$
$$x = \pm\sqrt{\frac{49}{9}}$$
$$x = \pm\frac{7}{3}$$

The solutions are $\frac{7}{3}$ and $-\frac{7}{3}$.

Check Your Work

$9x^2 - 49 = 0$	
$9\left(-\frac{7}{3}\right)^2 - 49$	0
$9\left(\frac{49}{9}\right) - 49$	0
$49 - 49$	0
0	0 ✓

$9x^2 - 49 = 0$	
$9\left(\frac{7}{3}\right)^2 - 49$	0
$9\left(\frac{49}{9}\right) - 49$	0
$49 - 49$	0
0	0 ✓

This equation is quadratic and it does not include a bx-term. So, to solve this equation first solve for x^2 and then take the square root of each side. Remember to include both the positive and negative root as solutions.

Since 49 is subtracted from the x^2-term, begin by adding 49 to each side.

Next, notice that the coefficient of x^2 is 9. Therefore, divide both sides by 9.

Now the equation is in the form $x^2 = c$, where $c = 49/9$. Take the square root of each side. The square root of 49 is 7 and the square root of 9 is 3, so the solutions are 7/3 and −7/3.

Example 3 Approximating Solutions

Again, this quadratic equation does not include a *bx*-term. So, square roots can be used to solve this equation.

Solve. Round to the nearest hundredth.

$0 = -3x^2 + 60$

$0 = -3x^2 + 60$	
$+3x^2 \quad +3x^2$	*Add $3x^2$ to both sides.*
$3x^2 = 60$	
$\frac{3x^2}{3} = \frac{60}{3}$	*Divide both sides by 3.*
$x^2 = 20$	
$x = \pm\sqrt{20}$	*Take the square root of both sides.*
$x = \pm 2\sqrt{5}$	
$x \approx \pm 4.47$	*Evaluate $\sqrt{20}$ on a calculator.*

Here, the solving begins by adding $3x^2$ to each side. Alternatively, 60 could be subtracted from each side. Either way the solutions will be the same.

The coefficient of the x^2-term in the equation $3x^2 = 60$ is 3. So, divide each side by 3.

Last, take the square root of each side. Remember to write the ± in front of the square root. The radicand, 20, is not a perfect square, so the solutions are not integers. In this case, find the exact answer by simplifying the square root. Since 20 has a perfect-square factor, 4, the root can be simplified.

The solutions can be approximated using a calculator.

Example 4 Consumer Application

A nursery manager is buying fencing to enclose three sides of a garden, with two short sides and one long side. The garden is a rectangle with sides 2*x* and 3*x*, where *x* is feet. The area of the garden is 3000 square feet. The manager can buy fencing in whole feet only. How many feet of fencing should he buy?

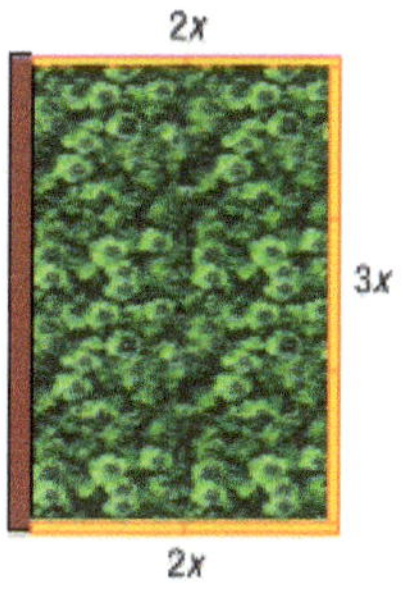

$\ell w = A$	*Use the formula for area of a rectangle.*
$(2x)(3x) = 3000$	
$\frac{6x^2}{6} = \frac{3000}{6}$	*Divide both sides by 6.*
$x^2 = 500$	
$x = \pm\sqrt{500}$	*Take the square root of both sides.*
$x = \pm 10\sqrt{5}$	
$x = 10\sqrt{5}$	*Negative numbers are not reasonable for length.*
$x \approx 22.4$ ft	*Evaluate $10\sqrt{5}$ on a calculator.*
$2x + 2x + 3x = P$	
$7x = P$	
$7(22.4) = P$	
$156.8 \text{ ft} = P$	

The nursery manager must buy 157 ft of fencing to enclose the garden.

This example involves the area of a rectangle. So, use the area formula $A = LW$ to write an equation. The area of the rectangle is 3000 square feet and the dimensions of the rectangle are represented by $2x$ and $3x$. Therefore, the equation is $3000 = (2x)(3x)$, which simplifies to $3000 = 6x^2$, or $6x^2 = 3000$. Notice that $6x^2 = 3000$ is a quadratic equation without a *bx*-term. Therefore, use square roots to solve the equation after solving for x^2.

The solutions to $6x^2 = 3000$ are approximately 22.4 and −22.4. However, since the question concerns length, the negative solution can be ignored.

The question—"How many feet of fencing should he buy?"—has not been answered yet. The nursery manager wants to enclose three sides of the garden with fence (two shorts sides and one long side), so the answer is the sum of the lengths of those three sides, $2x + 2x + 3x$. This expression simplifies to $7x$. Multiply 7 by 22.4 to find the total amount of fencing needing.

Sector Area and Arc Length

Key Objectives

- Find the area of sectors.
- Find arc lengths.

Key Terms

- A **sector of a circle** is a region bounded by two radii of the circle and their intercepted arc.
- A **segment of a circle** is a region bounded by an arc and its chord.
- **Arc length** is the distance along an arc measured in linear units.

Formulas

- **Sector Area** $A = \pi r^2 \left(\frac{m^\circ}{360^\circ}\right)$
- **Arc Length** $L = 2\pi r \left(\frac{m^\circ}{360^\circ}\right)$

Example 1 Finding the Area of a Sector

Find the area of sector *DAE*. Give your answer in terms of π and rounded to the nearest tenth.

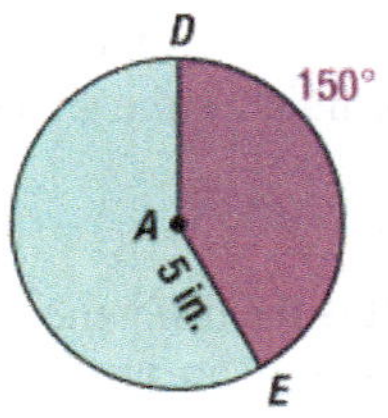

$A = \pi r^2 \left(\frac{m^\circ}{360^\circ}\right)$ Use the formula for area of a sector.

$= \pi (5)^2 \left(\frac{150^\circ}{360^\circ}\right)$ Substitute 5 for r and 150° for m.

$= \frac{125}{12}\pi \text{ in}^2 \approx 32.7 \text{ in}^2$ Simplify.

The area of a sector is determined in this example. The radius and the measure of the central angle are given.

To determine the area of the sector *DAE*, use the formula for the area of a sector. Substitute the given values for radius and the measure of the central angle into the formula. Solving for area yields $A \approx 32.7 \text{ in}^2$.

$A: \pi r^2$ $\frac{150}{360} = 41.6$
$= 25\pi$

Example 2 Agriculture Application

A circular plot with a 400 ft diameter is watered by a spray irrigation system. To the nearest square foot, what is the area that is watered as the sprinkler rotates through an angle of 60°?

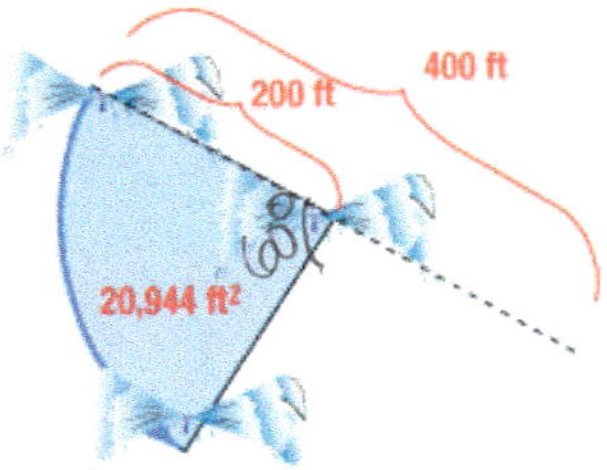

$A = \pi r^2 \left(\frac{m^\circ}{360^\circ}\right)$ Area of a sector

$= \pi (200)^2 \left(\frac{60^\circ}{360^\circ}\right)$ Substitute 200 for r and 60 for m.

$= \frac{20000}{3}\pi \approx 20{,}944 \text{ ft}^2$ Simplify.

The area of a sector is determined in this application example. The diameter and the measure of the central angle are given.

To determine the area of the sector covered by the irrigation system, use the formula for the area of a sector. Divide the diameter by 2 to find the radius. Substitute the values for radius and the measure of the central angle into the formula. Solving for area yields $A \approx 20{,}944 \text{ ft}^2$.

$A = \pi r^2$
$= 40{,}000\pi$
$= (40{,}000)\left(\frac{60}{360}\right)\pi$
$= 20{,}944 \text{ ft}^2$

Example 3 Finding the Area of a Segment

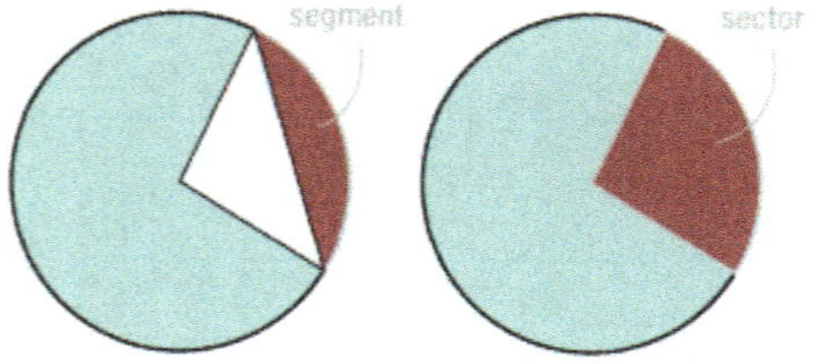

A segment is the area bounded by an arc and its chord.

The area of a segment is equal to the area of a sector minus the area of the triangle formed by the chord and the radii that define the sector.

Find the area of segment *BAC* to the nearest hundredth.

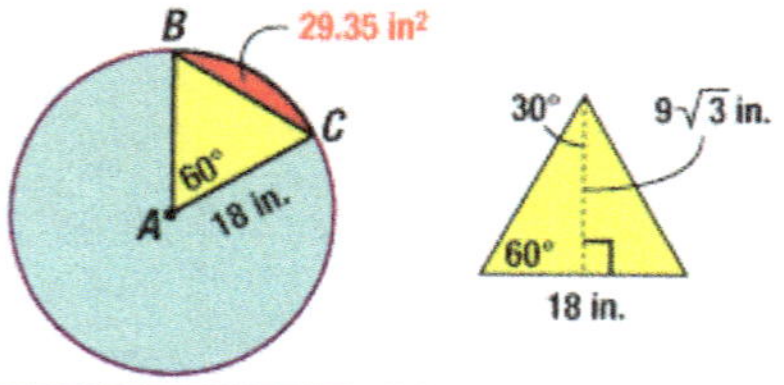

Find the area of sector *BAC*.

$A = \pi r^2\left(\frac{m^\circ}{360^\circ}\right)$ Area of a sector

$= \pi(18)^2\left(\frac{60^\circ}{360^\circ}\right)$ Substitute 18 for *r* and 60 for *m*.

$= 54\pi \text{ in}^2$ Simplify.

Find the area of $\triangle BAC$.

$A = \frac{1}{2}bh = \frac{1}{2}(18)(9\sqrt{3})$ $b = 18$ in. and $h = 9\sqrt{3}$ in.

$= 81\sqrt{3} \text{ in}^2$ Simplify.

area of segment *BAC* = area of sector *BAC* − area of $\triangle BAC$

$A = 54\pi - 81\sqrt{3} \approx 29.35 \text{ in}^2$

The area of a segment is determined in this example. The diameter and the measure of the central angle are given.

To determine the area of the segment, begin by finding the area of the sector defined by the central angle. Use the formula for sector area. Substitute the values for radius and the measure of the central angle into the formula. Solving for area yields $A = 54\pi \text{ in}^2$.

Find the area of the triangle by using the formula for the area of a triangle. The central angle of the triangle is 60° and the height of the triangle bisects the triangle into two 30°-60°-90° right triangles whose shorter legs are 9 inches long, (1/2)18 in., and whose longer leg, the height, is $9\sqrt{3}$. Substituting into the formula for the area of a triangle gives $81\sqrt{3} \text{ in}^2$.

The area of the segment is the difference between the two areas, which is approximately 29.35 in^2.

Example 4 Finding Arc Lengths

Find the arc length of an arc with measure 40° in a circle with radius 6 in. Give your answer in terms of π.

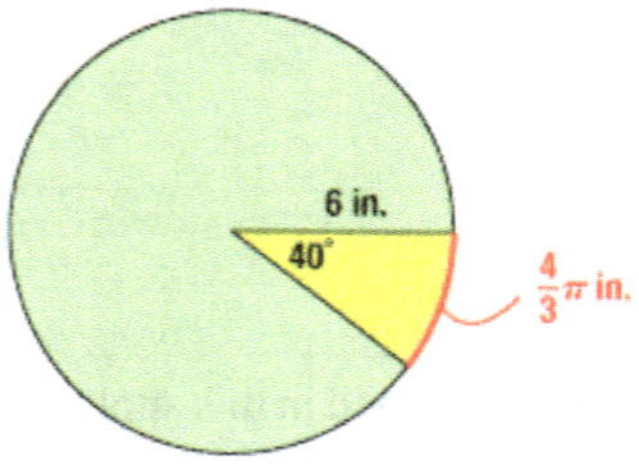

$L = 2\pi r\left(\frac{m^\circ}{360^\circ}\right)$ Use formula for arc length.

$L = 2\pi(6)\left(\frac{40^\circ}{360^\circ}\right)$ Substitute 6 for *r* and 40 for *m*.

$= \frac{4}{3}\pi \text{ in.}$ Simplify.

The length of an arc is determined in this example. The radius and the measure of the central angle are given.

To determine the length of the arc, use the formula for the length of an arc. Substitute the values for radius and the measure of the central angle into the formula. Solving for length yields $L = (4/3)\pi$ in.

$C = 2\pi r$

$2\pi r\left(\frac{40}{360}\right)$

$= 1.\bar{3}\pi$ in

Circles in the Coordinate Plane, Part 1

Key Objectives

- Write equations and graph circles in the coordinate plane.
- Use the equation and graph of a circle to solve problems.

Theorems, Postulates, Corollaries, and Properties

- **Equation of a Circle** The equation of a circle with center (h, k) and radius r is $(x - h)^2 + (y - k)^2 = r^2$.

Example 1 Writing the Equation of a Circle

Equation of a Circle Theorem

The equation of a circle with center (h, k) and radius r is $(x - h)^2 + (y - k)^2 = r^2$. (Equation of a Circle Thm.)

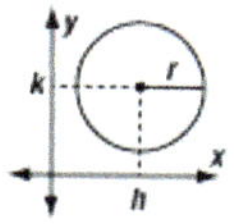

The equation of a circle with center (h, k) and radius r is $(x - h)^2 + (y - k)^2 = r^2$.

Write the equation of each circle.

⊙A with center $A(-1, 5)$ and radius 4

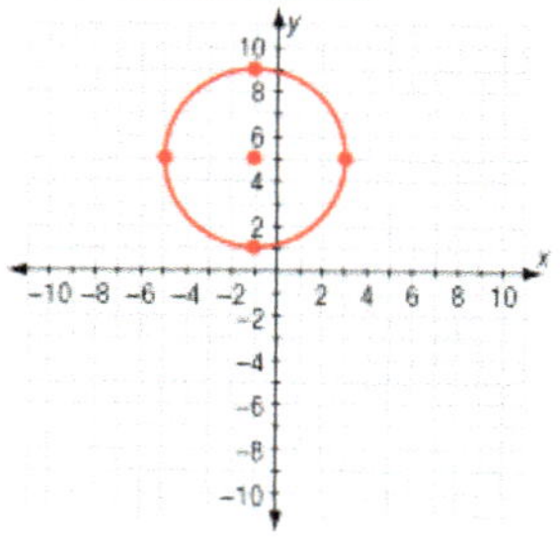

$(x - h)^2 + (y - k)^2 = r^2$ Equation of a circle

$(x - (-1))^2 + (y - 5)^2 = 4^2$ Substitute −1 for h, 5 for k, and 4 for r.

$(x + 1)^2 + (y - 5)^2 = 16$ Simplify.

The equation for a circle is determined in this example. The coordinates of the center of the circle and the radius are given.

Substitute the given values into the equation for a circle. Simplify the equation. The equation for the circle is $(x + 1)^2 + (y - 5)^2 = 16$.

Write the equation of each circle.

⊙B that passes through (−2, 3) and has a center $B(2, 0)$

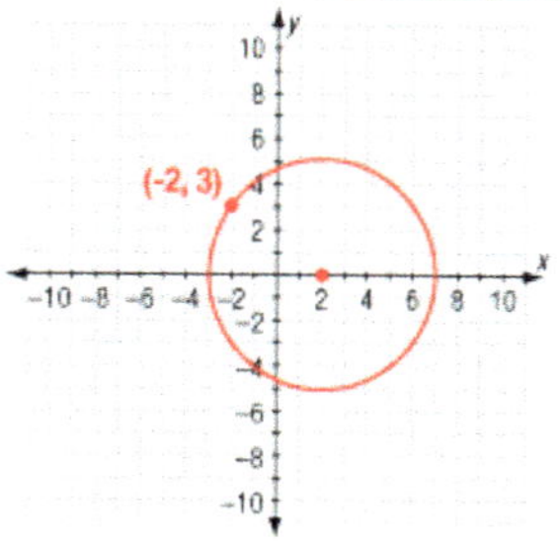

$r = \sqrt{(-2 - 2)^2 + (3 - 0)^2}$ Distance Formula

$= \sqrt{16 + 9} = \sqrt{25} = 5$

$(x - h)^2 + (y - k)^2 = r^2$ Equation of a circle

$(x - 2)^2 + (y - 0)^2 = 5^2$ Substitute 2 for h, 0 for k, and 5 for r.

$(x + 1)^2 + y^2 = 25$ Simplify.

The equation for a circle is determined in this example. The coordinates of the center of the circle and the coordinates of a point the circle passes through are given.

Begin by determining the length of the radius for the circle. A radius is a line segment from the center of the circle to any point on the circle. Find the length of the radius using the Distance Formula. Substitute the given values for the center of the circle and the point on the circle into the distance formula and solve for r. The length of the radius is 5 units. To find the equation for the circle, substitute the value for the radius and the coordinates of the center of the circle into the equation. Simplify the equation.The equation for the circle is $(x + 1)^2 + y^2 = 25$.

Example 3 Office Application

Two people walked from the water cooler to the location shown in the diagram below. How far apart are the two people? Each floor is 12 feet.

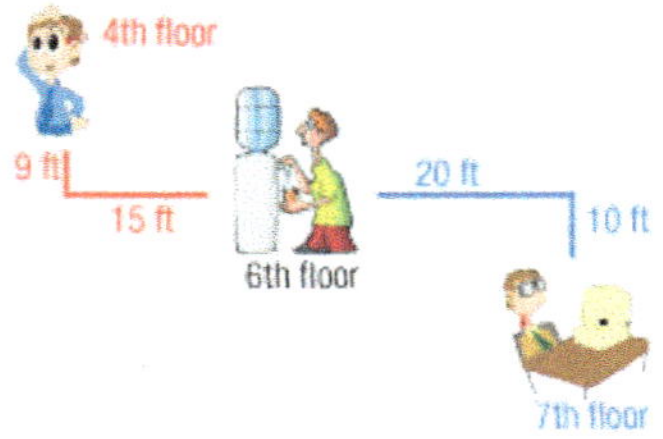

$$d = \sqrt{(15-(-20))^2 + (-9-10)^2 + (-24-12)^2}$$ Use the Distance Formula for three dimensions.

$$= \sqrt{35^2 + 19^2 + 36^2}$$ Substitute the given values.

$$= \sqrt{2882} \approx 53.68 \text{ ft.}$$ Simplify.

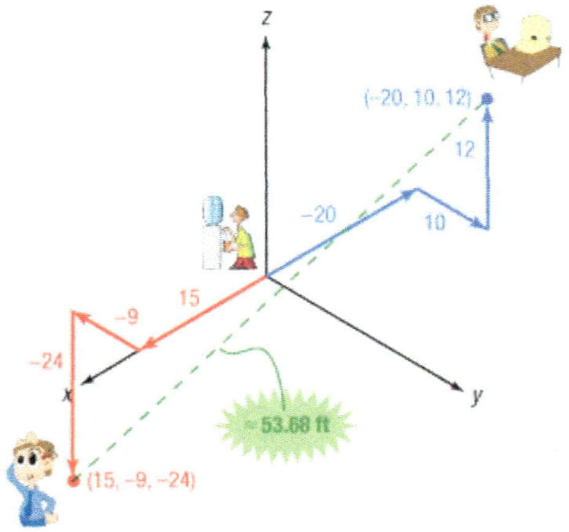

The straightline distance between two points is determined in this application example. The coordinates of the points are given in three-dimensional space.

Begin by determining the coordinates of the two people relative to the origin, which is the water cooler. A diagram is a useful way of visualizing the movements on the three axes and organizing the information. The x, y, z coordinates of the first person are $(-20, 10, 12)$. The coordinates of the second person are $(15, -9, -24)$.

To find the distance between the people, substitute the coordinates into the Distance Formula in Three Dimensions. Using a calculator, the distance is found to be approximately 53.68 feet.

Effects of Changing Dimensions Proportionally, Part 1

Key Objectives

- Describe the effect on perimeter and area when one or more dimensions of a figure are changed.
- Apply the relationship between perimeter and area in problem solving.

Effects of Changing Dimensions Proportionally		
Change in Dimensions	**Perimeter or Circumference**	**Area**
All dimensions multiplied by a	Changes by a factor of a	Changes by a factor of a^2

Changing dimensions proportionally means changing all dimensions of a figure by the same factor. This type of change has the effect of changing the perimeter or circumference of a figure by the same factor. However, the area of the figure changes by the square of the factor. This is expected because the area is a function of the square of the linear dimensions of a figure.

Example 1 Effects of Changing One Dimension

Describe the effect of each change on the area of the given figure.

the height of the triangle is doubled

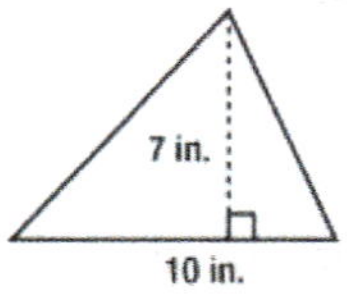

original dimensions: $A = \frac{1}{2}bh = \frac{1}{2} \cdot 10 \cdot 7 = 35 \text{ in}^2$

height doubled: $A = \frac{1}{2}bh = \frac{1}{2} \cdot 10 \cdot 14 = 70 \text{ in}^2$

Notice that $70 = 2(35)$, so when the height is doubled, the area is doubled.

The effect of doubling the height of a triangle on the area of the triangle is determined in this example.

To calculate the area of a triangle with height 7 inches and base 10 inches, substitute the values into the formula for area. $A = 1/2(10)(7) = 35 \text{ in}^2$.

To calculate the area of a triangle with height 14 inches and base 10 inches, substitute the values into the formula for area. $A = 1/2(10)(14) = 70 \text{ in}^2$.

When the height of the triangle is doubled, the area of the triangle also doubles. Note that this is not a proportional change in dimensions; only one dimension is changed.

Describe the effect of each change on the area of the given figure.

the base length of the parallelogram is multiplied by $\frac{1}{2}$

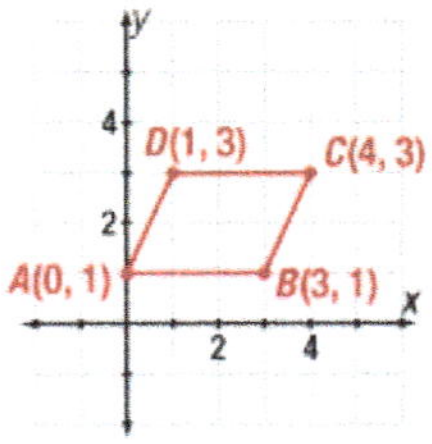

original dimensions: $A = bh = 3 \cdot 2 = 6 \text{ units}^2$

base length halved: $A = bh = \frac{1}{2}(3) \cdot 2 = 3 \text{ units}^2$

Notice that $3 = \frac{1}{2}(6)$, so when the base length is multiplied by $\frac{1}{2}$, the area is multiplied by $\frac{1}{2}$.

The effect of halving the base length of a parallelogram on the area of the parallelogram is determined in this example.

To calculate the area of a parallelogram with a base length of 3 units and height of 2 units, substitute the values into the formula for area. $A = (3)(2) = 6 \text{ units}^2$.

To calculate the area of a parallelogram with a base length of (1/2)3 units and height of 2 units, substitute the values into the formula for area. $A = (1/2)(3)(2) = 3 \text{ units}^2$.

When the length of the base of the parallelogram is halved, the area of the parallelogram also halves.

Example 2 Effects of Changing Dimensions Proportionally

Describe the effect of each change on the perimeter or circumference and the area of the given figure.

the length and width of the rectangle are both multiplied by 3

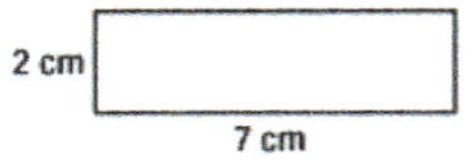

original dimensions:

$P = 2(7) + 2(2) = 14 + 4 = 18$ cm
$A = 2(7) = 14$ cm^2

length and width tripled:

$P = 2(21) + 2(6) = 42 + 12 = 54$ cm
$A = 6(21) = 126$ cm^2

Notice that $54 = 3(18)$, so when the length and width are multiplied by 3, the perimeter is multiplied by 3.

Notice that $126 = 9(14)$, so when the length and width are multiplied by 3, the area is mulitplied by 9 or 3^2.

The effect of multiplying both the length and width of a rectangle by 3 on the perimeter and area of the rectangle is determined in this example.

The perimeter of a rectangle with length 7 cm and width 2 cm is $P = 2 + 2 + 7 + 7 = 18$ cm. The area of the rectangle is $A = (2)(7) = 14$ cm^2.

The perimeter of a rectangle with length 21 cm and width 6 cm is $P = 6 + 6 + 21 + 21 = 54$ cm. The area of the rectangle is $A = (6)(21) = 126$ cm^2.

Tripling the length and width of the rectangle also triples the perimeter. But tripling the length and width (increasing the dimensions proportionally) increases the area by 3^2, or 9, times.

Describe the effect of each change on the perimeter or circumference and the area of the given figure.

the radius of $\odot B$ is multiplied by $\frac{1}{5}$

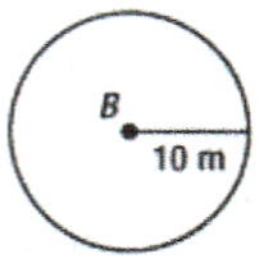

original dimensions:

$C = 2\pi r = 2\pi(10) = 20\pi$ m
$A = \pi r^2 = 100\pi$ m^2

radius multiplied by $\frac{1}{5}$:

$C = 2\pi r = 2\pi \cdot \frac{1}{5}(10) = 4\pi$ m
$A = \pi r^2 = \pi\left(\frac{1}{5}(10)\right)^2 = 4\pi$ m^2

Notice that $4\pi = \frac{1}{5}(20\pi)$, so when the radius is multiplied by $\frac{1}{5}$, the circumference is multiplied by $\frac{1}{5}$.

Notice that $4\pi = \frac{1}{25}(100\pi)$, so when the radius is multiplied by $\frac{1}{5}$, the area is multiplied by $\frac{1}{25}$, or $\left(\frac{1}{5}\right)^2$.

The effect of multiplying the radius of a circle by 1/5 on the circumference and area of the circle is determined in this example.

The circumference of a circle with radius 10 m is $C = 2\pi r = 20\pi$ m. The area of the circle is $A = \pi r^2 = 100\pi$ m^2.

The circumference of a circle with radius multiplied by 1/5 is $C = 2\pi r = 2\pi(1/5)(10) = 4\pi$ m. The area of the circle is $A = \pi r^2 = \pi[(1/5)(10)]^2 = 4\pi$ m^2.

The circumference has changed by a factor of 1/5 from 20π to 4π. The area has changed by a factor of $(1/5)^2 = (1/25)$ from 100π to 4π.

Effects of Changing Dimensions Proportionally, Part 2

Key Objectives

- Describe the effect on perimeter and area when one or more dimensions of a figure are changed.
- Apply the relationship between perimeter and area in problem solving.

Effects of Changing Dimensions Proportionally		
Change in Dimensions	**Perimeter or Circumference**	**Area**
All dimensions multiplied by a	Changes by a factor of a	Changes by a factor of a^2

Changing dimensions proportionally means changing all dimensions of a figure by the same factor. This type of change has the effect of changing the perimeter or circumference of a figure by the same factor. However, the area of the figure changes by the square of the factor. This is expected because the area is a function of the square of the linear dimensions of a figure.

Example 1 Effects of Changing Area

A circle has radius 7 ft. If the area is multiplied by 5, what happens to the radius?

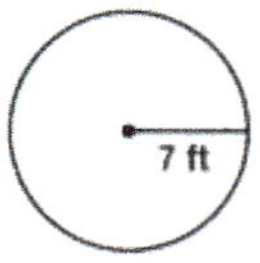

original dimensions:

$r = 7$ ft
$A = \pi r^2$
$A = \pi(7^2)$
$= 49\pi$ ft^2

area multiplied by 5:

$A = 5(49\pi) = 245\pi$ ft^2
$A = \pi r^2$
$245\pi = \pi r^2$
$245 = r^2$
$7\sqrt{5}$ ft $= r$

Notice that $7\sqrt{5} = \sqrt{5}(7)$, so when the area is multiplied by 5, the radius is multiplied by $\sqrt{5}$.

The effect of a change in area on the radius of a circle is determined in this example.

A circle has a radius of 7 ft. The area of the circle is $A = \pi r^2 = 49\pi$ ft^2.

Multiplying the area by a factor of 5 gives an area of 245π. Substitute this value into the formula for area to find the radius of a circle with this area. Solving for r gives $r = 7\sqrt{5}$.

Increasing the area by a factor of 5 has the effect of increasing the radius by a factor of $\sqrt{5}$.

A square has a side length of 5 m. If the area is tripled, what happens to the perimeter?

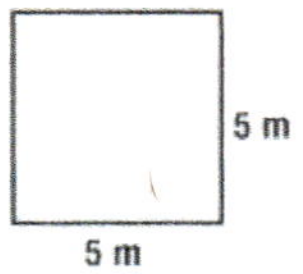

original dimensions:

$P = 4(5) = 20$ m
$A = s^2$
$A = 5^2 = 25$ m^2

area tripled:

$A = 3(25) = 75$ m^2
$A = s^2$
$75 = s^2$
$5\sqrt{3} = s$
$P = 4(5\sqrt{3}) = 20\sqrt{3}$ m

Notice that $20\sqrt{3} = \sqrt{3}(20)$, so when the area is multipled by 3, the perimeter is multiplied by $\sqrt{3}$.

The effect of a change in area on the perimeter of a square is determined in this example.

A square has sides of 5 m. The perimeter of the square is $4(5) = 20$ m. The area of the square is $A = s^2 = 5^2 = 25$ m^2.

Multiplying the area by a factor of 3 gives an area of 75 m^2. Substitute this value into the formula for area to find the length of the side of a square with this area. Solving for s gives $s = 5\sqrt{3}$. The perimeter of this square is $20\sqrt{3}$.

Increasing the area by a factor of 3 has the effect of increasing the perimeter by a factor of $\sqrt{3}$.

Example 2 Business Application

A tile manufacturer is exhibiting its tile sales in a graph that shows that sales totaled about 19 million in 2003, 38 million in 2004, and 76 million in 2005. The height of each tile is used to represent the number of tiles sold. Explain why the graph is misleading.

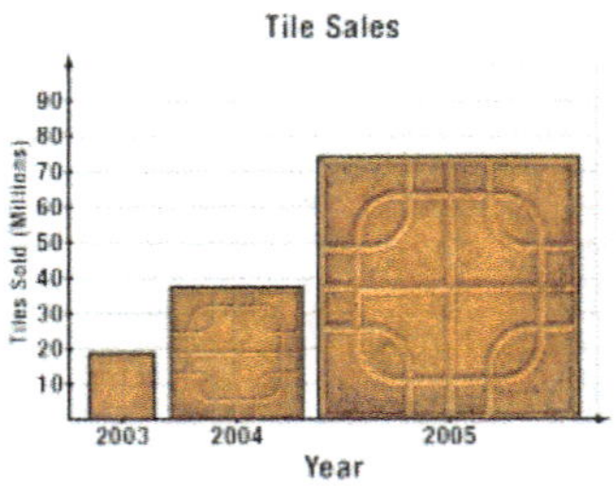

The height of the tile in 2005 is 4 times the height of the tile in 2003. This is a good representation of the sales in 2005 as compared to 2003.

But the tiles in the graph are squares and it seems like the sales are represented by the areas of the tiles. The area of the tile representing sales in 2005 is $4^2 = 16$ times the area of the tile representing sales in 2003.

The graph gives the misleading impression that the tile sales in 2005 were 16 times the tiles sales in 2003, when they were actually 4 times the 2003 sales.

The impact on a data presentation of a disproportionate change in area with a change in length of a figure is examined in this example.

The graph shows an increase in sales (the y-axis) over three years. From 2003 to 2005 the increase is from 19 million to 76 million, an increase of 4 times (19(4) = 76). But because the graph uses the device of a square tile to show the increase along the y-axis, it is also indirectly showing an increase in the area of the tiles. From 2003 to 2005, the areas of the tiles increase 16 times, since $A_{2003} = 19^2 = 361$ and $A_{2005} = 76^2 = 5776$, and 361(16) = 5776. This is a much greater factor of increase than the graph is supposed to represent, which makes it visually misleading, or at the very least unhelpful to the viewer.

Perimeter and Area in the Coordinate Plane, Part 1

Key Objectives

- Find the perimeters and areas of figures in a coordinate plane.

Example 1 Estimating Areas of Irregular Shapes in the Coordinate Plane

Estimate the area of the irregular shape.

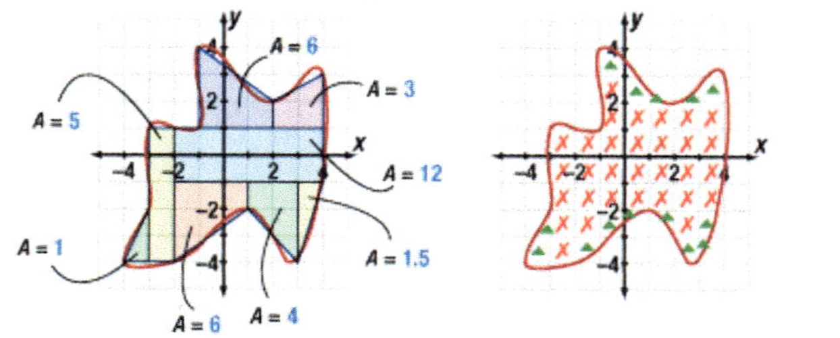

Method 1:

Approximate the area by adding the areas of the composite figures.

$A \approx 6 + 3 + 12 + 5 + 6 + 1 + 4 + 1.5$

$A \approx 38.5$ units^2 Simplify.

Method 2:

Approximate the area by counting the number of full and almost-full squares, and adding half the number of partial squares.

$A \approx 31 + \frac{1}{2}(14)$

$A \approx 38$ units^2 Simplify.

The area of an irregular shape in the coordinate plane is estimated using two methods.

In the first method, the irregular shape is approximated by filling it with smaller regular figures, such as triangles and rectangles, whose areas can be calculated. The area of the irregular shape is approximated by the area of the composite shape. This method yields an area estimate of 38.5 square units.

In the second method, the number of full square units enclosed by the irregular shape is counted. Then the number of square units with a portion lying within the irregular shape is counted. The area of the irregular shape is estimated as the number of full square units plus one-half the number of partial square units. This method yields an estimate of 38 square units.

Example 2 Finding Perimeter and Area in the Coordinate Plane

Draw and classify the polygon with vertices $A(0, 1)$, $B(3, 2)$, $C(2, -1)$, $D(-1, -2)$. Find the perimeter and area of the polygon.

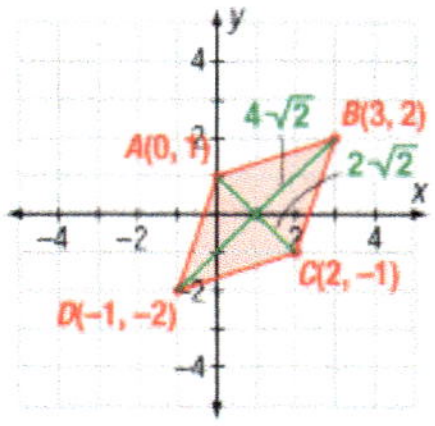

Find the length of each side.

$CD = AD = AB = BC = \sqrt{10}$

The four sides are congruent, so *ABCD* is a rhombus by definition.

Find the slope of each line.

slope of $\overline{CD}$: $\frac{\text{rise}}{\text{run}} = \frac{1}{3}$ slope of $\overline{AD}$: $\frac{\text{rise}}{\text{run}} = \frac{3}{1} = 3$

slope of $\overline{AB}$: $\frac{\text{rise}}{\text{run}} = \frac{1}{3}$ slope of $\overline{BC}$: $\frac{\text{rise}}{\text{run}} = \frac{3}{1} = 3$

Since the product of the slopes of two adjacent sides is not -1, the sides are not perpendicular. The rhombus is not a square.

Perimeter: $P = 4(\sqrt{10}) = 4\sqrt{10}$ units

The perimeter is $4\sqrt{10}$ units.

Area: $d_1 = \sqrt{2^2 + 2^2} = \sqrt{8} = 2\sqrt{2}$ $d_2 = \sqrt{4^2 + 4^2} = \sqrt{32} = 4\sqrt{2}$

$A = \frac{1}{2}d_1 d_2 = \frac{1}{2}(2\sqrt{2})(4\sqrt{2}) = 8$ units^2

The area is 8 square units.

The type of polygon drawn in the coordinate plane and its perimeter and area are determined in this example. The coordinates of the vertices of the polygon are given.

The lengths of the sides of the polygon are determined using the Distance Formula. All four sides are congruent, so the figure is a rhombus. The perimeter of the rhombus is $4\sqrt{10}$. The slopes of the sides can be determined from the formula for slope and the coordinates of the vertices. None of the products of the slopes of adjacent sides are -1, so none of the sides are perpendicular and this is not a square.

The area is one-half times the product of the lengths of the diagonals. The lengths of the diagonals can be calculated using the Distance Formula and the coordinates of opposite vertices. The area is found to be 8 square units.

